Time Series Analysis by Higher Order Crossings

IEEE Press
445 Hoes Lane, PO Box 1331
Piscataway, NJ 08855-1331

Technical Reviewers

Peter Brockwell
G. R. Dargahi-Noubary
Kurt Fristrup
Laveen Kanal
B. V. K. Vijaya Kumar
Mohsen Pourahmadi

Time Series Analysis by Higher Order Crossings

Benjamin Kedem

Department of Mathematics and Institute for Systems Research
University of Maryland, College Park

The Institute of Electrical and Electronics Engineers, Inc., New York

This book may be purchased at a discount from the publisher when ordered in bulk quantities. For more information contact:

IEEE PRESS Marketing
Attn: Special Sales
P.O. Box 1331
445 Hoes Lane
Piscataway, NJ 08855-1331
Fax: (908) 981–8062

Printed in the United States of America

10 9 8 7 6 5 4 3 2 1

ISBN 0-87942-299-8

IEEE Order Number: PC02949

Library of Congress Cataloging-in-Publication Data

Kedem, Benjamin
Time series analysis by higher order crossings / Benjamin Kedem.
p. cm.
Includes bibliographical references and index.
ISBN 0-87942-299-8
1. Time-series analysis. I. Title.
QA280.K43 1993
519.5′5—dc20 93-14967
CIP

In loving memory of our son Mattan

July 1971–August 1988

And I replied:
Ah, Lord God!
I do not know how to speak
For I am still a boy.
(Jeremiah 1:6)

". . . and they said—purple!"

What the caterpillar thought
Was the end of life . . .
The butterfly knew was
Just the beginning . . .
(The Compassionate Friends)

Contents

Preface

Imagine a bank of filters applied to a finite signal or time series, indeed a pervasive engineering practice. Typically, the application of each filter from the bank changes the signal oscillation pattern and alters the zero crossing count. Accordingly, the application of each member filter gives rise to a zero-crossing count, and the zero-crossing counts resulting from the application of some or all the filters are called *higher order crossings*, or HOC for short. This book is a treatise about HOC and related ideas. It explores, for the most part, an approach to time series analysis based on the surprisingly fruitful connection between filtering and zero-crossings.

In his important pioneering work in the 1920s, E. E. Slutsky used statistical and probabilistic tools to explain the effect of repeated linear filtering on the spectrum of a stationary time series. He recognized then what today is common knowledge, namely that conspicuous periodicities observed in a time record may be the result of linear filters applied to some underlying process, thus enhancing certain spectral bands. This program of repeated filtering, or more generally the application of filters from a designated family, is followed in this book, the extra feature being the involvement of zero-crossings observed in the filtered data. Our message is that in addition to the spectrum and autocorrelation, HOC families and sequences too constitute viable tools which can be used to summarize the oscillatory content of time series. The oscillatory information contained in

HOC families and sequences is useful for fast signal classification and spectral analysis, particularly frequency estimation in the presence of ambient noise.

It is impossible to write a book about zero-crossings without mentioning the name of another important pioneer, S. O. Rice, who in the 1940s derived, in addition to many other useful results, the celebrated formula for the expected number of level-crossings by a finitely long stationary Gaussian time series in continuous time. Through his formula, Rice was able to quantify the effect of a time-invariant linear filter on the expected zero-crossing rate, which is precisely the central idea of the present book. Interestingly, the extension of Rice's formula to general non-Gaussian stationary processes in continuous time is still an open problem, though some concrete particular extensions are possible (Chapter 4).

Two unrelated incidents in 1978 sparked my research on higher order crossings. First, or it might have been the other way around, Eric Slud asked me if I had ever thought of zero-crossings of differenced time series, which I never did, and second, the late Dr. Ismail Shimi of the Air Force Office of Scientific Research kindly suggested that I look into the "signature problem," having in mind reliability tests for vibrating engines. The subtle, however fortunate, connection between the inquiry and suggestion becomes more apparent when one thinks of the very first few zero-crossing counts from repeated differences as a means for summarizing "all the oscillatory information" in a signature. From this to a more systematic consideration of zero-crossing counts in filtered time series the distance is rather short, the proof being the present work.

The book in its present form is an outgrowth of my article "Spectral Analysis and Discrimination by Zero-Crossings," *Proceedings of the IEEE*, 1986, and is meant to be accessible to signal processing practitioners and time series analysts whose work might lead them to consider zero-crossing-based statistical inference. Since zero-crossings convey a strong graphical appeal, most of the ideas presented are sufficiently intuitive and can be grasped quite easily. In general, however, familiarity with basic statistical concepts is quite essential.

The book can also serve as a text for a graduate seminar or a selected topics course in time series analysis. I have taught parts of the book in various graduate time series courses at the University of Maryland, College Park, as well as at Tsinghua University, Beijing, China, in the winter of 1990–91. In addition, some of the material was presented in a special five hour broadcast via satellite from the University of Maryland, College Park, to industrial and research outfits in the United States on January 21, 1988.

In addition to "applied" material, there is also enough here for the mathematically inclined to ponder over. For example, is there a closed form relationship between the expected number of zero-crossings in a finite stationary time series and the first order autocorrelation (Chapter 4)? Can we expect to see more zero-crossings as a result of high-pass filtering (Chapter 4)? Does the asymptotic empirical zero-crossing rate of a second order autoregressive stationary Gaussian process with unit roots differ from that of a pure random sinusoid

(Chapter 4)? To what extent do HOC sequences determine the spectrum of a stationary time series (Chapter 5)? To what value does the zero-crossing rate of a repeatedly differenced stationary time series converge (Chapter 5)? Does the observed zero-crossing rate of a stationary mixed spectrum Gaussian time series converge to a constant (Chapter 6)? Can the spectrum of a clipped process have a jump at the origin even when the spectrum of the original unclipped process is continuous there (Chapter 6)? At what rate should the bandwidth of a bandpass filter contract to produce almost surely convergent HOC sequences in frequency estimation without ever loosing the frequency of interest (Chapter 7)? Regarding these and many other zero-crossing problems, the book purports to provide information not easily accessible in most time series and signal processing books. The information is brought forward in the form of copious examples (some with real data), algorithms, theorems, problems, figures, discussions, and references. With regard to the theorems about HOC, I have tried to give at least an indication of proof in most cases. Appropriate references are indicated in the exceptional cases.

Chapter 1 introduces our agenda. For self-containment, Chapters 2 and 3 introduce some basic material from probability theory and the theory of stationary processes, emphasizing those concepts needed for the development of the next five chapters. Chapters 4, 5, 6, 7, and 8, where the theory and applications of HOC are developed, comprise the main body of this book. Chapter 9 deals with the problem of level-crossing prediction given time dependent covariates, and is more statistical in nature. The chapter introduces the useful notion of partial likelihood in connection with logistic regression applied to time series. There, new concepts are defined as needed. Each chapter contains a problems section and a list of relevant references. Some data sets used in the book are compiled in Appendix A.

This work is the result of cooperation and interaction with many colleagues, co-workers, and students, who collaborated with me in various forms, and who enthusiastically shared their talents and time. Each and every one of them contributed significantly to my research, and the joint references at the end of each chapter bear witness to some of this appreciated collaboration. I take this opportunity to thank them all wholeheartedly. The names that should be mentioned in particular include John Barnett, Mordechai Berger, Phineas Dickstein, Shuyuan He, Christian Houdré, Ta-Hsin Li, Avraham Tal, and Sid Yakowitz. Above all, I owe a great deal of deep gratitude to Eric Slud for numerous influential ideas, enlightening discussions, and words of wisdom regarding HOC. His invaluable contribution is manifested throughout the book. For the record, the term "higher order crossings," was coined in our joint 1979 technical reports, "Higher order crossings in the discrimination of time series, I, II," TR 79-66,81, Mathematics Department, University of Maryland, College Park.

It is also a pleasure to acknowledge the help of several individuals who have read large parts of the manuscript and made corrections as well as very helpful suggestions. Among them were John Barnett, Reza Dargahi-Noubary, Christian Houdré, Ta-Hsin Li, Chun-Kuo Li, Silvia Lopes, Donald Martin, Harry Pavlopoulos, and James Troendle. The advice and help of Alfred Gray regarding LaTeX is much appreciated.

Some of my research on higher order crossings was conducted while I was a visitor at the Institute for Physical Science and Technology at the University of Maryland, College Park (1978–79, 1988–89), RAFAEL, Israel (1982–84), and Institute for Systems Research at the University of Maryland, College Park (1991–present). I would like to thank these institutions for their hospitality and interest in my work.

Finally, I would like to acknowledge the generous and much appreciated financial support I received over the years from the Air Force Office of Scientific Research and the Office of Naval Research. In retrospect, I could not have carried on my research program without their support and encouragement.

Benjamin Kedem
College Park, Maryland

1

Introduction

In its most ordinary sense, a time series $Z_1, Z_2, \ldots, Z_N$, is a sequence of observations or measurements ordered in time. Examples are abundant. Speech records, sampled vibration signatures, daily variation of latitude, weekly sales of a certain commodity, hourly rain rate averaged over a given area, sampled electrical noise, are all examples of time series. The list of examples goes on and on indefinitely and runs across the natural and social sciences as well as engineering. This large diversity of fields of study where one encounters data in the form of time series, fields ranging from economics to geophysics and systems engineering, makes time series analysis an interdisciplinary field that draws heavily on mathematical, statistical, and engineering ideas and methods. This fact can be fully appreciated by simply glancing at the list of references given at the end of each chapter in this book. It is not a surprise then that the interdisciplinary nature of time series analysis has been responsible for the advancement of a number of approaches encompassing a wide variety of analytical tools appropriate for data in which the observations are ordered in time. But whatever the approaches and techniques are, they all cater to the fundamental fact that *almost all observed time series are oscillatory, displaying local and global up and down movements as time progresses*. In recognizing this fact, we echo the opening sentence of Slutsky's pioneering paper of 1927 [24]: "Almost all of the phenomena of economic life, like many other processes, social, meteorological, and others, occur in sequences of rising and falling movements, like waves." In this book we shall be concerned with this type of oscillation. Our primary objective is to explore a certain approach to time series analysis and signal processing suitable for "capturing," summarizing, and processing the oscillation observed in time series records. This is done, for the most part, by higher order crossings (HOC).

To explain the notion of HOC, and hint at possible extensions, we must first turn to filtering.

Time ordering makes it possible to apply to a time series certain operations, treating it as a function of time. Of these operations, linear filtering is one of the most indigenous elements of time series analysis. Now, in general, when a filter is applied to a time series it changes its oscillation. The change in oscillation, however, can be expressed by the change in the zero-crossing count effected by the filter. Much of this book is simply an elaboration of this basic observation.

Consider a finite time series oscillating about level zero, and count the number of observed zero-crossings. Apply to the time series a filter, and count the number of zero-crossings in the filtered time series. Apply to the original time series yet another filter, and again observe the resulting number of zero-crossings, and so on—filter and count, filter and count. The resulting (higher order) zero-crossing counts are referred to as *higher order crossings* or HOC for short. By applying a specific sequence of filters to a time series, we obtain the corresponding sequence of zero-crossing counts, and call the sequence of counts a HOC sequence. By appropriate filter design, we can construct many different types of HOC sequences, interesting for their own sake, but also useful for the purpose of spectral analysis and discrimination.

We can now be more precise about the goals set up in this book. To a high degree, the book's purpose is to explore theoretically as well as empirically the fruitful connection between time-invariant linear filtering and zero-crossing counts, for the purpose of time series analysis and signal processing. As we shall see, this connection sheds light on interesting and surprisingly useful properties of zero-crossings appropriate for fast time series analysis and signal processing. Furthermore, it will become evident that HOC constitute a domain by itself which "sits" between the time and spectral domains. Although each of these domains maintains its own peculiar identity, in many respects all three domains are equivalent.

The preceding definition of HOC should be viewed in a narrow sense only. A more suitable definition, as well as a certain generalization, will be discussed later. For the time being, the given definition suffices for bringing across some essential ideas.

To help motivate some key ideas presented in the book, it is instructive, at this early stage, to consider in the remainder of the chapter some typical examples of HOC sequences and their applications.

Before plunging into the examples, recall that

$$X_t = A\cos(\omega t + \phi)$$

is a *sinusoid* fluctuating around level 0 with *amplitude* A, *angular frequency* ω, and *phase* ϕ. The variable t stands for *time* and could be *discrete* (e.g., $t = \ldots, -2, -1, 0, 1, 2, \ldots$) or *continuous* (e.g., $-\infty < t < \infty$). The angular frequency ω is measured in radians per unit time, and $2\pi/\omega$ is the *period* which is the length of time needed for the sinusoid to complete one cycle. Thus, the period of $\cos t$ is 2π. The reciprocal of the period, $f = \omega/2\pi$, is the *frequency* measured in cycles per unit time. Except for very few exceptions, we always use angular frequencies but usually drop "angular" for brevity. From the trigonometric identity

$$\cos(\alpha + \beta) = \cos\alpha\cos\beta - \sin\alpha\sin\beta$$

it follows that

$$X_t = A'\cos(\omega t) + B'\sin(\omega t)$$

where $A' = A\cos\phi$ and $B' = -A\sin\phi$. We often assume that A' and B' are random, in which case the sinusoid becomes a *random sinusoid*. When A' and B' are random, the source of randomness can be a random phase ϕ and/or a random amplitude A. Colored noise can often be represented as a sum of random sinusoids with different frequencies.

Example 1.1: Superposition of Two Sinusoids ([1], [10], [16])

Consider the superposition of two sinusoids in discrete time,

$$Z_t = 3.86211\sin(0.5t) + 1.11689\sin(2t) \tag{1.1}$$

$t = 1, 2, 3, \ldots$, with angular frequencies $\omega_1 = 0.5$, and $\omega_2 = 2$ radians per unit time. The graph of Z_t, $t = 1, 2, \ldots, 50$, is shown in Figure 1.1(a). With only a few exceptions, as a general rule we adhere to discrete time, but it is useful to portray the graphs of time series on a continuous scale so as to convey the impression of "zero-crossings." This is achieved by connecting the discrete time observations by line segments. Later we shall give an exact definition of what we really mean by zero-crossing counts in discrete time.

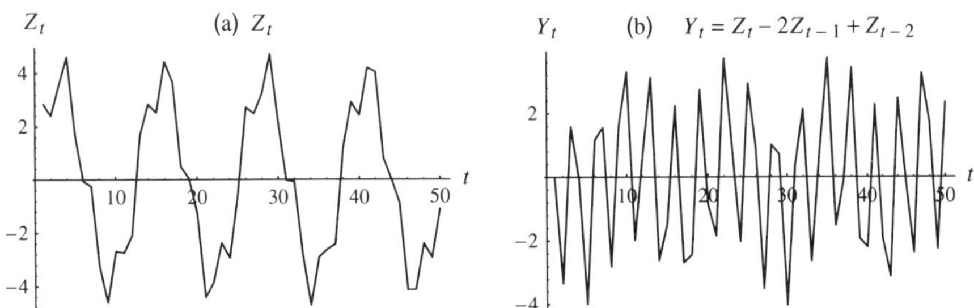

Figure 1.1: *(a) Superposition of two sinusoids, Z_t. (b) Second difference, Y_t.*

Suppose it is desired to estimate the frequencies ω_1, ω_2. Let D_1 denote the number of zero-crossings observed in a section of length $N = 1000$ from Z_t, $t = 51, 52, \ldots, 1050$, say. At a later stage we will argue that the number of zero-crossings, after normalization, tends to "land" on, or very near, the dominant frequency when one exists. For the moment accept this intuitive principle, referred to as the *dominant frequency principle*, without proof. Now in (1.1), $\omega_1 = 0.5$ is the dominant frequency since its amplitude is more than three times as large as the other amplitude corresponding to $\omega_2 = 2$. By the dominant frequency principle, then, ω_1 can be estimated from D_1. Indeed, in the time series $Z_{51}, Z_{52}, \ldots, Z_{1050}$ one finds that $D_1 = 159$, and by proper normalization, our estimate becomes,

$$\hat{\omega}_1 = \frac{\pi \times D_1}{N-1} = \frac{\pi \times 159}{999} = 0.500013$$

which is a fairly precise estimate of 0.5 by any account.

The same procedure can be applied to estimate ω_2, provided ω_2 is rendered dominant. This can be done by operating on Z_t with a high-pass filter that amplifies high frequency components. Recall that a difference defined by $Z_t - Z_{t-1}, t = 2, 3, 4, \ldots$, is a high-pass filter. A second difference $Z_t - 2Z_{t-1} + Z_{t-2}, t = 3, 4, 5, \ldots$, obtained by differencing $Z_t - Z_{t-1}$, is even a more pronounced high-pass filter. By applying the second difference to Z_t, the second frequency ω_2 becomes dominant, and hence detectable from the resulting zero-crossings in the filtered data. The dominancy of ω_2 in the twice differenced data can be perceived clearly from Figure 1.1(b). The figure shows that an appreciable amount of power was shifted away from ω_1 towards ω_2.

Let Y_t be the second difference,

$$Y_t = Z_t - 2Z_{t-1} + Z_{t-2}$$

$t = 3, 4, 5, \ldots$, and consider the time series of length $N = 1000$, $Y_{51}, Y_{52}, \ldots, Y_{1050}$. Denote by D_3 the corresponding zero-crossing count. Then we observe $D_3 = 636$, and our estimate is,

$$\hat{\omega}_2 = \frac{\pi \times D_3}{N-1} = \frac{\pi \times 636}{999} = 2.00005$$

which is again, fairly precise. Apparently, the HOC pair (D_1, D_3) contains useful spectral information.

We have thus demonstrated two crucial points. First, that a dominant frequency is detectable from zero-crossings, and second, that relative dominancy among frequencies can be achieved by proper filtering.

■

Example 1.1 continued: Weighted Average

Suppose none of the frequencies in (1.1) is sufficiently dominant, where does the zero-crossing count D_1 land then? The answer is that the normalized zero-crossing count $\pi D_1/999$ tends to land somewhere between ω_1 and ω_2. To illustrate this, suppose the two amplitudes in (1.1) are equal to 1,

$$Z_t = \sin(0.5t) + \sin(2t)$$

The zero-crossing count D_1, in a series of length $N = 1000$,

$$Z_{51}, Z_{52}, \ldots, Z_{1050}$$

is equal to 448, and

$$\omega_1 < \frac{\pi \times 448}{999} = 1.40884 < \omega_2$$

This is a manifestation of the fact, to be discussed in Chapters 4 and 6, that for large N, *the normalized zero-crossing rate $\pi D_1/(N-1)$ tends to fall between the lowest and highest frequencies in the spectral support.*

■

Example 1.2: Convergence to the Highest Frequency ([9], [11])

In the previous illustration, we have seen the simplest example of the fact that the zero-crossing count can be "pushed around" by appropriate filtering. In the specific case discussed there, the zero-crossing count was pushed upward by two sequential differences. What happens to the zero-crossing count if we keep differencing the time series again and again? To answer this question, we note that the repeated differencing means the sequential application of a high-pass filter. Accordingly, by applying higher order differences sequentially to our sum of two sinusoids in (1.1), we keep enhancing and amplifying the dominancy of $\omega_2 = 2$, so that by the dominant frequency principle *all* the (normalized) zero-crossing counts from a certain point on should "land" on ω_2. This is demonstrated in Figure 1.2, which gives the graph of $\pi D_j/999$ versus j, $j = 1, 2, \ldots, 30$, where D_j is the zero-crossing count in the $(j-1)$th difference. From the figure we can see that the $\pi D_j/999$ converge to ω_2, it being the highest frequency in the spectrum. *This is an illustration of the general fact that the expected normalized HOC sequence, $\pi E[D_j]/(N-1)$, $j = 1, 2, \ldots$, from repeated differencing, always converges to the highest frequency in the spectrum regardless of its type.* This will be proved rigorously in Chapter 5.

■

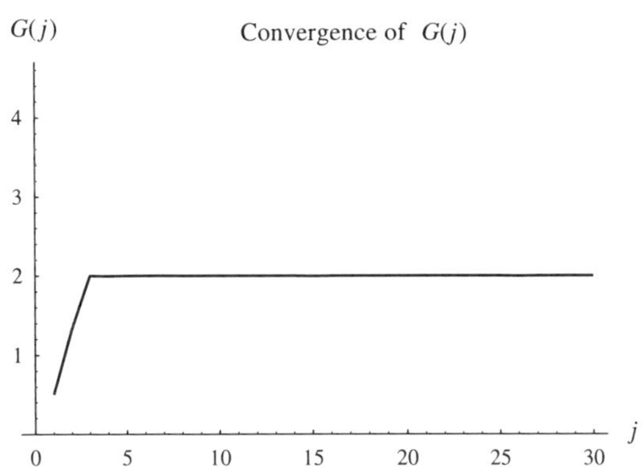

Figure 1.2: *Convergence of $G(j) = \pi D_j/999$ to the highest frequency* 2.

Example 1.3: Signal Detection ([12])

Figure 1.3 displays sections from two time series observed at instants $t = 1, 2, \ldots, 1000$. Although one of these time series is white noise, the two time series appear to be quite similar. The problem is to tell which of the two is *not* white noise. This can be revealed very quickly from HOC sequences obtained by repeated differencing.

Let D_1 be as in the previous two examples, the zero-crossing count in the unfiltered series, and let D_2, D_3, D_4, D_5, D_6 be the zero-crossing counts (HOC) corresponding to the first, second, third, fourth, and fifth differences of the time series, respectively. From series (a) we have,

$$(D_1, D_2, D_3, D_4, D_5, D_6) = (506, 679, 727, 763, 785, 799)$$

Likewise, series (b) gives,

$$(D_1, D_2, D_3, D_4, D_5, D_6) = (491, 713, 761, 799, 817, 833)$$

We can see that the two HOC sequences increase, a typical behavior of HOC from repeated differencing. The plot of D_j versus j gives a curve or a path when the points (j, D_j) are connected by line segments. From Figure 1.3 we can see that the first HOC path corresponding to series (a) (thin curve) enters certain probability limits (dark solid curves) derived under the hypothesis of white noise, while the second HOC path (dashed curve) lies partially outside the limits. From this one can conclude with a *high probability* that the first series oscillates as white noise, while the second is not quite white and in fact contains a signal, as is indeed the case. In statistical terms, the hypothesis that series (b) is white noise is rejected. Notice that *had we based our decision on the number of zero-crossings D_1 alone, the hypothesis would have erroneously been accepted.*

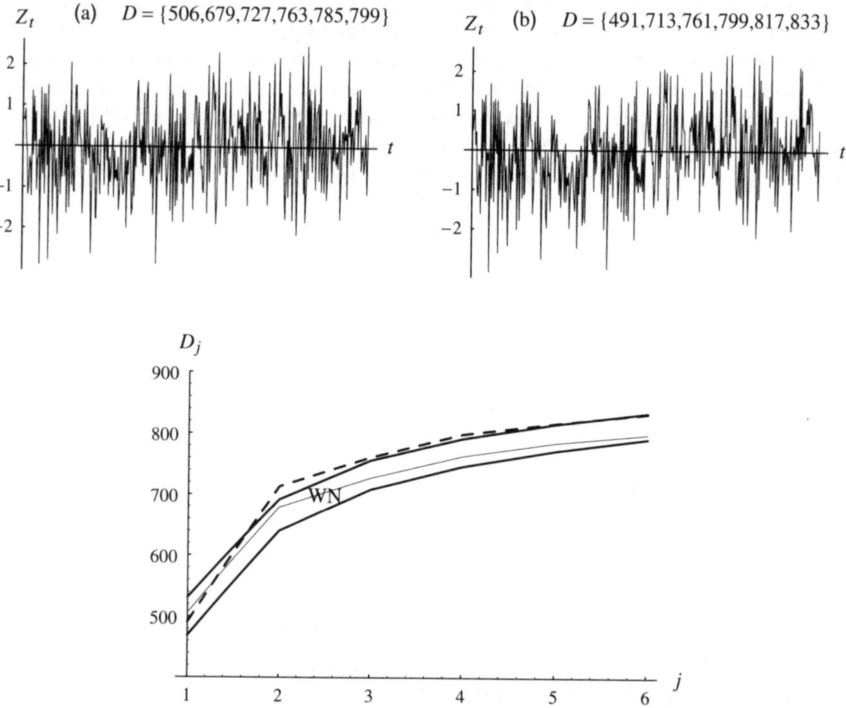

Figure 1.3: *HOC paths depicting $\{D_1, \cdots, D_6\}$ from two time series. The path corresponding to series (a) (thin curve) is within the white noise bounds (dark curves)—the hypothesis of white noise is accepted. The path from series (b) (dashed) is not completely inside the bounds—the hypothesis of white noise is rejected.*

In Chapter 8 we will show how to derive the probability limits. Similar limits can be obtained for many different types of HOC sequences produced by different families of filters.

Another useful fact is that the difference or dissimilarity between the two graphs can be expressed very economically by the HOC sequence

$$\{D_1, D_2, D_3, D_4, D_5, D_6\}$$

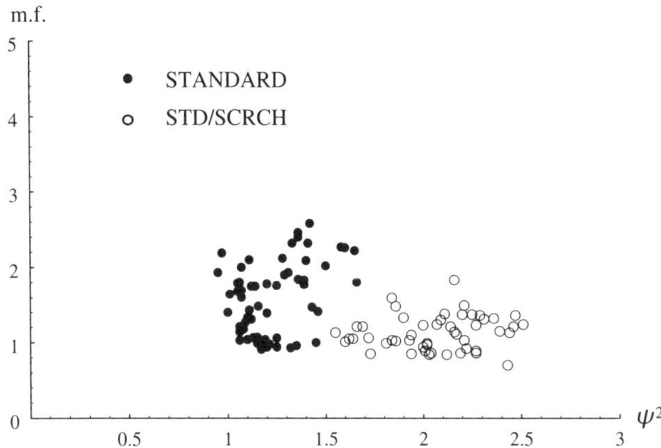

Figure 1.4: *Separation between standard (dot) and scorched (o) adhesive joints by a scatterplot of ψ^2 versus mean frequency f_4. (Source: P. Dickstein in private communication.)*

This points to the discrimination potential of HOC.

■

Example 1.4: Nondestructive Testing ([2], [3], [4])

The discrimination potential of HOC can be further illustrated by the following real engineering application. This application is also one of our most compelling examples for the use of HOC in discrimination between time series.

Various limitations of metal bonding by welding and riveting have led to the widely used method of adhesive bonding. This, however, demands the use of strictly nondestructive testing of the adhesion quality. Nondestructive testing of bonded metal adherents can be carried out by the analysis of ultrasonic echo signals. A recent effort in this direction has been directed at the study of aluminum-to-aluminum adhesive bonding [2], [3], [4]. As it turns out, a certain function (quadratic form) of HOC from repeated differencing, as defined in the previous example, plays an important role in discriminating between various kinds of adhesives. The quadratic form, called the ψ^2 statistic, will be discussed in Chapter 8. For the time being, it is instructive to look at Figure 1.4 and realize the discrimination power of ψ^2. In this application, the ψ^2 is used as a measure of distance from white noise.

The figure refers to the following experiment. Ultrasonic echo signals were obtained, using a 1 MHz ultrasonic probe, from two types of specimens of aluminum-to-aluminum adhesive bonds immersed in water [2]. One type of specimen corresponds to standard adhesion, the other to the same adhesion except that the adhesive layer was scorched by exposure to elevated temperatures. The figure shows the pairs (ψ^2, f_4) obtained from ultrasonic echos corresponding to the two types of adhesion. The values of the first component, ψ^2, are normalized. The second component, f_4, is the mean frequency of the estimated power spectrum of the ultrasonic signal. A quick look at Figure 1.4 shows that the ψ^2 is able to give almost perfect separation between the standard and scorched adhesives. The mean frequency, on the other hand, is much less potent in this case.

It is interesting to note that a similar use of the ψ^2 statistic revealed differences between different surface treatments prior to adhesion when using an ultrasonic probe operating at 1 MHz [2]. We shall return to this application again in Chapter 8.

■

Example 1.5: Frequency Detection by HOC Sequences [5], [13]

A novel feature of this book is the construction of convergent (normalized) HOC sequences applied in frequency detection and estimation in the presence of noise. The following special case is typical and gives the gist of the method: the iterative tuning of parametric filters by HOC sequences. The method itself is discussed more fully in Chapter 7.

Suppose the time series $Z_1, Z_2, \ldots, Z_N$, consists of a sinusoid (signal) with frequency ω_1 plus noise. We illustrate how to construct an expected (normalized) HOC sequence that converges to ω_1, using a certain iterative procedure.

Consider the "$\alpha - filter$"

$$\mathcal{L}_\alpha(Z)_t = Z_t + \alpha Z_{t-1} + \alpha^2 Z_{t-2} + \cdots \tag{1.2}$$

where $-1 < \alpha < 1$. For each α, $\mathcal{L}_\alpha$ defines a new filter. The number of zero-crossings in the filtered data

$$\mathcal{L}_\alpha(Z)_1, \mathcal{L}_\alpha(Z)_2, \ldots, \mathcal{L}_\alpha(Z)_N$$

is denoted by D_α, and the family $\{D_\alpha\}$, $-1 < \alpha < 1$, is a HOC family. Under certain conditions, whenever $\alpha \leq \cos(\omega_1)$, then

$$\alpha \leq \cos\left(\frac{\pi E[D_\alpha]}{N-1}\right) \leq \cos(\omega_1) \tag{1.3}$$

The important thing to observe is that $\cos\left(\pi E[D_\alpha]/(N-1)\right)$ is *closer* to $\cos(\omega_1)$ than α (and hence also $\pi E[D_\alpha]/(N-1)$ is closer to ω_1 than $\cos^{-1}(\alpha)$! See Figure 1.5.) Therefore, from (1.3), if we choose β such that

$$\beta = \cos\left(\frac{\pi E[D_\alpha]}{N-1}\right)$$

then $\beta \leq \cos(\omega_1)$, and

$$\alpha \leq \beta \leq \cos\left(\frac{\pi E[D_\beta]}{N-1}\right) \leq \cos(\omega_1)$$

We are now closer to $\cos(\omega_1)$ than before and can get even closer if we choose

$$\gamma = \cos\left(\frac{\pi E[D_\beta]}{N-1}\right)$$

because then (1.3) implies

$$\alpha \leq \beta \leq \gamma \leq \cos\left(\frac{\pi E[D_\gamma]}{N-1}\right) \leq \cos(\omega_1)$$

By this iterative process, each iteration brings us closer to $\cos(\omega_1)$ until convergence occurs. Under some conditions spelled out in Chapter 7, convergence to $\cos(\omega_1)$, and not to any other point before it, is guaranteed. It follows that the normalized expected HOC sequence,

$$\frac{\pi E[D_\alpha]}{N-1}, \frac{\pi E[D_\beta]}{N-1}, \frac{\pi E[D_\gamma]}{N-1}, \ldots$$

converges to ω_1.

What really was illustrated above is a certain *contraction mapping*—formed by a parametric filter together with the corresponding HOC family—along with its *fixed point iteration*. As it turns out, similar contraction mappings can be obtained by parametrizing—through parametric filtering—the first order autocorrelation or other quantities such as certain least squares estimates. As the reader may suspect, herein lies a possible generalization of the HOC idea! For more on this, see the discussion in Chapter 7. ■

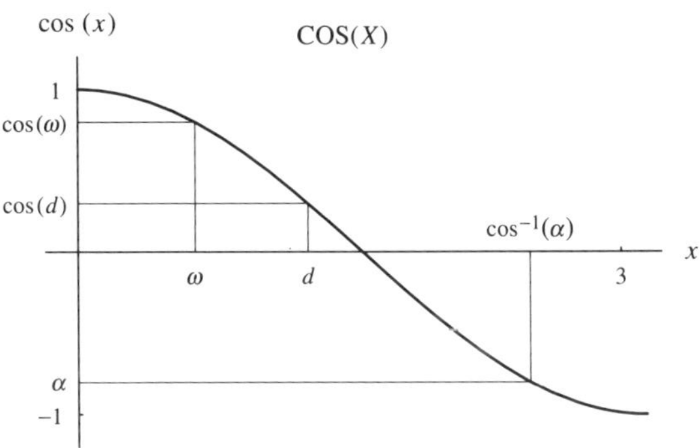

Figure 1.5: *Define* $d \equiv \pi D_\alpha/(N-1)$. *When* $\cos(d)$ *is closer to* $\cos(\omega)$ *than* α, *then* d *is closer to* ω *than* $\cos^{-1}(\alpha)$.

Example 1.6: Tracking Vocal Sound of a Humpback Whale [14]

The vocal sound of cetaceans is a subject of interest to marine biologists who consider the sound of these animals as a feature of their adaptation to aquatic existence [22]. Tracking and identification of distinctive whistle patterns of whales and porpoises is thus an important element of the research on cetaceans. The actual business of recording sound at sea presents some difficulties because of ambient sea noise due in part to machinery on board ships and to the movement of waves against bodies such as ship hulls. In what follows we apply HOC analysis to a vocal sound series uttered by a humpback whale in ambient sea noise. The series[1] was recorded 45 miles east of Boston. More information about the series and the recording device can be found in [22].

The series consists of 3 seconds worth of data sampled at the rate of 20 KHz, and recorded with 12 bit resolution in the form of integers in the range 0–4095. Because we are interested in tracking as well as identification, the data were partitioned into nonoverlapping stretches of 0.1 second containing 2048 observations each.

The data are analyzed by *HOC-grams* defined by [14],

[1]Thanks are due to K. Fristrup of Woods Hole Oceanographic Institution for the data.

$$\frac{D_\alpha}{N-1} - \frac{1}{\pi}\cos^{-1}(\alpha), \qquad \alpha \in (-1, 1)$$

for 19 α-values, $\alpha = -0.9, -0.8, \ldots, 0, 0.1, \ldots, 0.9$, and for nonoverlapping stretches of data each corresponding to 0.1 second. The HOC D_α are defined as in the previous example using the "$\alpha - filter$."

In the HOC-gram, the ordinate gives the α-values while the abscissa refers to time. The values of the HOC-gram are given in the form of grey levels. Darker levels correspond to more negative values of $(D_\alpha/N-1) - (1/\pi)\cos^{-1}(\alpha)$, while lighter levels correspond to more positive values. We can see from the HOC-gram given in Figure 1.6(a) that the entire data set contains three distinct pronounced utterances. From [22] we know that this is what was in fact expected.

It is interesting to see what a spectral-gram from the same stretches of data gives. Figure 1.6(b) gives the logspectral-gram obtained from AR-spectral estimates by fitting to each subseries of 0.1 second an AR(20) model. Here the ordinate refers to time, and the abscissa to frequency. It is possible to see some correspondence between the HOC-gram and the logspectral-gram, but the HOC-gram gives a much clearer tracking of the vocal sound series.

■

Example 1.7: Connection with the Autocorrelation ([7], [8])

The first order autocorrelation of a stationary time series is closely related to the (normalized) zero-crossing rate. Because this connection plays an important role in the book, readers will find themselves better off being acquainted with this fact as early as possible. We note that the precise relationship in general is still a challenging open problem. Fortunately enough, however, in the Gaussian case (as well as in some other cases) there is a neat compact formula that connects the first order autocorrelation with the expected zero-crossing rate. Accordingly, for a zero-mean stationary Gaussian process $\{Z_t\}$, $t = 0, \pm1, \pm2, \ldots$, we have

$$\rho_1 = \cos\left(\frac{\pi E[D_1]}{N-1}\right) \tag{1.4}$$

where ρ_1 is the first order autocorrelation and D_1 is the number of zero-crossings in a time series $Z_1, Z_2, \ldots, Z_N$. In Chapter 4 we shall study this formula and its ramifications in detail.

As an application of the formula, consider the first order autoregressive process

$$Z_t = \phi Z_{t-1} + \epsilon_t$$

where $|\phi| < 1$, and $\{\epsilon_t\}$, $t = 0, \pm1, \pm2, \ldots$, is Gaussian white noise. Then $\{Z_t\}$ is a zero-mean stationary Gaussian process whose first order autocorrelation is the parameter ϕ itself. Then the formula suggests

$$\hat{\phi} = \hat{\phi}_N = \cos\left(\frac{\pi D_1}{N-1}\right) \tag{1.5}$$

as an estimate of ϕ. See [8, p. 50]. The statistical properties of this estimate can be studied with the help of the theory presented in Chapter 6.

■

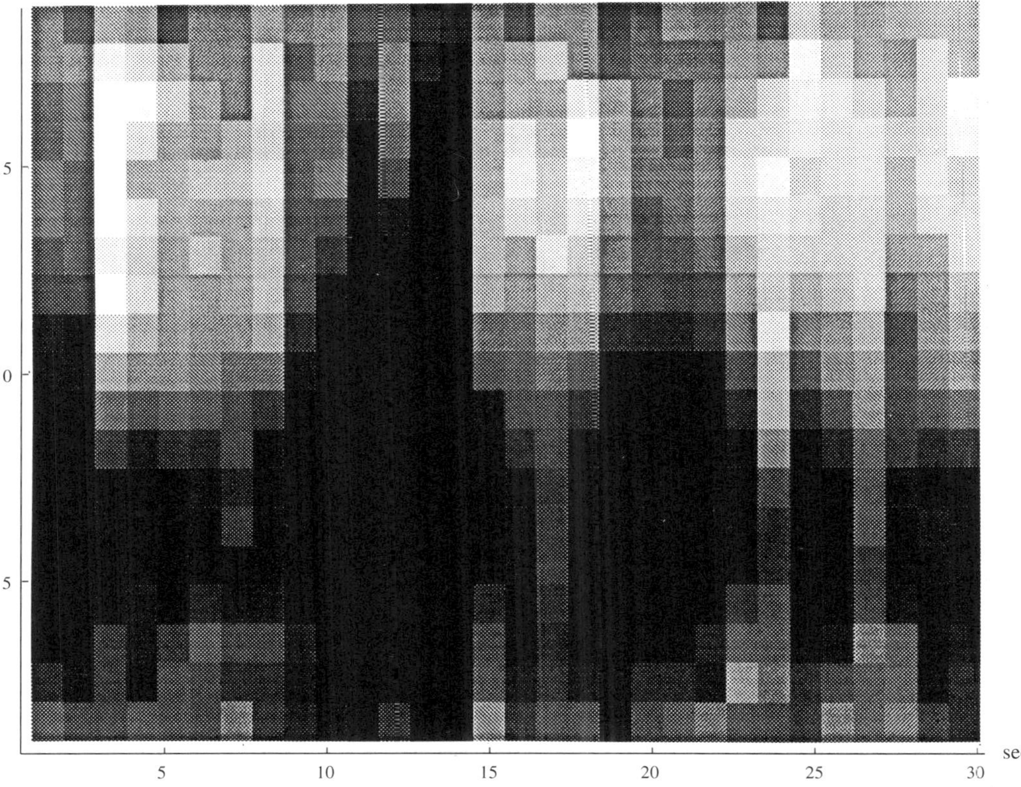

Figure 1.6(a): *HOC-gram from the vocal sound of a humpback whale.*

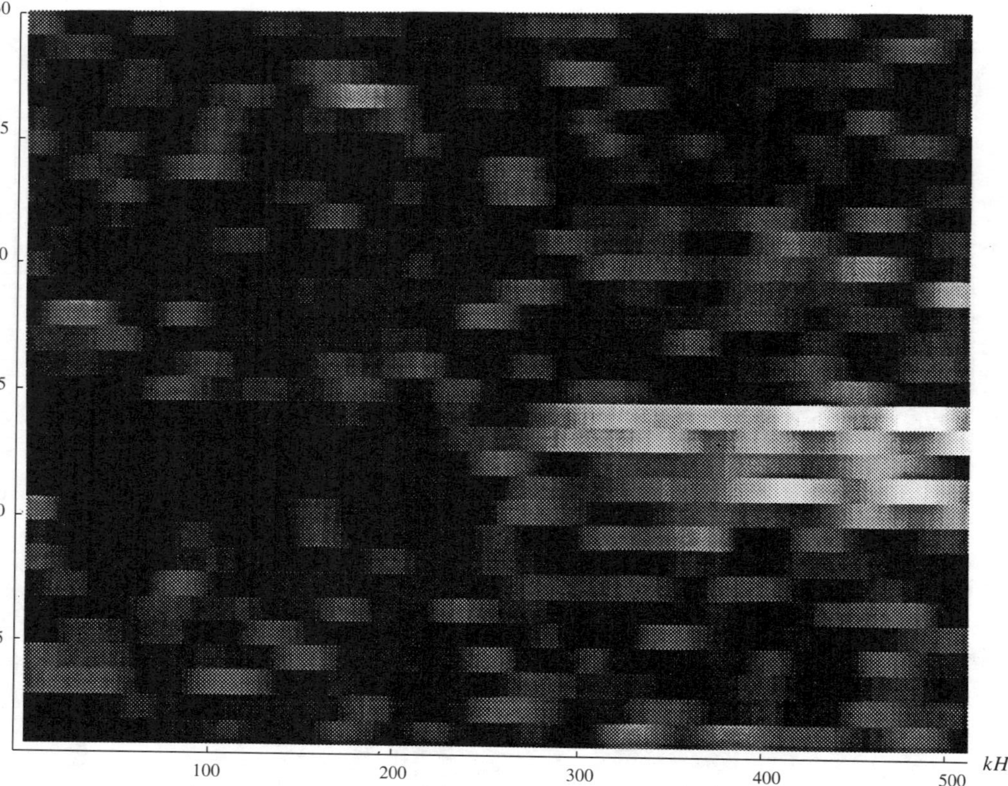

Figure 1.6(b): *The corresponding logspectral-gram.*

Example 1.8: Likelihood Considerations ([7], [8])

It is interesting to note that the expression for $\hat{\phi}$ in (1.5) can also be obtained from the following argument concerning clipped data [7]. Let $X_t = 1$ when $Z_t \geq 0$, and 0 otherwise. Then a zero-crossing in Z_t corresponds to a symbol change in X_t, and D_1 can be defined as the number of symbol changes in the binary time series $X_1, X_2, \ldots, X_N$. Now define

$$\lambda_1 \equiv P(X_t = 1 | X_{t-1} = 1)$$

Then our formula (1.4) takes the form (why?),

$$\phi = -\cos(\pi\lambda_1)$$

Thus, any estimate for λ_1 gives automatically also an estimate for ϕ. However, λ_1 can now be estimated from the binary clipped data. A way to approach the problem is to see what a $0-1$ Markov chain with parameters $(p = 1/2, \lambda_1)$ would *suggest* as an estimate for λ_1. This procedure allows us to bring in maximum likelihood considerations that can lead to the *construction* of useful estimates, even though our clipped binary process $\{X_t\}$ is *not* Markov [7], [23]. Define: $s = \sum_{t=1}^{N} x_t$, $r = \sum_{t=2}^{N} x_t x_{t-1}$, and $h = x_1 + x_N$. For a binary Markov chain, the log-likelihood function of λ_1 is given by

$$\log L(\lambda_1) = [2r - 2s + h + (N-1)]\log(\lambda_1) + [2s - 2r - h]\log(1-\lambda_1) - \log(2)$$

and this is maximized for

$$\hat{\lambda}_1 = \frac{2r - 2s + h + (N-1)}{N-1}$$

A closer look reveals, however, that $2s - 2r - h$ is precisely the *number of symbol changes* in $X_1, X_2, \ldots, X_N$, so that in fact

$$\hat{\lambda}_1 = 1 - \frac{D_1}{N-1}$$

and our estimate for ϕ becomes

$$-\cos\left(\pi - \frac{\pi D_1}{N-1}\right) = \cos\left(\frac{\pi D_1}{N-1}\right)$$

which is exactly what we had before. Once an estimate has been constructed by some method we can study its properties rigorously. In the present case $\{Z_t\}$ has a continuous spectrum in addition to being stationary and Gaussian, and so the process is ergodic as well. It follows that,

$$\lim_{N\to\infty} \hat{\phi} = \phi$$

with probability one.

The preceding argument brings to light the importance of the notion of likelihood and maximum likelihood estimation in time series analysis. In general, to obtain the likelihood, it is necessary to know or make assumptions concerning the joint probability distribution of the data. For stationary binary data this can be achieved by assuming a specific dependence structure such as Markov dependence as was done above. In recent years the notion of likelihood has been generalized and extended (or shall we say stretched) in various ways. For a certain type of *conditional inference* it is sufficient to model the conditional probability law of an observation given the past data and past auxiliary information in order to execute a "pseudolikelihood" analysis, and this carries over to nonstationary binary data as well. We shall see more of this in Chapter 9 where we discuss *partial likelihood* inference in the prediction of level-crossings.

■

Example 1.9: Spectral Moments and HOC in Continuous Time

There is a wide literature on random vibration of mechanical and physical systems in which the notion of HOC from repeated stochastic differentiation is essentially synonymous with spectral moments of stationary Gaussian processes. To understand why this is so, we turn to some of the more traditional level- and zero-crossing results concerning the expected zero-crossing rate of a zero-mean continuous-time stationary Gaussian process. The results are credited to Rice [18].

First we establish some moment relations. Suppose $\{Z(t)\}$, $-\infty < t < \infty$, is a real-valued zero-mean stationary process, not necessarily Gaussian, with autocovariance $R(\tau)$ and spectral distribution function $F(\omega)$ (Chapter 3). Denote by $Z^{(k)}(t)$ the kth derivative (in mean square) of $Z(t)$, and by $R_{Z^{(k)}}(\tau)$ the corresponding autocovariance. Under the assumption that all the relevant spectral moments exist,

$$R_{Z^{(k)}}(\tau) = (-1)^k R^{(2k)}(\tau)$$

from which we derive,

$$\mathrm{Var}[Z^{(k)}(t)] = R_{Z^{(k)}}(0) = (-1)^k R^{(2k)}(0) = \int_{-\infty}^{\infty} \omega^{2k}\, dF(\omega) \tag{1.6}$$

where the last integral is referred to as the $2k$th spectral moment of $F(\omega)$. Note that because of the symmetry of $dF(\omega)$, all the odd spectral moments are equal to 0.

Let $p(z, \dot{z})$ be the joint probability density of $(Z(t), Z'(t))$, and u a fixed level. From Problem 11 in Chapter 4 we know that the expected level-crossing rate, that is the expected number of u-level crossings (upcrossings and downcrossings of level u) per unit time, is given by the general formula,

$$\begin{aligned} E(\text{Level-Crossing Rate}) &= \int_{-\infty}^{\infty} |\dot{z}| p(u, \dot{z})\, d\dot{z} \\ &= p_Z(u) E[|Z'(t)| \,\big|\, Z(t) = u] \end{aligned} \tag{1.7}$$

where $p_Z(u)$ is the probability density of $Z(t)$ evaluated at u, and $E[|Z'(t)| \,\big|\, Z(t) = u]$ is the conditional expectation of $|Z'(t)|$ given that $Z(t) = u$. In particular, when $\{Z(t)\}$ is also Gaussian, $Z(t)$ and $Z'(t)$ are *independent*, and upon noting that $Z'(t) \sim \mathcal{N}(0, -R''(0))$, the general formula (1.7) reduces to

$$\begin{aligned} E(\text{Level-Crossing Rate}) &= p_Z(u) E[|Z'(t)|] \\ &= \frac{1}{\pi}\sqrt{-\frac{R''(0)}{R(0)}} \exp\left\{\frac{-u^2}{2R(0)}\right\} \\ &= \frac{1}{\pi}\sqrt{\frac{\mathrm{Var}[Z'(t)]}{\mathrm{Var}[Z(t)]}} \exp\left\{\frac{-u^2}{2\mathrm{Var}[Z(t)]}\right\} \end{aligned} \tag{1.8}$$

This is Rice's famous formula obtained in his well-known 1944 article [18].

Now let $D_1(T)$ be the number of zero-crossings (i.e., $u = 0$) by $\{Z(t)\}$ in the time interval $(0, T]$. Then from (1.7),

$$E(\text{Zero-Crossing Rate}) = \frac{E[D_1(T)]}{T} = p_Z(0)E[|Z'(t)| \,\big|\, Z(t) = 0] \tag{1.9}$$

By substituting $u = 0$ in Rice's formula (1.8), this reduces in the Gaussian case to $p_Z(0)E$ $[|Z'(t)|]$, or

$$\frac{E[D_1(T)]}{T} = \frac{1}{\pi}\sqrt{-\frac{R''(0)}{R(0)}} = \frac{1}{\pi}\sqrt{\frac{\text{Var}[Z'(t)]}{\text{Var}[Z(t)]}} = \frac{1}{\pi}\left\{\frac{\int_{-\infty}^{\infty}\omega^2\, dF(\omega)}{\int_{-\infty}^{\infty} dF(\omega)}\right\}^{1/2} \tag{1.10}$$

Denote by $D_2(T)$ the number of zero-crossings by $\{Z'(t)\}$ in $(0, T]$, then obviously $D_2(T)$ is also the number of extrema (minima and maxima) observed in $\{Z(t)\}$ in $(0, T]$, and so from (1.6) and (1.10), the expected number of peaks and troughs per unit time is given by

$$\frac{E[D_2(T)]}{T} = \frac{1}{\pi}\sqrt{-\frac{R^{(4)}(0)}{R''(0)}} = \frac{1}{\pi}\sqrt{\frac{\text{Var}[Z''(t)]}{\text{Var}[Z'(t)]}} \tag{1.11}$$

In general, if we let $D_k(T)$ be the number of zero-crossings by the $(k-1)$th derivative $\{Z^{(k-1)}(t)\}$ in $(0, T]$, then the Gaussian assumption gives

$$\frac{E[D_{k+1}(T)]}{T} = \frac{1}{\pi}\sqrt{-\frac{R^{(2k+2)}(0)}{R^{(2k)}(0)}} = \frac{1}{\pi}\sqrt{\frac{\text{Var}[Z^{(k+1)}(t)]}{\text{Var}[Z^{(k)}(t)]}} \tag{1.12}$$

or, in terms of spectral moments, the expected simple HOC from repeated differentiation is given by

$$E[D_{k+1}(T)] = \frac{T}{\pi}\left\{\frac{\int_{-\infty}^{\infty}\omega^{2k+2}\, dF(\omega)}{\int_{-\infty}^{\infty}\omega^{2k}\, dF(\omega)}\right\}^{1/2} \tag{1.13}$$

We point out that (1.10) is the continuous time analog of (1.4) and that the two are quite close, a fact revealed by a Taylor series expansion of $\cos(x)$ [10]. In Chapter 4 we shall present certain concrete extensions of Rice's formula by considering monotone functions of Gaussian processes.

An immediate consequence of (1.13), is an expression for the normalized $2k$th spectral moment in terms of a product of expected HOC,

$$\begin{aligned} E[D_1(T)]\cdots E[D_K(T)] &= \left[\frac{T}{\pi}\right]^K \left\{\frac{\text{Var}[Z^{(K)}(t)]}{\text{Var}[Z(t)]}\right\}^{1/2} \\ &= \left[\frac{T}{\pi}\right]^K \left\{\frac{\int_{-\infty}^{\infty}\omega^{2K}\, dF(\omega)}{\int_{-\infty}^{\infty} dF(\omega)}\right\}^{1/2} \end{aligned} \tag{1.14}$$

Unfortunately, no such simple formula exists for $E[D_1(T)\cdots D_K(T)]$.

As a typical application of these zero-crossing formulas, we consider fluctuating random stress of mechanical systems, $\{Z(t)\}, 0 < t < \infty$, modeled as a stationary zero-mean

Gaussian process. A simple parameter describing the fatigue damage due to random loading is the average rise and fall per unit time defined as $E[|Z'(t)|]$, divided by the expected number of extrema per unit time $E[D_2(T)/T]$ [19], [20]. Accordingly, from (1.9) and (1.11) we have,

$$\bar{h} \equiv \frac{E[|Z'(t)|]}{E[D_2(T)/T]} = \frac{1}{p_Z(0)} \frac{E[D_1(T)]}{E[D_2(T)]} = -R''(0)\sqrt{\frac{2\pi}{R^{(4)}(0)}} \leq \sqrt{2\pi R(0)}$$

and equality is approached for narrow band processes.

Another application is from the field of seismology. An analysis of earthquake records made in [6] shows that the number of zero-crossings and extrema in a given interval are useful quantities in modeling the stationary component of earthquake accelerograms. Ground acceleration $X(t)$ is modeled as the product

$$X(t) = V(t) \cdot Z(t)$$

where $V(t)$ is a modulating deterministic function, and $Z(t)$ a stationary Gaussian process with mean zero and unit variance, and spectral density

$$f(\omega) = A\exp(-\omega^2c^2) + B\omega^2\exp(-4\omega^2c^2)$$

where c is a positive parameter. Then, since $\text{Var}[Z(t)] = 1$,

$$\int_{-\infty}^{\infty} f(\omega)\, d\omega = 1$$

A second equation is obtained from (1.10),

$$\frac{1}{\pi}\left\{\int_{-\infty}^{\infty} \omega^2 f(\omega)\, d\omega\right\}^{1/2} = \frac{E[D_1(T)]}{T}$$

Integration with respect to $f(\omega)$—done easily by recognizing that $f(\omega)$ is a "mixture" of two probability densities—yields two equations,

$$\frac{\sqrt{\pi}}{c}\left[A + \frac{B}{16c^2}\right] = 1$$

$$\frac{\sqrt{\pi}}{2c^3}\left[A + \frac{3B}{64c^2}\right] = \pi^2\left[\frac{E[D_1(T)]}{T}\right]^2$$

Solving for A, B, we get

$$A = \frac{c}{\sqrt{\pi}}\left[8c^2\pi^2\left(\frac{E[D_1(T)]}{T}\right)^2 - 3\right]$$

$$B = \frac{64c^3}{\sqrt{\pi}}\left[1 - 2c^2\pi^2\left(\frac{E[D_1(T)]}{T}\right)^2\right]$$

This procedure, which suggests a way for parametrizing a spectral density by the expected zero-crossing rate, was found adequate for observed earthquake data in [6]. Note that c must be chosen so that the spectral density is non-negative.

Recall that (1.10) was derived under the Gaussian assumption. In applications, however, the Gaussian assumption is not guaranteed and the common wisdom is to use (1.10) as a *definition* of the expected zero-crossing rate in the hope that the true expected rate is sufficiently close. An example of this practice is an application to ground motion produced by earthquakes [17]. Let $A(f,r)$ be the displacement amplitude spectrum for the main ground motion at frequency $f = \omega/2\pi$ and epicentral distance r within a time period of duration T. It is assumed that within this time period the motion constitutes a stationary random process. Let

$$m_k = \frac{1}{\pi}\int_0^\infty \omega^k |A(f,r)|^2 \, d\omega$$

and *define* the expected number of zero-crossings by

$$N \equiv \frac{T}{\pi}\sqrt{\frac{m_2}{m_0}}$$

Then the mean peak ground motion is approximated by [17]

$$a_0 = \sqrt{\frac{m_0}{T}}\left[\sqrt{2\ln N} + \frac{0.57722}{\sqrt{2\ln N}}\right]$$

■

Example 1.10: The Distribution of Peak Magnitude

In the literature on random vibration of mechanical structures an important problem is to determine the probability distribution of peak load where the load is viewed as a stationary random process in continuous time. It turns out that the previous development contains the desired answer under the restriction that the process is a narrow band stationary Gaussian process with mean zero. When the band is sufficiently narrow, the narrow band assumption ensures that the peaks are nonnegative, and that the expected upcrossing rate of level zero is equal to the peak rate per unit time. Conversely, when the expected upcrossing rate of level zero is equal to the peak rate per unit time, the Gaussian assumption guarantees a narrow band. A precise mathematical statement of this is Corollary 4.2 and the ensuing discussion in Chapter 4.

In deriving the peak magnitude distribution we follow a well-known standard argument (e.g., see [25, p. 209]). Let $\{Z(t)\}$, $-\infty < t < \infty$, be a real-valued zero-mean stationary Gaussian process, and denote by $N_u \equiv N_u(0,T]$ the number of upcrossings of level u by $\{Z(t)\}$ in the time interval $(0,T]$. Then from (1.8)—we divide by 2—the expected upcrossing rate per unit time of level u is given by

$$\frac{E[N_u]}{T} = \frac{1}{2\pi}\sqrt{\frac{\text{Var}[Z'(t)]}{\text{Var}[Z(t)]}}\exp\left\{\frac{-u^2}{2\text{Var}[Z(t)]}\right\} \tag{1.15}$$

Let X be the random peak magnitude, and $p(x)$ its probability density. Our problem is to determine $p(x)$ from (1.15) under the narrow band assumption.

The trick is to note that,

$$\begin{aligned} p(x)\,dx &\approx \frac{\text{Expected rate per unit time of peaks lying in } (x, x+dx)}{\text{Expected rate per unit time of peaks}} \\ &\approx \frac{(N_x - N_{x+dx})/T}{N_0/T} \approx \frac{-dN_x/T}{N_0/T} \end{aligned}$$

or

$$p(x)dx = \frac{x}{\text{Var}[Z(t)]} \exp\left\{\frac{-x^2}{2\text{Var}[Z(t)]}\right\} dx$$

Since for narrow band X is nonnegative the complete statement is

$$p(x) = \begin{cases} \frac{x}{\text{Var}[Z(t)]} \exp\left\{\frac{-x^2}{2\text{Var}[Z(t)]}\right\}, & x \geq 0 \\ 0, & x < 0 \end{cases} \tag{1.16}$$

which we recognize as the Rayleigh probability density function shown in Figure 1.7 for $\text{Var}[Z(t)] = 1$.

■

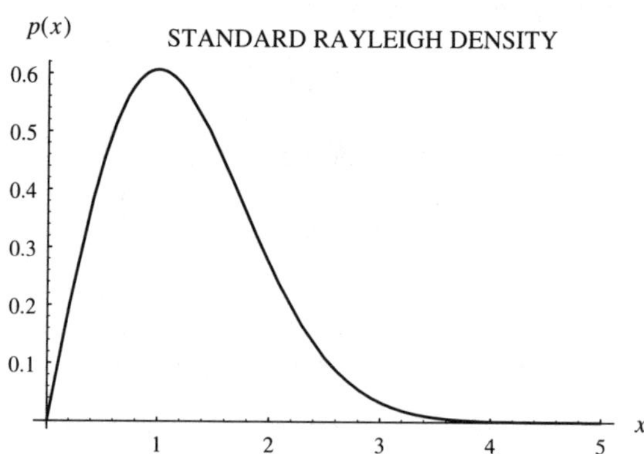

Figure 1.7: *Graph of the standard Rayleigh probability density.*

In the rest of the book, the ideas and concepts illustrated in the preceding examples will be studied in detail, and further developed by means of theory and quite a few additional examples. This program is carried out within the rich theory of stationary stochastic processes. This provides a convenient framework that leads to a natural connection between HOC and familiar concepts and tools such as the spectrum and autocorrelation, entities that are indigenous to the theory of stationary processes. However, the notion of HOC is general enough, and need not be tied down to stationary processes.

1.1 PROBLEMS AND COMPLEMENTS

1. Consider the time series

$$Z_t = \cos(t) + 2\cos(2t), \qquad t = 1, 2, \cdots, 1000$$

   (a) Let D_1 be the number of zero-crossings observed in the series. Provide lower and upper bounds for $\pi D_1/999$.

   (b) Suppose the component $5\cos(0.8t)$ is added to Z_t. Where then will $\pi D_1/999$ tend to land?

   (c) Generate artificial data to verify your answers in (a) and (b).

2. For a time series $\{Z_t\}$, $t = 0, \pm 1, \pm 2, \ldots$, the (backward) difference operator $\bigtriangledown$ is defined by

$$\bigtriangledown Z_t \equiv Z_t - Z_{t-1}$$

and $\bigtriangledown^k = \bigtriangledown(\bigtriangledown^{k-1})$. Show that

$$\bigtriangledown^k Z_t = \sum_{j=0}^{k} \binom{k}{j} (-1)^j Z_{t-j}$$

(Hint: $\bigtriangledown = 1 - \mathcal{B}$, where $\mathcal{B} Z_t = Z_{t-1}$.)

3. Consider the nth degree polynomial $Q(t) = a_0 + a_1 t + a_2 t^2 + \cdots + a_n t^n$. Show that for $k \geq n + 1$,

$$\bigtriangledown^k Q(t) = 0$$

For most time series, the kth difference only enhances the observed oscillation. Hence conclude that a polynomial oscillates in a fundamentally different way than most oscillatory time series observed in nature.

4. Let $X_1, X_2, \ldots, X_N$ be a stationary binary (0-1) time series such that $P(X_t = 1) = P(X_t = 0) = 1/2$, and $\lambda_1 = P(X_t = 1 | X_{t-1} = 1)$. Define the statistics $S = \sum_{t=1}^{N} X_t$, $R = \sum_{t=2}^{N} X_t X_{t-1}$, and $H = X_1 + X_N$.

(a) Verify that the number of symbol changes, D_1, in the binary series is given by

$$D_1 = 2S - 2R - H$$

(b) Show that $\text{Cov}(|X_t - X_{t-1}|, X_t) = 0$.

(c) If in addition $\{X_t\}$ is a Markov chain, show that [7]

$$\lim_{N \to \infty} N \times \text{Var}\left(\frac{S}{N}\right) = \frac{\lambda_1}{4(1 - \lambda_1)}$$

and

$$\lim_{N \to \infty} (N-1) \times \text{Var}\left(\frac{R}{N-1}\right) = \frac{\lambda_1[(1 - \lambda_1)(2 - \lambda_1) + \lambda_1]}{4(1 - \lambda_1)}$$

5. *Inference for periodically correlated sequences [15].* A process $\{Z_t\}$, $t = 0, \pm 1, \pm 2, \ldots$, is called periodically correlated (PC) with minimal period T, if T is the smallest positive integer such that

$$E[Z_u Z_v] \equiv R(u, v) = R(u + T, v + T)$$

$$E[Z_u] \equiv m(u) = m(u + T)$$

Let $X_t = 0.3 X_{t-1} + \epsilon_t$, $t = 0, \pm 1, \pm 2, \ldots$, where $\{\epsilon_t\}$ is white Gaussian noise, and define

$$Z_t = \cos\left(\frac{2\pi t}{10}\right) \times X_t$$

(a) Show that $\{Z_t\}$ is PC with period $T = 5$.

(b) Let

$$s(j) \equiv \frac{1}{N-j} \sum_{k=1}^{N-j} |Z_k - Z_{k+j}|$$

$j = 0, 1, 2, \ldots, N-1$, and $\overline{s} \equiv (1/N) \sum_{j=0}^{N-1} s(j)$. Argue that T can be estimated from the zero-crossings of the sequence

$$s^*(j) \equiv s(j) - \overline{s}, \qquad j = 0, 1, 2, \ldots, M$$

where $M << N-1$. See Figure 1.8.

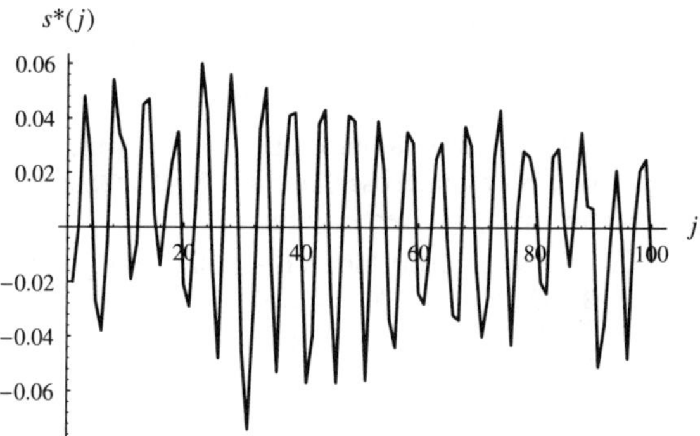

Figure 1.8: *Graph of* $s^*(j)$. *The period* T *is estimated from* $2 \times 100/($*Number zero-crossings by* $s^*(j))$.

6. *Properties of the autocovariance (e.g., see [21]).* Suppose $\{X(t)\}$, $-\infty < t < \infty$, is a real-valued zero-mean stationary process with autocovariance $R(\tau) = E[X(t)X(t+\tau)] = E[X(t-\tau)X(t)] = R(-\tau)$. Use formal differentiation with respect to τ inside the expectation to show that

(a) $R'(\tau) = E[X(t)X'(t+\tau)] = -E[X'(t-\tau)X(t)]$. Hence, by putting $\tau = 0$, conclude that $R'(0) = E[X(t)X'(t)] = -E[X'(t)X(t)] = 0$.

(b) $R''(\tau) = -E[X'(t-\tau)X'(t)]$. Hence, $R''(0) = -E[(X'(t))^2]$.

(c) $R'''(0) = 0$.

7. Let X be a normal random variable with mean 0 and variance σ^2. Show that $E[|X|] = \sqrt{2/\pi}\sigma$.

REFERENCES

[1] Becker, A. and R.S. Mackay, "Maximum frequency indication: Minima counting versus zero-crossing counting," *IEEE Trans. Biomed. Eng.*, Vol. BME-29, No. 3, pp. 213–215, Mar. 1982.

[2] Dickstein, P.,"Spectral analysis of ultrasound from multi-layered media," Ph.D. Thesis, Department of Nuclear Engineering, Technion-Israel Institute of Technology, Haifa, Jan. 1988.

[3] Dickstein, P., Y. Segal, E. Segal, and A.N. Sinclair, "Statistical pattern recognition techniques: A sample problem of ultrasonic determination of interfacial weakness in adhesive joints," *J. Nondestr. Eval.*, Vol. 8, pp. 27–35, 1989.

[4] Dickstein, P.A., S. Girshovich, Y. Sternberg, A.N. Sinclair, and H. Leibovich, "Ultrasonic feature-based classification of the interfacial condition in composite adhesive joints," *J. Res. Nondestr. Eval.*, Vol. 2, No. 4, pp. 207–224, 1990.

[5] He, S. and B. Kedem, "Higher order crossings of an almost periodic random sequence in noise," *IEEE Trans. Inform. Theory*, Vol. IT-35, pp. 360–370, Mar. 1989.

[6] Iyengar, R.N. and K.T.S.R. Iyengar, "A nonstationary random process model for earthquakes accelerograms," *Bull. Seismol. Soc. Am.*, Vol. 59, No. 3, pp. 1163–1188, June 1969.

[7] Kedem, B., "Exact maximum likelihood estimation of the parameter in the AR(1) process after hard limiting," *IEEE Trans. Inform. Theory*, Vol. IT-22, No. 4, pp. 491–493, July 1976.

[8] Kedem, B., *Binary Time Series*, New York: Dekker, 1980.

[9] Kedem, B., "Search for periodicities by axis-crossings of filtered time series," *Signal Process.*, Vol. 10, No. 2, pp. 129–144, March 1986.

[10] Kedem, B., "Spectral analysis and discrimination by zero-crossings," *Proceedings of the IEEE*, Vol. 74, No. 11, pp. 1477–1493, Nov. 1986.

[11] Kedem, B., "Detection of periodicities by higher order crossings," *J. Time Ser. Anal.*, Vol. 8, No. 1, pp. 39–50, 1987.

[12] Kedem, B., "A fast graphical goodness of fit test for time series models" in *Applied Probability, Stochastic Processes, and Sampling Theory*, I.B. MacNeill and G.J. Umphry, eds., Reidel: Dordrecht, the Netherlands, pp. 65–76, 1987.

[13] Kedem, B., "Contraction mappings in mixed spectrum estimation," in *New Directions in Time Series Analysis*, Part I, D. Brillinger et al. eds., pp. 169–191, New York: Springer-Verlag, 1992.

[14] Kedem, B. and T. Li, "Higher order crossings from a parametric family of linear filters," Tech. Rep. TR-89-47, Mathematics Dept., Univ. of Maryland, College Park, 1989.

[15] Martin, E.K.D. and B. Kedem, "Estimation of the Period of Periodically Correlated Sequences", *J. Time Ser. Anal.*, Vol. 14, No. 2, pp. 193–205, 1993.

[16] Offner, F.F., "Comments on *Maximum frequency indication: Minima counting versus zero-crossing counting, IEEE Trans. Biomed. Eng.*, Vol. BME-30, No. 1, p. 78, Jan. 1983.

[17] Ou, G.-B. and R.B. Herrmann, "A statistical model for ground motion produced by earthquakes at local and regional distances," *Bull. Seismol. Soc. Am.*, Vol. 80, pp. 1397–1417, Dec. 1990.

[18] Rice, S.O., "Mathematical analysis of random noise," *Bell Syst. Tech. J.*, Vol. 23, pp. 282–332, 1944, and Vol. 24, pp. 46–156, 1945, reprinted in *Selected Papers on Noise and Stochastic Processes*," N. Wax, ed., New York: Dover, 1954.

[19] Rice, J.R. and F.P. Beer, "On the distribution of rises and falls in a continuous random process," *J. Basic Eng., Trans. ASME*, Series D, Vol. 87, pp. 398–404, June 1965.

[20] Rice, J.R., F.P. Beer, and P.C. Paris, "On the prediction of some random loading characteristics relevant to fatigue," in *Acoustical Fatigue in Aerospace Structures*, Proc. Second Int. Conf., Dayton, Ohio, Apr. 29–May 1, 1964, W.J. Trapp and D.M. Forney, Jr., eds., Syracuse, N.Y.: Syracuse Univ. Press, pp. 121-144, 1965.

[21] Robson, J.D., *An Introduction to Random Vibration*, Edinburgh: Edinburgh Univ. Press, 1964.

[22] Schevill, W.E. and W.A. Watkins, *Whale and Porpoise Voices*, Woods Hole Oceanographic Institution, Woods Hole, Mass., 1962.

[23] Slud, E., "Clipped Gaussian processes are never m-step Markov," *J. Multivariate Anal.*, Vol. 29, pp. 1–14, 1989.

[24] Slutsky, E.E., "The summation of random causes as the source of cyclic processes," *Prob. Econ. Cond.*, Vol. 3, No. 1, 1927 (in Russian). English translation in *Econometrica*, Vol. 5, pp. 105–146, Apr. 1937.

[25] Yang, C.Y., *Random Vibration of Structures*, New York: Wiley-Interscience, 1986.

2

Fundamentals of Probability Theory

In this chapter we review some fundamental concepts and notions from the theory of probability that are essential for the rest of the book. One of the goals of this review is to go over some particular facts that are peculiar to the ideas expressed in later chapters. Thus, for example, we emphasize the notion of a mixed distribution because it will help in understanding the meaning of a mixed spectrum, and we discuss certain stochastic ordering because it helps to explain the action of high-pass and low-pass filters in a neat and novel way. Another special feature that we discuss in this chapter is a certain orthant probability formula that is of fundamental importance from our point of view because under some conditions it is instrumental in obtaining a spectral representation for the expected number of zero-crossings. Also, we discuss the notion of correlation between random variables and make it a point to emphasize a certain aspect of it regarding the general appearance of time series. Detailed introductory accounts on the subject of probability with emphasis on applications can be found in [14], [16], and [17].

2.1 RANDOM VARIABLES

A time series is a family of random variables, most often ordered in time, so that it is appropriate to begin with a brief description of random variables and their different types.

Examples of random variables are numerous and can be found in nearly every scientific and technological endeavor. Any physical measurement corrupted by noise, the value of a noisy signature at time t, the intensity of a speech sound uttered by an individual, the amount of rainfall at a certain location during the month of August, the number of pedestrians on a section of a certain sidewalk at noon, foot size of an individual, are all examples of random variables. The term "random" refers to our inability to predict the values of these variables with certainty. Thus in the case of the pedestrians, their number at noon could be either 0 or 1 or 2, ..., but beyond this sure event any prediction is uncertain. The real interest in random variables lies in their *values* and in certain functions defined on these values such as *distribution functions* and *density functions*. The set of all possible values of a random variable is called the *sample space* associated with the random variable. Random variables will be denoted by capital letters such as X, Y, Z, and their values by lower case letters such as x, y, z, respectively.

Random variables are classified by their values. When the possible values of a random variable are distinct values of the form $x_1, x_2, x_3, \ldots$, the random variable is called *discrete*. When the set of possible values is an interval or a union of intervals on the real line, the random variable is called *continuous*. In this case the values are admitted on a continuum. There are cases, however, when the random variable admits distinct values with positive probability and also values on a continuum. Such a random variable is said to be of a *mixed type*. A more precise definition of these three types of random variables will be given shortly. For the time being, we note that in the pedestrians example the random variable is discrete, foot size is a continuous random variable, but the amount of rainfall in August is a mixed random variable because this random variable may admit the value 0, when it does not rain at all during August, with positive probability. Otherwise the amount of rain is observed on a continuum.

2.2 THE DISTRIBUTION FUNCTION

The probabilistic behavior of a random variable is determined by its distribution function. A clear understanding of the meaning of a distribution function of a random variable will prove useful later when we introduce the spectral distribution function of a stationary process.

If A is an event, the probability of A is denoted by $P(A)$. Recall that the probability of any event is between 0 and 1, the probability of a sure event is 1, and if A, B are two events that cannot occur together, then

$$P(A \text{ or } B) = P(A) + P(B)$$

Clearly, if A^c is the complement of the event A, then either A or A^c, but not both, must occur so that $P(A) + P(A^c) = 1$, and this implies that the probability of an event that cannot occur is 0.

Consider a random variable X and the event that the value of X is less than or equal to x, $\{X \le x\}$. The probability of this event is denoted by $P(X \le x)$, and we define a function $F(x)$ by the equation

$$F(x) = P(X \le x)$$

for all real x. $F(x)$ is called the *cumulative distribution function*, or simply the *distribution function* of X. The distribution function has some interesting properties, but its most characteristic property is that it is monotone increasing for all real x. In general the distribution function is not continuous, but it is always continuous from the right, a fact that is signified by the equation $F(x) = F(x+)$. To show that $F(x)$ is monotone, note that for $x \leq y$, the events $\{X \leq x\}$ and $\{x < X \leq y\}$ cannot occur together, and by the axioms of probability

$$P(X \leq y) = P(X \leq x) + P(x < X \leq y)$$

Therefore, by the definition of $F(x)$, we obtain, after rearranging terms,

$$F(y) - F(x) = P(x < X \leq y) \geq 0$$

and we see that $F(x)$ is a *nondecreasing function*. The right continuity property of $F(x)$ can also be obtained from the last equation. Because probability is always between 0 and 1, and assigns the value 0 to an impossible event and the value 1 to a sure event, $F(x)$ is sandwiched between 0 and 1, approaching 0 when $x \to -\infty$ and approaching 1 when $x \to \infty$, respectively. Thus we have

1. $0 \leq F(x) \leq 1$
2. $\lim_{x \to -\infty} F(x) = 0$
3. $\lim_{x \to \infty} F(x) = 1$
4. $x \leq y \Rightarrow F(x) \leq F(y)$
5. $F(x) = F(x+)$

Properties 1–5 are some of the basic properties of $F(x)$. However, what is well known in addition but not always emphasized is the fact that F can be expressed as a sum of two[1] components, a "continuous" component and a "discrete" component. Let α be a number between 0 and 1, $0 \leq \alpha \leq 1$. Then $F(x)$ can be expressed as a convex combination of two other *distribution functions*,

$$F(x) = \alpha F_c(x) + (1 - \alpha) F_d(x)$$

where F_c is absolutely continuous and F_d is a step function. The absolute continuity of F_c means that there is a function f, called the *probability density function* of X, such that

$$F_c(x) = \int_{-\infty}^{x} f(y)\, dy$$

or equivalently,

$$f(x) = \frac{dF_c(x)}{dx}$$

[1]There is a third component, referred to as "singular," that is somewhat pathological. In this book the singular component is assumed missing.

The step function F_d is a function with jumps at discrete points $x_1, x_2, x_3, \dots$, where the magnitude of the jump at x_i is given by $p(x_i)$. The $p(x)$ is called a *probability function* and plays the role of a discrete density relative to $F_d(x)$:

$$F_d(x) = \sum_{x_i \le x} p(x_i)$$

or

$$p(x_i) = F_d(x_i) - F_d(x_{i-1})$$

We can now be more precise in classifying random variables. When $\alpha = 1$ the random variable is called continuous and is said to have a continuous distribution. In this case $F = F_c$. When $\alpha = 0$ the random variable is called discrete and is said to have a discrete distribution. In this case $F = F_d$. But when $0 < \alpha < 1$, both F_c and F_d are present and the random variable and its distribution are said to be of a mixed type. In this general case we compute probabilities by combining an integral and a sum:

$$P(a \le X \le b) = \alpha \int_a^b f(y)\, dy + (1 - \alpha) \sum_{a \le x_i \le b} p(x_i)$$

It is worth noting that mixed distributions arise very naturally in applications whenever a continuous variable undergoes quantization that lumps together certain values but leaves other values untouched.

Before going to some examples that will help clarify the decomposition of $F(x)$, it is helpful to introduce a simplifying notation. First observe that we can use the differential notation

$$dF_c(x) = f(x)dx$$

We *define* a similar differential notation for the step function $F_d(x)$ by

$$dF_d(x) = 0, \qquad x \ne x_i$$

but

$$dF_d(x) = p(x_i), \qquad x = x_i$$

for $i = 1, 2, 3, \dots$. With this convention we define the symbolism

$$dF(x) = \alpha\, dF_c(x) + (1 - \alpha)\, dF_d(x)$$

to gain the technical convenience of writing

$$\begin{aligned} P(X \le y) &= \int_{-\infty}^{y} dF(x) \\ &= \alpha \int_{-\infty}^{y} dF_c(x) + (1 - \alpha) \int_{-\infty}^{y} dF_d(x) \\ &= \alpha \int_{-\infty}^{y} f(x)\, dx + (1 - \alpha) \sum_{x_i \le y} p(x_i) \end{aligned}$$

and

$$\int_{-\infty}^{\infty} g(x)\, dF(x) = \alpha \int_{-\infty}^{\infty} g(x)\, dF_c(x) + (1-\alpha) \int_{-\infty}^{\infty} g(x)\, dF_d(x)$$

$$= \alpha \int_{-\infty}^{\infty} g(x) f(x)\, dx + (1-\alpha) \sum_{i=1}^{\infty} g(x_i) p(x_i)$$

The reader may recognize the last expression as the *mathematical expectation*, or simply the *expectation* of $g(X)$. Usually the expectation of $g(X)$ is abbreviated by

$$E(g(X))$$

In particular the expectation of $g(X) = X^k$ is the kth order moment of X, and by the above rule

$$E(X^k) = \int_{-\infty}^{\infty} x^k\, dF(x)$$

$$= \alpha \int_{-\infty}^{\infty} x^k f(x)\, dx + (1-\alpha) \sum_{i=1}^{\infty} x_i^k p(x_i)$$

The *variance* of a random variable X is defined by

$$\mathrm{Var}(X) = E(X - E(X))^2$$
$$= E(X^2) - (E(X))^2$$

The variance of a random variable is a measure of dispersion. The larger the variance, the larger is the dispersion of the values of the random variable.

2.2.1 Some Order Relations

With the help of the complementary function

$$P(X > x) = 1 - F(x)$$

we can obtain inequalities between expected values that will be very helpful in understanding the action of certain linear operations on time series. Of special interest to us is to establish a certain inequality between the expected values of nonnegative random variables.

By a nonnegative random variable we mean a random variable whose possible values are nonnegative. The expectation of a nonnegative random variable is related to its distribution function in an interesting way. To see that, let X be a nonnegative random variable with distribution function F. Then by changing the order of integration

$$E(X) = \int_0^{\infty} x\, dF(x)$$

$$= \int_0^{\infty} \left[\int_0^x dy \right] dF(x)$$

$$= \int_0^\infty \left[\int_y^\infty dF(x) \right] dy$$

$$= \int_0^\infty P(X > y)\, dy$$

$$= \int_0^\infty [1 - F(y)]\, dy$$

Now let X, Y be two nonnegative random variables such that for all z

$$P(X > z) \geq P(Y > z)$$

Then from the previous result

$$E(X) = \int_0^\infty P(X > z)\, dz \geq \int_0^\infty P(Y > z)\, dz = E(Y)$$

If in addition $g(x)$ is a nonnegative increasing function, then[2]

$$P(g(X) > z) = P(X > g^{-1}(z)) \geq P(Y > g^{-1}(z)) = P(g(Y) > z)$$

and by the last result

$$E(g(X)) \geq E(g(Y))$$

We have proved the following fact.

Theorem 2.1. *Let X, Y be two nonnegative random variables. Suppose that $P(X > z) \geq P(Y > z)$ for all z. Then*

$$E(X) \geq E(Y)$$

Further, if in addition $g(x)$ is a nonnegative increasing function, then also

$$E(g(X)) \geq E(g(Y))$$

For an extension of Theorem 2.1 see Example 2.5 in the next section.

2.3 EXAMPLES OF DISTRIBUTIONS

In the literature the term "distribution" refers to the general probabilistic behavior of a random variable as expressed by its distribution function or equivalently its density and probability functions. The examples of distributions to be discussed next reflect the particular needs of this book. For more examples the reader should consult the references at the end of this chapter.

---

[2]The term $g^{-1}(z)$ is defined as the smallest x such that $g(x) > z$.

Example 2.1: The Normal Distribution

A continuous random variable X that takes values in the interval $(-\infty, \infty)$ and has density

$$f(x) = \frac{1}{\sigma\sqrt{2\pi}} e^{-(x-\mu)^2/2\sigma^2}, \qquad -\infty < x < \infty$$

is said to have a normal, or Gaussian, distribution with mean (expected value) μ and variance σ^2, and we write $X \sim N(\mu, \sigma^2)$. The square root of σ^2, σ, is called the *standard deviation* of X. The standardized random variable

$$Z = \frac{X - \mu}{\sigma}$$

has density

$$\phi(z) = \frac{1}{\sqrt{2\pi}} e^{-z^2/2}, \qquad -\infty < z < \infty$$

The density $\phi(z)$ is called the *standard normal density*, and Z is said to have the standard normal distribution with mean 0 and variance 1. By the newly adopted notation we write $Z \sim N(0, 1)$. It is customary to denote the distribution function of Z by $\Phi(z)$,

$$\Phi(z) = \int_{-\infty}^{z} \phi(u)\, du = \int_{-\infty}^{z} \frac{1}{\sqrt{2\pi}} e^{-u^2/2}\, du$$

Both $\phi(z)$ and $\Phi(z)$ are plotted in Figure 2.1. Note that by symmetry, $\Phi(-z) = 1 - \Phi(z)$ for any z.

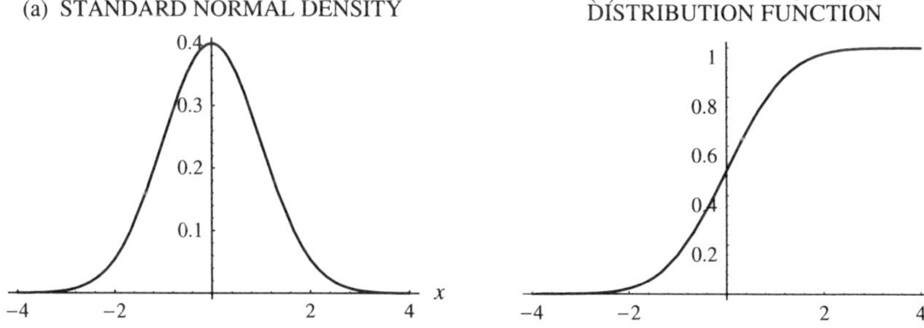

Figure 2.1: *(a) The standard normal density* $\phi(x)$. *(b) The standard normal distribution function* $\Phi(x)$.

To show that $E(Z) = 0$ and $\text{Var}(Z) = 1$, we have

$$E(Z) = \int_{-\infty}^{\infty} \frac{z}{\sqrt{2\pi}} e^{-z^2/2}\, dz = -\frac{e^{-z^2/2}}{\sqrt{2\pi}} \Big|_{-\infty}^{\infty} = 0$$

and by integration by parts,

$$\begin{aligned} \text{Var}(Z) = E(Z^2) &= \int_{-\infty}^{\infty} \frac{z^2}{\sqrt{2\pi}} e^{-z^2/2}\, dz \\ &= -\frac{ze^{-z^2/2}}{\sqrt{2\pi}} \Big|_{-\infty}^{\infty} + \int_{-\infty}^{\infty} \phi(z)\, dz = 0 + 1 = 1 \end{aligned}$$

By expressing X in terms of Z, $X = \sigma Z + \mu$, we can verify that

$$E(X) = \sigma E(Z) + \mu = \mu$$

$$\text{Var}(X) = \sigma^2 \text{Var}(Z) = \sigma^2$$

If $F(x)$ is the distribution function of X, then F and Φ are related by the equation

$$\begin{aligned} F(x) &= P(X \leq x) \\ &= P(Z \leq \frac{x-\mu}{\sigma}) \\ &= \Phi(\frac{x-\mu}{\sigma}) \end{aligned}$$

Thus all normal probabilities can be obtained from $\Phi(z)$ whose values are given in normal tables.

■

Example 2.2: Clipped Random Variable

Let Z be a continuous random variable such that $P(Z \geq 0) = p$, and $P(Z < 0) = 1 - p$. Define a new random variable X by "clipping" Z,

$$X = \begin{cases} 1 & \text{if } Z \geq 0 \\ 0 & \text{if } Z < 0 \end{cases}$$

Because X admits only two distinct values 0 and 1, it is a discrete random variable. Clearly, $p(1) = P(X = 1) = p$, and $p(0) = P(X = 0) = 1 - p$. The distribution function of X is the step function

$$\begin{aligned} F(x) &= 0, & x &< 0 \\ &= 1-p, & 0 &\leq x < 1 \\ &= 1, & 1 &\leq x \end{aligned}$$

The distribution function for $p = 0.6$ is plotted in Figure 2.2. Because $X^2 = X$,

$$E(X) = E(X^2) = 0 \cdot (1-p) + 1 \cdot p = p$$

and

$$\text{Var}(X) = E(X^2) - E^2(X) = p - p^2 = p(1-p)$$

If $Z \sim N(0,1)$, then $p = 1 - \Phi(0) = 0.5$.

■

Example 2.3: Construction of a Mixed Distribution

A mixed distribution can be constructed by "mixing" a continuous distribution and a discrete distribution. Consider the continuous distribution defined by the density

$$\begin{aligned} f(x) &= \frac{1}{4}, & 0 &\leq x \leq 4 \\ &= 0, & &\text{otherwise} \end{aligned}$$

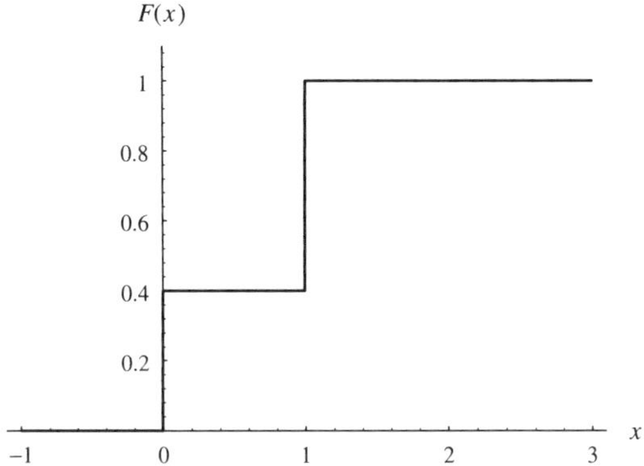

Figure 2.2: *The distribution function of a clipped random variable;* $p = 0.6$.

The corresponding distribution function is

$$\begin{aligned} F_c(x) &= 0, & x < 0 \\ &= \frac{x}{4}, & 0 \le x < 4 \\ &= 1, & 4 \le x \end{aligned}$$

Next consider the discrete distribution with probability function

$$p(1) = p(2) = p(3) = \frac{1}{3}$$

The corresponding distribution function is

$$\begin{aligned} F_d(x) &= 0, & x < 1 \\ &= \frac{1}{3}, & 1 \le x < 2 \\ &= \frac{2}{3}, & 2 \le x < 3 \\ &= 1, & 3 \le x \end{aligned}$$

The distribution function given by the convex combination

$$F(x) = \frac{1}{2}F_c(x) + \frac{1}{2}F_d(x)$$

defines a mixed distribution where

$$\begin{aligned} F(x) &= 0, & x < 0 \\ &= \frac{x}{8}, & 0 \le x < 1 \\ &= \frac{1}{6} + \frac{x}{8}, & 1 \le x < 2 \\ &= \frac{1}{3} + \frac{x}{8}, & 2 \le x < 3 \\ &= \frac{1}{2} + \frac{x}{8}, & 3 \le x < 4 \\ &= 1, & 4 \le x \end{aligned}$$

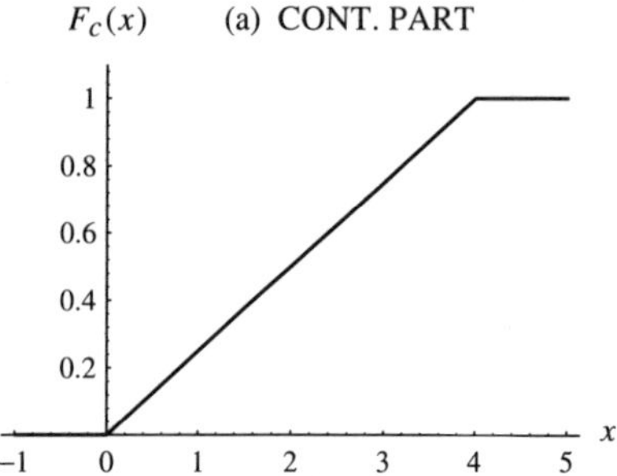

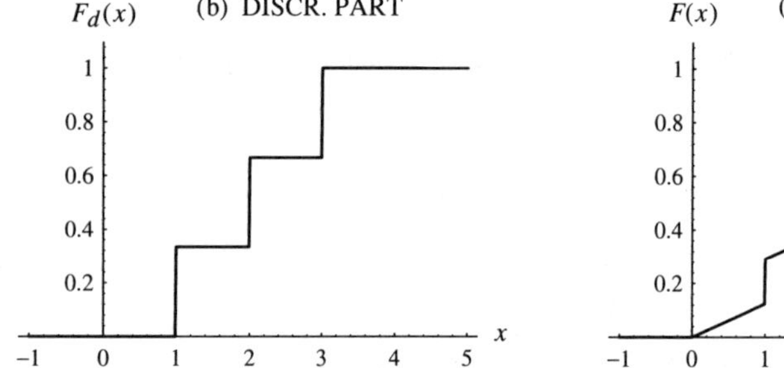

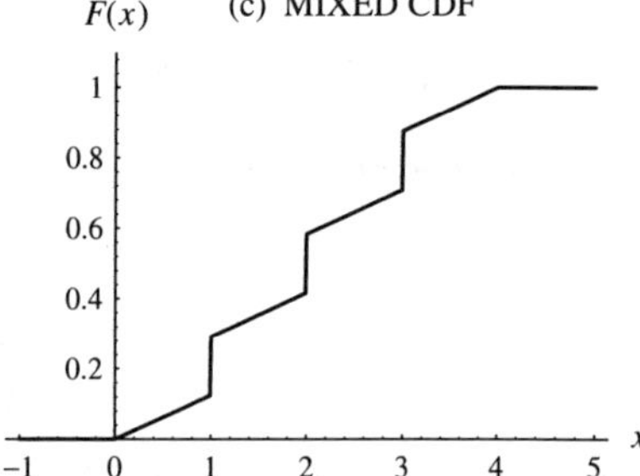

Figure 2.3: *Components of a mixed distribution. (a) Continuous $F_c(x)$. (b) Discrete $F_d(x)$. (c) Mixed $F(x) = 0.5F_c(x) + 0.5F_d(x)$.*

Figure 2.3 portrays $F(x)$ and its two components. If X is the random variable with distribution function $F(x)$, then it is a random variable of a mixed type and its first two moments are

$$\begin{aligned}
E(X) &= \int_{-\infty}^{\infty} x\, dF(x) \\
&= \frac{1}{2}\int_{-\infty}^{\infty} x\, dF_c(x) + \frac{1}{2}\int_{-\infty}^{\infty} x\, dF_d(x) \\
&= \frac{1}{2}\left[\int_0^4 \frac{x}{4}\, dx + \frac{1}{3}\sum_{k=1}^{3} k\right] = \frac{1}{2}(2+2) = 2
\end{aligned}$$

$$\begin{aligned}
E(X^2) &= \int_{-\infty}^{\infty} x^2\, dF(x) \\
&= \frac{1}{2}\int_{-\infty}^{\infty} x^2\, dF_c(x) + \frac{1}{2}\int_{-\infty}^{\infty} x^2\, dF_d(x) \\
&= \frac{1}{2}\left[\int_0^4 \frac{x^2}{4}\, dx + \frac{1}{3}\sum_{k=1}^{3} k^2\right] = \frac{1}{2}\left(\frac{64}{12} + \frac{14}{3}\right) = 5
\end{aligned}$$

and $\text{Var}(X) = 5 - 4 = 1$.

■

Example 2.3′: Construction of Mixed Lognormal Distribution

The problem of constructing a mixed model does come up in real applications. Consider the random variable R that gives the values of rain intensity at a fixed location. Because it does not rain all the time R admits the value 0 with positive probability, but otherwise when it rains R admits values in the interval $(0, \infty)$. Thus the meteorologist is faced with the problem of modeling a mixed distribution containing an "atom" at 0. If the rain intensity is measured by radar, the reflectivity also has a mixed distribution. A distribution that is often used to model rain intensity is the *mixed lognormal distribution* [12]. The continuous component is modeled by the lognormal density

$$f(r) = \begin{cases} \frac{1}{\sqrt{2\pi}\sigma r} \exp(-\frac{(\log r - \mu)^2}{2\sigma^2}), & r > 0 \\ 0, & r \leq 0 \end{cases} \tag{2.1}$$

where the parameters satisfy $-\infty < \mu < \infty$, and $\sigma > 0$. The discrete component is given by (no modeling is needed here. Why?)

$$F_d(r) = \begin{cases} 0 & \text{if } r < 0 \\ 1 & \text{if } r \geq 0 \end{cases}$$

Let $G(r)$ be the distribution function of R. The mixed lognormal distribution is defined by

$$G(r) = pF_c(r) + (1-p)F_d(r)$$

where $0 < p < 1$ is the probability of rain. The kth moment of R is [1]

$$E(R^k) = p\int_0^\infty r^k f(r)\, dr = p \cdot \exp(k\mu + \frac{1}{2}k^2\sigma^2)$$

from which the mean and variance can be easily obtained.

In general, if a random variable Y has a continuous distribution with a density f as in (2.1), then $\log Y \sim N(\mu, \sigma^2)$ (this is the reason for the name "lognormal"), and we say that Y has a lognormal distribution with parameters μ, σ^2. This is signified by the symbolism: $Y \sim \Lambda(\mu, \sigma^2)$. Observe that the mixed lognormal distribution contains an additional parameter p, the "success probability."

■

Example 2.4: Components of a Limiter

In Example 2.3 it was shown how to construct $F(x)$ from complete knowledge of $\alpha(= \frac{1}{2})$, F_c, F_d. Conversely, from a given F we can recover these three components explicitly.

Suppose Z is a standard normal random variable, $Z \sim N(0, 1)$. Define X by

$$X = \begin{cases} 1 & \text{if } Z \geq 1 \\ Z & \text{if } -2 < Z < 1 \\ -2 & \text{if } Z \leq -2 \end{cases}$$

The values of X range from -2 to 1, and its distribution function is given by

$$\begin{aligned} F(x) &= 0, & x &< -2 \\ &= \Phi(x), & -2 &\le x < 1 \\ &= 1, & 1 &\le x \end{aligned}$$

The jumps in $F(x)$ occurring at $-2, 1$, and seen very well in Figure 2.4, render the distribution of X mixed. By differentiating $F(x)$ wherever possible we have

$$\begin{aligned} \alpha f(x) &= \phi(x), & -2 < x < 1 \\ &= 0, & \text{otherwise} \end{aligned}$$

and $f(x)$ is known apart from a scale factor. Similarly, by differencing $F(x) - F(x-)$ at the jump points we obtain

$$(1-\alpha)p(-2) = \Phi(-2) = 0.0228$$

$$(1-\alpha)p(1) = 1 - \Phi(1) = 0.1587$$

and so, the probability function is also known up to a constant. To find the constant α we integrate (remembering that $f(x)$ must integrate to 1):

$$\alpha \int_{-2}^{1} f(x)\, dx = \alpha \cdot 1 = \int_{-2}^{1} \phi(x)\, dx$$

Therefore

$$\alpha = \Phi(1) - \Phi(-2) = 0.8185$$

By substituting 0.8185 for α we can solve for the density and probability functions explicitly. For example,

$$\begin{aligned} p(-2) &= 0.0228/(1 - 0.8185) = 0.1256 \\ p(1) &= 0.1587/(1 - 0.8185) = 0.8744 \end{aligned}$$

■

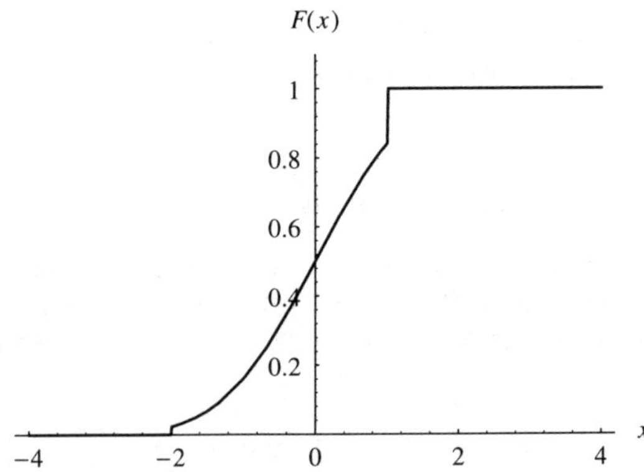

Figure 2.4: *The distribution function of a limiter.*

Example 2.5: Extension of Theorem 2.1

In Theorem 2.1 all references to "nonnegative" can be dropped as we shall show next by using two useful limiters [20, p. 252].

Let X be a random variable with distribution function $F(x)$. Define two new random variables X^+, X^- by limiting X:

$$X^+ = \begin{cases} X, & \text{if } X \geq 0 \\ 0, & \text{if } X < 0 \end{cases}$$

$$X^- = \begin{cases} 0, & \text{if } X \geq 0 \\ -X, & \text{if } X < 0 \end{cases}$$

Then X^+, X^- are nonnegative and are referred to as the *positive* and *negative parts* of X, respectively. Their distribution functions are

$$\begin{aligned} F_{X^+}(x) &= 0, & x < 0 \\ &= F(0), & x = 0 \\ &= F(x), & x > 0 \end{aligned}$$

$$\begin{aligned} F_{X^-}(x) &= 0, & x < 0 \\ &= 1 - F(0-), & x = 0 \\ &= P(X \geq -x), & x > 0 \end{aligned}$$

When $0 < F(0) < 1$ and X is continuous, X^+, X^- are of mixed type. We can now write

$$X = X^+ - X^-$$

Let Y be another random variable with the corresponding positive and negative parts Y^+, Y^-. Suppose that for all a, $P(X > a) \geq P(Y > a)$. If $a < 0$, then since X^+, Y^+ are never negative $P(X^+ > a) = P(Y^+ > a) = 1$. But if $a \geq 0$, then by our hypothesis

$$P(X^+ > a) = P(X > a) \geq P(Y > a) = P(Y^+ > a)$$

and we can conclude that for all a

$$P(X^+ > a) \geq P(Y^+ > a)$$

Therefore by Theorem 2.1

$$E(X^+) \geq E(Y^+)$$

In the same way we can show that for all a

$$P(X^- > a) \leq P(Y^- > a)$$

and again from Theorem 2.1

$$E(X^-) \leq E(Y^-)$$

Thus by the linearity of "E"

$$E(X) = E(X^+) - E(X^-) \geq E(Y^+) - E(Y^-) = E(Y)$$

regardless of the type of X, Y. Hence, in the hypothesis of Theorem 2.1 we can relax the requirement that X, Y be nonnegative.

■

Example 2.6: A Population Shift [18, p. 232]

To give an illustration that clarifies the foregoing example, consider two random variables X, Y, not necessarily nonnegative, with distribution functions F_X, F_Y, respectively. Assume that for some fixed $b > 0$

$$F_Y(x) = F_X(x + b)$$

for all x. Then for all x

$$P(Y > x) = P(X > x + b) \leq P(X > x)$$

and by the foregoing discussion

$$E(Y) \leq E(X)$$

But this can also be verified directly:

$$E(X) = \int_{-\infty}^{\infty} x \, dF_X(x) = \int_{-\infty}^{\infty} (v + b) \, dF_Y(v) = E(Y) + b$$

and the inequality holds due to a shift to the right in the mean of X. On the other hand, a shift in the mean should not affect the dispersion:

$$\begin{aligned}
\text{Var}(X) &= \int_{-\infty}^{\infty} (x - E(X))^2 \, dF_X(x) \\
&= \int_{-\infty}^{\infty} (v + b - (E(Y) + b))^2 \, dF_Y(v) \\
&= \int_{-\infty}^{\infty} (v - E(Y))^2 \, dF_Y(v) \\
&= \text{Var}(Y)
\end{aligned}$$

■

2.4 MULTIVARIATE CONCEPTS

To be able to speak intelligently about time series, familiarity with the basic notions of joint distributions of several random variables is needed first. Of the many topics that come under the general heading of multivariate analysis, we wish to review in particular the notions of independence and dependence, covariance and correlation, and orthant probabilities of elliptically symmetric bivariate distributions. We specialize the exposition to the continuous case only since the discrete case is entirely analogous except that densities are replaced by probability functions and integrals are replaced by sums.

The *joint distribution function* of the random variables $X_1, \ldots, X_N$ is defined by

$$F(x_1, \ldots, x_N) = P(X_1 \leq x_1, \ldots, X_N \leq x_N)$$

If $F_{X_j}(x_j)$ denotes the *marginal* distribution function of X_j, then

$$\begin{aligned} F_{X_j}(x_j) &= \lim_{x_i \to \infty, i \neq j} F(x_1, \ldots, x_N) \\ &\equiv F(\infty, \ldots, \infty, x_j, \infty, \ldots, \infty) \end{aligned}$$

The random variables $X_1, \ldots, X_N$ are said to be *independent* if

$$F(x_1, \ldots, x_N) = F_{X_1}(x_1) \cdots F_{X_N}(x_N)$$

for all $x_1, \ldots, x_N$. Otherwise the random variables $X_1, \ldots, X_N$ are said to be *dependent*. In the continuous case there exists a *joint density* $f(x_1, \ldots, x_N)$ such that

$$F(x_1, \ldots, x_N) = \int_{-\infty}^{x_1} \cdots \int_{-\infty}^{x_N} f(u_1, \ldots, u_N)\, du_1 \cdots du_N$$

and probabilities are obtained by integration. For $A \in \mathcal{R}^n$,

$$P((X_1, \ldots, X_N) \text{ in } A) = \int_A \cdots \int f(x_1, \ldots, x_N) dx_1 \cdots dx_N$$

The *marginal* density of X_j can be obtained by integrating $f(x_1, \ldots, x_N)$ with respect to all its arguments except x_j

$$f_{X_j}(x_j) = \int_{-\infty}^{\infty} \cdots \int_{-\infty}^{\infty} f(x_1, \ldots, x_N)\, dx_1 \cdots dx_{j-1}\, dx_{j+1} \cdots dx_N$$

In this case the condition of independence is expressed by the product of densities

$$f(x_1, \ldots, x_N) = f_{X_1}(x_1) \cdots f_{X_N}(x_N)$$

It should be noted that the notion of independence is a matter of a definition that very often agrees with our intuitive notion of independence, but not necessarily so.

Example 2.7: Uniform Distribution on the Unit Disc

Let X, Y be uniformly distributed in the unit disc. By this we mean that their joint density is given by

$$\begin{aligned} f(x, y) &= \frac{1}{\pi}, \qquad x^2 + y^2 \leq 1 \\ &= 0, \qquad \text{otherwise} \end{aligned}$$

For fixed x, y varies from $-\sqrt{1 - x^2}$ to $\sqrt{1 - x^2}$, and therefore

$$f_X(x) = \int_{-\sqrt{1-x^2}}^{\sqrt{1-x^2}} \frac{1}{\pi}\, dy = \frac{2}{\pi}\sqrt{1 - x^2}, \qquad -1 < x < 1$$

Similarly

$$f_Y(y) = \frac{2}{\pi}\sqrt{1 - y^2}, \qquad -1 < y < 1$$

Therefore

$$f_X(0)f_Y(0) = (\frac{2}{\pi})^2$$

but

$$f(0,0) = \frac{1}{\pi} \neq f_X(0)f_Y(0)$$

and so X, Y are dependent. This agrees with our intuition since large values of X near ± 1 imply that Y must be close to 0.

■

Dependent random variables can be transformed to produce independent random variables and vice versa.

Example 2.8: Polar Coordinates

Consider again X, Y in Example 2.7 and define two new random variables:

$$R = \sqrt{X^2 + Y^2}, \qquad \Theta = \tan^{-1}\frac{Y}{X}$$

For every pair (x, y) there is exactly one pair (r, θ). The density $g(r, \theta)$ of (R, Θ) can be obtained from the density $f(x, y)$ of (X, Y) by the formula

$$g(r, \theta) = f(x, y)|J|$$

where now $x = r\cos(\theta)$ and $y = r\sin(\theta)$ are considered functions of (r, θ) and

$$|J| = r\cos^2(\theta) + r\sin^2(\theta) = r$$

is the absolute value of the determinant

$$J = \begin{vmatrix} \frac{\partial x}{\partial r} & \frac{\partial x}{\partial \theta} \\ \frac{\partial y}{\partial r} & \frac{\partial y}{\partial \theta} \end{vmatrix}$$

Therefore the joint density of (R, Θ) is

$$g(r, \theta) = \frac{1}{\pi}r = \frac{r}{\pi}, \qquad 0 \leq r \leq 1, \quad 0 \leq \theta < 2\pi$$

and the two marginal densities are

$$f_R(r) = \int_0^{2\pi} \frac{r}{\pi}\, d\theta = 2r, \qquad 0 \leq r \leq 1$$

and

$$f_\Theta(\theta) = \int_0^1 \frac{r}{\pi}\, dr = \frac{1}{2\pi}, \qquad 0 \leq \theta < 2\pi$$

It follows that R and Θ are independent. This again is rather expected on intuitive grounds since a completely random direction says nothing about the distance from the origin.

Clearly, we can go in the reverse direction and obtain X, Y, two dependent random variables, from R, Θ, two independent random variables.

■

2.4.1 Moments and Correlation

Let $f(x_1, \ldots, x_N)$ be the joint density of $X_1, \ldots, X_N$ and consider the function $g(X_1, \ldots, X_N)$. The expectation of $g(X_1, \ldots, X_N)$ can be evaluated by integration with respect to $f(x_1, \ldots, x_N)$:

$$E[g(X_1, \ldots, X_N)] = \int_{-\infty}^{\infty} \cdots \int_{-\infty}^{\infty} g(x_1, \ldots, x_N) f(x_1, \ldots, x_N)\, dx_1 \cdots dx_N$$

This implies that

$$E(\sum_{j=1}^{N} X_j) = \sum_{j=1}^{N} E(X_j)$$

Thus, the expectation of a sum of random variables is the sum of their expectations regardless of dependence. This linear property, that we already invoked earlier, does not hold in general for the variance of a sum of random variables unless they are uncorrelated as we explain below.

The *covariance* of two random variables X, Y is defined by

$$\begin{aligned}\text{Cov}(X, Y) &\equiv E[(X - E(X))(Y - E(Y))] \\ &= E(XY) - E(X)E(Y)\end{aligned}$$

When

$$\text{Cov}(X, Y) = 0$$

we say that X, Y are *uncorrelated*. In particular, when X, Y are independent, then they are also uncorrelated because independence implies

$$E(XY) = E(X)E(Y)$$

The converse is not necessarily true. Two dependent random variables may nonetheless be uncorrelated as the next example shows.

Example 2.9: Dependent Yet Uncorrelated Random Variables

Consider the random variables in Examples 2.7 and 2.8, and observe that

$$E(\cos(\Theta)) = \frac{1}{2\pi} \int_0^{2\pi} \cos(\theta)\, d\theta = 0$$

and

$$E(\sin(\Theta)\cos(\Theta)) = \frac{1}{2\pi} \int_0^{2\pi} \sin(\theta)\cos(\theta)\, d\theta = \frac{1}{4\pi} \sin^2(\theta)|_0^{2\pi} = 0$$

Therefore by the independence of R and Θ,

$$E(X) = E(R\cos(\Theta)) = E(R)E(\cos(\Theta)) = 0$$

and

$$E(XY) = E(R^2 \sin(\Theta)\cos(\Theta)) = E(R^2)E(\sin(\Theta)\cos(\Theta)) = 0$$

and so

$$\text{Cov}(X, Y) = 0$$

and we see that X, Y are uncorrelated. But X, Y were found in Example 2.7 to be dependent.

■

The covariance is bilinear:

$$\text{Cov}(\sum_{i=1}^{M} a_i X_i, \sum_{j=1}^{N} b_j Y_j) = \sum_{i=1}^{M}\sum_{j=1}^{N} a_i b_j \text{Cov}(X_i, Y_j)$$

This and the simple observation that $\text{Var}(X) = \text{Cov}(X, X)$ yield

$$\begin{aligned}\text{Var}(\sum_{i=1}^{N} X_i) &= \text{Cov}(\sum_{i=1}^{N} X_i, \sum_{j=1}^{N} X_j) = \sum_{i=1}^{N}\sum_{j=1}^{N} \text{Cov}(X_i, X_j)\\ &= \sum_{j=1}^{N} \text{Var}(X_j) + 2\sum_{i<j} \text{Cov}(X_i, X_j)\end{aligned}$$

It follows that when the random variables $X_1, \ldots, X_N$ are (pairwise) uncorrelated, then the variance of a sum is the sum of the variances,

$$\text{Var}(\sum_{i=1}^{N} X_i) = \sum_{i=1}^{N} \text{Var}(X_i)$$

a fact we alluded to earlier.

2.4.2 The Correlation Coefficient

The covariance is a measure of association between random variables, a fact that is better manifested through the normalized quantity

$$\rho_{xy} = \frac{\text{Cov}(X, Y)}{\sqrt{\text{Var}(X)}\sqrt{\text{Var}(Y)}}$$

We refer to ρ_{xy} as the *correlation coefficient* between X and Y. By definition $\rho_{xy} = 0$ if and only if X, Y are uncorrelated. We will show that the correlation coefficient is a measure of the degree of linearity between two random variables and that it satisfies the inequality $|\rho_{xy}| \leq 1$. Values of ρ_{xy} near $+1$ or near -1 mean a high degree of linearity between jointly distributed X and Y, while

$\rho_{xy} = 0$ indicates that the relationship between X and Y cannot be linear. To demonstrate these important facts, we digress momentarily to discuss a famous inequality.

Consider the random variables U, V and $(U - tV)^2$ where t is a real number and where we assume $E(U^2) > 0$, $E(V^2) > 0$ to avoid trivialities. Recall that the variance of a random variable X is never negative because we average (by taking the expectation) the nonnegative quantity $(X - E(X))^2$. Similarly, because $(U - tV)^2$ is nonnegative for all t, so is its expectation. Therefore for all real t

$$0 \leq E(U - tV)^2 = t^2 E(V^2) - 2tE(UV) + E(U^2)$$

This is a nonnegative quadratic function *in t* whose minimum occurs at

$$t_0 = \frac{E(UV)}{E(V^2)} = \frac{E(UV)}{\sqrt{E(U^2)}\sqrt{E(V^2)}} \cdot \frac{\sqrt{E(U^2)}}{\sqrt{E(V^2)}}$$

By substituting t_0 for t and rearranging terms we obtain

$$0 \leq E(U - t_0 V)^2 = \frac{E(U^2)E(V^2) - (E(UV))^2}{E(V^2)}$$

and therefore

$$[E(UV)]^2 \leq E(U^2)E(V^2)$$

This inequality is the well-known *Cauchy–Schwarz* inequality. We can see that the Cauchy–Schwarz inequality becomes an equality if and only if

$$U = t_0 V$$

with probability one. With this result in mind we go back to ρ_{xy}.

By making the substitution $U = X - E(X)$ and $V = Y - E(Y)$ it follows from the Cauchy–Schwarz inequality that

$$[\text{Cov}(X, Y)]^2 \leq \text{Var}(X) \cdot \text{Var}(Y)$$

or that

$$|\rho_{xy}| \leq 1$$

with equality holding (that is $|\rho_{xy}| = 1$) if and only if

$$(X - E(X)) = \rho_{xy} \frac{\sigma_x}{\sigma_y} (Y - E(Y))$$

with probability one where $\sigma_x = \sqrt{\text{Var}(X)}$ and $\sigma_y = \sqrt{\text{Var}(Y)}$. We have thus established the following important fact.

Theorem 2.2. *Let X, Y be jointly distributed random variables with correlation coefficient ρ_{xy}. Then*

$$-1 \leq \rho_{xy} \leq 1$$

Further, $\rho_{xy} = 1$ *if and only if*

$$(X - E(X)) = \frac{\sigma_x}{\sigma_y}(Y - E(Y))$$

with probability one, and $\rho_{xy} = -1$ *if and only if*

$$(X - E(X)) = -\frac{\sigma_x}{\sigma_y}(Y - E(Y))$$

with probability one.

From Theorem 2.2 we expect the graphs of long sequences of random variables to be rather "smooth" or slowly varying when the correlation between neighboring random variables is high (that is close to 1), but otherwise when the correlation between neighboring random variables is low (that is close to -1) we expect the graphs to display very rapid oscillation. The following example is an illustration of this effect.

Example 2.10: Correlation in a Moving Average

Suppose $\ldots, X_{-2}, X_{-1}, X_0, X_1, X_2, \ldots$ is an infinite sequence of uncorrelated random variables with $E(X_i) = 0$ and $\text{Var}(X_i) = \sigma^2$ for all i. Define the *moving average* sequence for $t = \ldots, -2, -1, 0, 1, 2, \ldots$ and for $m \geq 1$ by

$$Y_t = \frac{X_t + X_{t-1} + \cdots + X_{t-(m-1)}}{m}$$

Then for all t, $E(Y_t) = 0$, $\text{Var}(Y_t) = \sigma^2/m$. The assumption that the random variables are uncorrelated with zero expectations means that

$$\text{Cov}(X_i, X_j) = E(X_i X_j) = 0 \qquad \text{for } i \neq j$$

and so by a simple manipulation

$$\text{Cov}(Y_t, Y_{t-1}) = E(Y_t Y_{t-1}) = \frac{\sigma^2(m-1)}{m^2}$$

Therefore

$$\rho_{y_t y_{t-1}} = \frac{m-1}{m}$$

Now set $m = 1000$ and consider the finite subsequence $Y_1, Y_2, \ldots, Y_N$. Then

$$\rho_{y_2 y_1} = \rho_{y_3 y_2} = \cdots = \rho_{y_N y_{N-1}} = 0.999$$

and by Theorem 2.2 $Y_1 \simeq Y_2 \simeq \cdots \simeq Y_N$ with high probability. In fact the sequence $Y_1, \ldots, Y_N$ will appear very "smooth" as opposed to the sequence $Y_1, -Y_2, Y_3, -Y_4, Y_5, \ldots,$ whose correlation between neighboring points is close to -1 so that its appearance is very oscillatory. We have plotted $Y_t, t = 1, 2, \ldots, 100$, for various values of m in Figure 2.5.

■

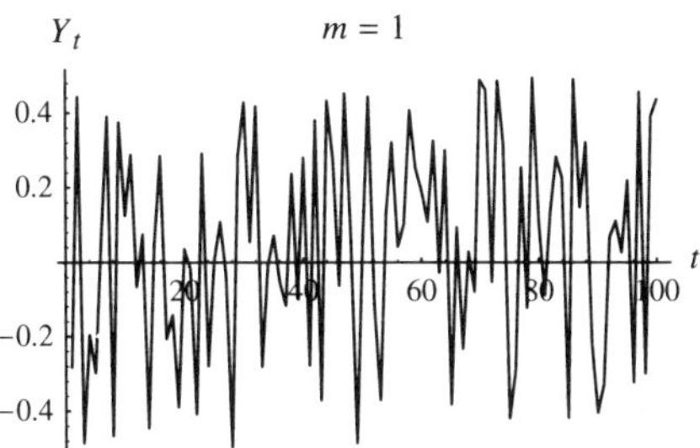

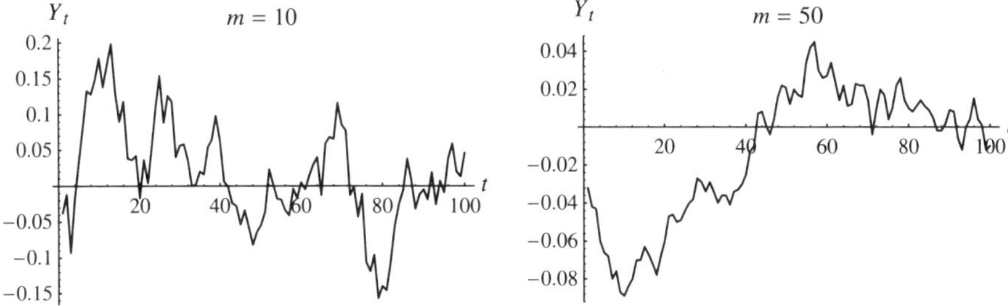

Figure 2.5: *Moving averages corresponding to different values of m.*

2.4.3 The Joint Moment-Generating Function

A useful device for obtaining moments and determining distributions is the *moment-generating function* (MGF). For a single random variable X the moment-generating function is defined by

$$M_X(t) = E(e^{tX})$$

provided it exists in the interval $-a < t < a$ for some positive a. In advanced probability theory it is proved that $M_X(t)$ completely determines the distribution of X. The moments of X are obtained by repeated differentiation of $M_X(t)$ at $t = 0$:

$$E(X^k) = M_X^{(k)}(0)$$

Similarly, the *joint* moment-generating function of two random variables X, Y is defined by

$$M(s,t) = E\left(e^{sX+tY}\right)$$

provided it exists for $-a < s < a, -b < t < b$ for some positive a, b. Again, $M(s,t)$ completely determines the joint distribution of X, Y and moments are obtained by differentiation at $(0,0)$. For example:

$$E(X) = \frac{\partial}{\partial s}M(0,0) \qquad\qquad E(Y) = \frac{\partial}{\partial t}M(0,0)$$

$$E(X^2) = \frac{\partial^2}{\partial s^2}M(0,0) \qquad E(Y^2) = \frac{\partial^2}{\partial t^2}M(0,0)$$

and

$$E(XY) = \frac{\partial^2}{\partial s \partial t}M(0,0)$$

Observe that from $M(s,t)$ we can determine the marginal distributions of X and Y since

$$M(s,0) = M_X(s) \qquad M(0,t) = M_Y(t)$$

In general the joint moment-generating function of a vector of random variables $X_1, \ldots, X_N$ is defined by

$$M_{X_1 \ldots X_N}(t_1, \ldots, t_N) = E\left(e^{t_1 X_1 + \cdots t_N X_N}\right)$$

In the next section we apply the moment-generating function in the important special case of the two-dimensional normal distribution. Here, however, we note the following fact.

Example 2.11: MGF of the Normal Distribution

Suppose $X \sim N(\mu, \sigma^2)$. The moment-generating function of X is determined by integration:

$$\begin{aligned} M_X(t) &= \int_{-\infty}^{\infty} e^{tx} \frac{1}{\sqrt{2\pi}\sigma} \exp\left[-\frac{(x-\mu)^2}{2\sigma^2}\right] dx \\ &= \exp\left(\mu t + \frac{\sigma^2 t^2}{2}\right) \int_{-\infty}^{\infty} \phi(z)\, dz \\ &= \exp\left(\mu t + \frac{\sigma^2 t^2}{2}\right) \end{aligned}$$

This is a well-known formula that is very useful in identifying the normal distribution because the normal distribution $N(\mu, \sigma^2)$ is the only distribution with moment-generating function given by $\exp(\mu t + (\sigma^2 t^2/2))$. We can now verify that μ and σ^2 are indeed the mean and variance of X. We have

$$E(X) = M'(0) = \mu$$

and

$$E(X^2) = M''(0) = \sigma^2 + \mu^2$$

so that $\text{Var}(X) = \sigma^2 + \mu^2 - \mu^2 = \sigma^2$

■

2.4.4 The Bivariate Normal Distribution

Two random variables X, Y whose joint density function is given by

$$f(x,y) = \frac{1}{2\pi\sigma_x\sigma_y\sqrt{1-\rho^2}} \exp\left(-\frac{q}{2}\right), \qquad -\infty < x, y < \infty$$

where

$$q = \frac{1}{1-\rho^2}\left[\left(\frac{x-\mu_X}{\sigma_x}\right)^2 - 2\rho\left(\frac{x-\mu_X}{\sigma_x}\right)\left(\frac{y-\mu_Y}{\sigma_y}\right) + \left(\frac{y-\mu_Y}{\sigma_y}\right)^2\right]$$

are said to have a *bivariate normal distribution* with parameters

$$\mu_X, \mu_Y, \sigma_x^2, \sigma_y^2, \rho$$

and the density $f(x, y)$ is called the *bivariate normal density function*. Here we assume that σ_x, σ_y are both positive and that $|\rho| < 1$ to avoid division by 0. The special case with $\mu_X = \mu_Y = 0, \sigma_x^2 = \sigma_y^2 = 1$ is referred to as the standard form of the distribution. The bivariate normal distribution serves as a model in a wide variety of applications, and it plays an important role in our development. It is therefore important to be familiar with some of the basic properties of this distribution and to be able to interpret its parameters. In the following we shall make use of the moment-generating function in identifying various normal distributions.

The joint moment-generating function of X, Y is given by

$$M(t_1, t_2) = \exp\left(t_1\mu_X + t_2\mu_Y + \frac{t_1^2\sigma_x^2 + 2\rho\sigma_x\sigma_y t_1 t_2 + t_2^2\sigma_y^2}{2}\right)$$

The manner in which the last formula is obtained is very similar to the univariate case considered in Example 2.11. Quite a few results can now be derived from $M(t_1, t_2)$. Our first observation is that

$$M_X(t) = M(t, 0) = \exp(t\mu_X + \frac{t^2\sigma_x^2}{2})$$

which we recognize from Example 2.11 to be the moment-generating function of a normal distribution with mean μ_X and variance σ_x^2. Therefore

$$X \sim N(\mu_X, \sigma_x^2)$$

Similarly

$$M_Y(t) = M(0, t) = \exp(t\mu_Y + \frac{t^2\sigma_y^2}{2})$$

and thus also

$$Y \sim N(\mu_Y, \sigma_y^2)$$

We see that the two marginal distributions of X, Y are normal with the indicated parameters. To interpret the fifth parameter ρ note that

$$E(XY) = \frac{\partial^2}{\partial s \partial t}M(0, 0) = \mu_X\mu_Y + \rho\sigma_x\sigma_y$$

and the covariance of X, Y is

$$\text{Cov}(X,Y) = E(XY) - E(X)E(Y) = \mu_X\mu_Y + \rho\sigma_x\sigma_y - \mu_X\mu_Y = \rho\sigma_x\sigma_y$$

Therefore ρ is the correlation coefficient between X and Y. In particular we see immediately that if $\rho = 0$, then

$$f(x, y) = f_X(x)f_Y(y)$$

and X, Y are independent. But since we already know that independence implies zero correlation we see that in the case of the bivariate normal distribution X, Y are independent if and only if they are uncorrelated. This fact is one of the attractive features of the bivariate normal distribution.

Next we consider the linear combination

$$U = aX + bY$$

for real a, b. To find the distribution of U it is sufficient to find its moment-generating function. Accordingly, we have

$$\begin{aligned} M_U(t) &= E\left[e^{tU}\right] = E\left[e^{t(aX+bY)}\right] = M(at, bt) \\ &= \exp[(a\mu_X + b\mu_Y)t + \frac{1}{2}(a^2\sigma_x^2 + b^2\sigma_y^2 + 2ab\rho\sigma_x\sigma_y)t^2] \end{aligned}$$

Therefore, by inspection U has the univariate normal distribution

$$U \sim N(a\mu_X + b\mu_Y, a^2\sigma_x^2 + b^2\sigma_y^2 + 2ab\rho\sigma_x\sigma_y)$$

We leave it as a simple exercise to show that in fact the mean and variance of this normal distribution coincide with $E(aX + bY)$ and $\text{Var}(aX + bY)$, respectively. Just remember that

$$E(aX + bY) = aE(X) + bE(Y)$$

and

$$\text{Var}(aX + bY) = a^2\text{Var}(X) + b^2\text{Var}(Y) + 2ab\text{Cov}(X, Y)$$

So, any linear combination of X, Y is itself normally distributed. It is convenient to collect all our new results in the form of a theorem.

Theorem 2.3. *Let X, Y be distributed according to a bivariate normal distribution with parameters $\mu_X, \mu_Y, \sigma_x^2, \sigma_y^2, \rho$. Then we have:*

1. *μ_X, σ_x^2 are the mean and variance of X, respectively.*
   μ_Y, σ_y^2 are the mean and variance of Y, respectively.
   ρ is the correlation between X, Y.
2. *The marginal distribution of X is $N(\mu_X, \sigma_x^2)$.*
   The marginal distribution of Y is $N(\mu_Y, \sigma_y^2)$.

3. *X and Y are independent if and only if $\rho = 0$.*
4. *Any linear combination of X, Y has a normal distribution. If $U = aX + bY$, then*

$$U \sim N(a\mu_X + b\mu_Y, a^2\sigma_x^2 + b^2\sigma_y^2 + 2ab\rho\sigma_x\sigma_y)$$

Example 2.12: Independent Normal Random Variables Are Jointly Normal

If X, Y are two independent normal random variables, then they have a joint bivariate normal distribution. For example, let X, Y be independent standard normal random variables. Then their joint density

$$f(x, y) = \phi(x)\phi(y) = \frac{1}{2\pi} \exp \frac{-(x^2 + y^2)}{2}$$

is the bivariate normal density with parameters $\mu_X = \mu_Y = 0$, $\sigma_x^2 = \sigma_y^2 = 1$, and $\rho = 0$. However, if $X \sim N(0, 1)$ and $Y \sim N(0, 1)$ and if in addition X and Y are dependent, then their joint density is not necessarily that of the bivariate normal distribution. An example is furnished by the joint density [10]

$$\begin{aligned} f_a(x, y) &= \phi(x)\phi(y) + (a^2/2\pi e)xy, && |x|, |y| < 1 \\ &= \phi(x)\phi(y), && \text{otherwise} \end{aligned}$$

where a is sufficiently small so that $f_a(x, y) \geq 0$. Then X and Y are both standard normal random variables, but their joint distribution is not bivariate normal. Observe that X, Y are dependent because

$$\text{Cov}(X, Y) = \frac{a^2}{2\pi e} \int_{-1}^{1} \int_{-1}^{1} x^2 y^2 \, dx \, dy = \frac{2a^2}{9\pi e}$$

The lesson to be learned from this example is that in general marginal distributions do not determine joint distributions.

■

2.4.5 An Orthant Probability

Suppose X_1, X_2 have a bivariate normal distribution with zero means and variances equal to 1. Then X_1, X_2 are distributed according to the standard bivariate normal distribution with density

$$f(x_1, x_2) = \frac{1}{2\pi\sqrt{1 - \rho^2}} \exp\left[-\frac{1}{2(1 - \rho^2)}(x_1^2 - 2\rho x_1 x_2 + x_2^2)\right]$$

defined for all x_1, x_2 such that, $-\infty < x_1, x_2 < \infty$. The probability

$$P(X_1 \geq 0, X_2 \geq 0) = \int_0^\infty \int_0^\infty f(x_1, x_2) \, dx_1 \, dx_2$$

is referred to as an *orthant probability* [15]. We would like to derive an expression or formula for this probability. Because ρ is the only parameter of the standard normal distribution, any expression for the orthant probability must be a function of ρ. In fact, from Theorems 2.2 and 2.3, we can already see that when $\rho = 0$ the orthant probability is equal to $1/4$, when ρ approaches -1 the probability approaches 0, and when ρ approaches 1 the probability approaches $1/2$. This suggests that as a function of ρ the orthant probability in question is monotone increasing.

We shall derive a somewhat more general result. For this end it is useful to reparametrize $f(x_1, x_2)$ using matrix notation. Define the vector $\mathbf{x} = (x_1, x_2)'$, where $\mathbf{x}'$ is the transpose of the vector $\mathbf{x}$, and define the matrix

$$\mathbf{\Sigma} = \begin{pmatrix} 1 & \rho \\ \rho & 1 \end{pmatrix}$$

Then we have

$$f(\mathbf{x}) = \frac{1}{2\pi}|\mathbf{\Sigma}|^{-1/2}\exp(-\frac{1}{2}\mathbf{x}'\mathbf{\Sigma}^{-1}\mathbf{x})$$

where we note that $\mathbf{\Sigma}$ is symmetric and also positive definite when $|\rho| < 1$. We can still go a bit further and define the function

$$\psi(u) = \frac{1}{2\pi}\exp(-\frac{1}{2}u)$$

Then we can express the standard bivariate normal density in the more general form

$$f(\mathbf{x}) = |\mathbf{\Sigma}|^{-1/2}\psi(\mathbf{x}'\mathbf{\Sigma}^{-1}\mathbf{x})$$

Any distribution whose density admits the last form for some function $\psi(u)$ defined on $[0, \infty)$ and parametrized by a symmetric positive definite matrix $\mathbf{\Sigma}$ is called an *ellipsoidal distribution* [9] or *elliptically symmetric distribution* [13]. Thus the standard normal distribution is a member of the class of ellipsoidal distributions, and any general result that holds for this class also covers automatically the standard normal case. This is true in particular for the bivariate orthant probability, the subject of our next result.

Theorem 2.4. *Assume that X_1, X_2 have zero means, variances equal to 1, and correlation ρ. Suppose X_1, X_2 have a bivariate ellipsoidal distribution with density*

$$f(\mathbf{x}) = |\mathbf{\Sigma}|^{-1/2}\psi(\mathbf{x}'\mathbf{\Sigma}^{-1}\mathbf{x})$$

for some positive definite matrix $\mathbf{\Sigma}$ and suitable function $\psi(u)$ on $[0, \infty)$. Then

$$P(X_1 \geq 0, X_2 \geq 0) = \frac{1}{4} + \frac{1}{2\pi}\sin^{-1}\rho$$

Proof. [7] It can be shown that there exists a constant $c > 0$ such that

$$\mathbf{\Sigma} = c \begin{pmatrix} 1 & \rho \\ \rho & 1 \end{pmatrix}$$

By switching to polar coordinates and then making the transformation

$$u = \frac{r^2(1 - \rho \sin 2\theta)}{c(1 - \rho^2)}$$

we have

$$\begin{aligned} P(X_1 \geq 0, X_2 \geq 0) &= \int_0^\infty \int_0^\infty \frac{1}{\sqrt{c^2(1-\rho^2)}} \psi(\frac{x_1^2 + x_2^2 - 2\rho x_1 x_2}{c(1-\rho^2)})\, dx_1\, dx_2 \\ &= \int_0^{\pi/2} \int_0^\infty \frac{1}{\sqrt{c^2(1-\rho^2)}} \psi(\frac{r^2(1-\rho\sin 2\theta)}{c(1-\rho^2)}) r\, dr\, d\theta \\ &= \sqrt{1-\rho^2} \int_0^{\pi/2} \frac{d\theta}{2(1-\rho\sin 2\theta)} \int_0^\infty \psi(u)\, du \\ &= (\frac{\pi}{4} + \frac{1}{2}\sin^{-1}\rho) \int_0^\infty \psi(u)\, du \end{aligned}$$

To evaluate the remaining integral note that

$$\int_{-\infty}^\infty \int_{-\infty}^\infty \frac{1}{\sqrt{c^2(1-\rho^2)}} \psi(\frac{x_1^2 + x_2^2 - 2\rho x_1 x_2}{c(1-\rho^2)})\, dx_1\, dx_2 = 1$$

and by the same transformation as above

$$\int_0^\infty \psi(u)\, du = [\sqrt{1-\rho^2} \int_0^{2\pi} \frac{d\theta}{2(1-\rho\sin 2\theta)}]^{-1} = \frac{1}{\pi}$$

In evaluating the last integral it helps if we recognize that

$$\int_0^{2\pi} \frac{d\theta}{2(1-\rho\sin 2\theta)} = \int_0^{\pi/2} \frac{d\theta}{1-\rho\sin 2\theta} + \int_0^{\pi/2} \frac{d\theta}{1+\rho\sin 2\theta}$$

Collecting our results we obtain

$$\begin{aligned} P(X_1 \geq 0, X_2 \geq 0) &= (\frac{\pi}{4} + \frac{1}{2}\sin^{-1}\rho)(\frac{1}{\pi}) \\ &= \frac{1}{4} + \frac{1}{2\pi}\sin^{-1}\rho \end{aligned}$$

□

We can now state the following useful result.

Corollary 2.1. *Assume X_1, X_2 have a bivariate normal distribution with zero means, arbitrary variances, and correlation ρ:*

$$E(X_1) = E(X_2) = 0$$

$$\mathrm{Var}(X_1) = \sigma_{x_1}^2, \qquad \mathrm{Var}(X_2) = \sigma_{x_2}^2$$

$$\mathrm{Corr}(X_1, X_2) = \rho$$

Then

$$P(X_1 \geq 0, X_2 \geq 0) = \frac{1}{4} + \frac{1}{2\pi} \sin^{-1} \rho$$

Proof. Define $Y_1 = X_1/\sigma_{x_1}$ and $Y_2 = X_2/\sigma_{x_2}$. From their joint moment-generating function Y_1, Y_2 have a standard bivariate normal distribution with correlation ρ. Then from Theorem 2.4

$$P(X_1 \geq 0, X_2 \geq 0) = P(Y_1 \geq 0, Y_2 \geq 0) = \frac{1}{4} + \frac{1}{2\pi} \sin^{-1} \rho$$

□

We shall refer to the last expression as the *arcsine formula*. The origin of this formula can be traced back to Stieltjes [23] and Sheppard [21] who published their respective works a century ago. At a later stage we will show that the arcsine formula is the basis for a spectral representation for (expected) zero-crossing counts and that it has an interesting formal counterpart that holds quite generally. The formula has a rather clever extension to the case of three jointly normal random variables but the extension stops short of simple generalization when four or more random variables are present. This will be discussed in the next section.

Remark. In the proofs of Theorem 2.4 and Corollary 2.1 we assume that $|\rho| < 1$. But the arcsine formula also holds for the degenerate case $|\rho| = 1$, as is evident from Theorem 2.2. In the normal case, when $\rho = 1$ we are effectively dealing with the probability that a single normal random variable with zero mean exceeds zero. This probability is $1/2$, which is also the value obtained from the arcsine formula by plugging in $\rho = 1$. When $\rho = -1$ we have an event that cannot occur and the arcsine formula yields the value 0 for the orthant probability as it should.

Example 2.13: Other Probabilities

Let X, Y be jointly normal with zero means, arbitrary variances, and correlation ρ. Then

$$\tfrac{1}{2} = P(Y \geq 0) = P(X \geq 0, Y \geq 0) + P(X < 0, Y \geq 0)$$

Therefore

$$P(X < 0, Y \geq 0) = \frac{1}{2} - \frac{1}{4} - \frac{1}{2\pi} \sin^{-1} \rho = \frac{1}{4} - \frac{1}{2\pi} \sin^{-1} \rho$$

Similarly

$$\tfrac{1}{2} = P(X < 0) = P(X < 0, Y < 0) + P(X < 0, Y \geq 0)$$

and

$$P(X < 0, Y < 0) = \frac{1}{2} - \frac{1}{4} + \frac{1}{2\pi} \sin^{-1} \rho = \frac{1}{4} + \frac{1}{2\pi} \sin^{-1} \rho$$

■

Example 2.14: The Uniform Distribution on the Unit Disc Is Ellipsoidal

Let X_1, X_2 be jointly uniformly distributed on the unit disc as in Example 2.7: $f(x_1, x_2) = 1/\pi$ for $x_1^2 + x_2^2 \leq 1$, but $f(x_1, x_2) = 0$ otherwise. For u in $[0, \infty)$ define

$$\psi(u) = \begin{cases} \frac{1}{\pi} & \text{if } u \leq 1 \\ 0 & \text{otherwise} \end{cases}$$

Then

$$f(\mathbf{x}) = \psi(\mathbf{x}'\mathbf{x})$$

This defines an ellipsoidal distribution with $\boldsymbol{\Sigma}$ being the identity matrix. From Example 2.9 we know that X_1, X_2 are uncorrelated. The standardized variables are also jointly ellipsoidal and uncorrelated, and by Theorem 2.4,

$$P(X_1 \geq 0, X_2 \geq 0) = \tfrac{1}{4}$$

as is well expected.

■

2.4.6 The Multivariate Normal Distribution

The (univariate) normal and the bivariate normal distributions are special cases of the *multivariate normal distribution*, a distribution that plays an important role in applications. There is a vast literature on the multivariate normal distribution, its properties, and its numerous applications in estimation, detection, hypothesis testing, and control, to mention only a few areas of applications. Very detailed accounts of general properties of the multivariate normal distribution can be found in [19, ch. 8] and [2, ch. 2]. We will be content with a very concise introduction to this topic.

We treat all vectors as column vectors. Accordingly, we let

$$\mathbf{X} = (X_1, \ldots, X_N)'$$

be a vector of random variables such that

$$\mu_i = E(X_i)$$

and

$$\sigma_{ij} = \text{Cov}(X_i, X_j)$$

The *mean* of X is the $N \times 1$ vector of means

$$\boldsymbol{\mu} = (\mu_1, \ldots, \mu_N)'$$

and the *covariance matrix* is defined by the $N \times N$ square matrix

$$\boldsymbol{\Sigma} = \begin{pmatrix} \sigma_{11} & \sigma_{12} & \cdot & \cdot & \cdot & \sigma_{1N} \\ \sigma_{21} & \sigma_{22} & \cdot & \cdot & \cdot & \sigma_{2N} \\ \cdot & \cdot & \cdot & \cdot & \cdot & \cdot \\ \cdot & \cdot & \cdot & \cdot & \cdot & \cdot \\ \cdot & \cdot & \cdot & \cdot & \cdot & \cdot \\ \sigma_{N1} & \sigma_{N2} & \cdot & \cdot & \cdot & \sigma_{NN} \end{pmatrix}$$

Because $\sigma_{jj} = \text{Var}(X_j)$ the matrix $\boldsymbol{\Sigma}$ is sometimes also called the *variance-covariance* matrix. Such a matrix is always symmetric, having the variances on the main diagonal and the covariances everywhere else, and it can be shown to be nonnegative definite.

We say [19] that X has a multivariate normal distribution if and only if every linear combination

$$a_1X_1 + \cdots + a_NX_N$$

has a (univariate) normal distribution. In particular, when $\boldsymbol{\Sigma}$ is positive definite it is nonsingular and X has the density

$$f(\mathbf{x}) = (2\pi)^{-N/2}|\boldsymbol{\Sigma}|^{-1/2} \exp[-\frac{1}{2}(\mathbf{x} - \boldsymbol{\mu})'\boldsymbol{\Sigma}^{-1}(\mathbf{x} - \boldsymbol{\mu})]$$

If $\boldsymbol{\Sigma}$ is singular (has no inverse), the density does not exist but the *characteristic function* always exists and is given by

$$\begin{aligned} E(e^{i\mathbf{t}'\mathbf{X}}) &\equiv E(\cos(\mathbf{t}'\mathbf{X})) + iE(\sin(\mathbf{t}'\mathbf{X})) \\ &= \exp(i\boldsymbol{\mu}'\mathbf{t} - \frac{1}{2}\mathbf{t}'\boldsymbol{\Sigma}\mathbf{t}) \end{aligned}$$

where $\mathbf{t} = (t_1, \ldots, t_N)'$ is a real vector and $i = \sqrt{-1}$. The characteristic function plays the same role as the moment-generating function, but whereas the moment-generating function need not exist, the characteristic function always exists and can be used in generating moments and in identifying distributions. Because the characteristic function of the multivariate normal distribution depends on $\boldsymbol{\mu}$ and $\boldsymbol{\Sigma}$ only, so does the distribution itself. This is significant for applications because means, variances, and covariances can be estimated quite accurately from data, and hence the distribution itself can be approximated once data become available. The notation $N(\boldsymbol{\mu}, \boldsymbol{\Sigma})$ is used to denote the multivariate normal distribution with mean vector $\boldsymbol{\mu}$ and covariance matrix $\boldsymbol{\Sigma}$, and when a random vector X follows this distribution, we write

$$X \sim N(\boldsymbol{\mu}, \boldsymbol{\Sigma})$$

We now note the following useful facts (compare with Theorem 2.3).

1. From the definition it follows that the marginal distribution of any subset of components of X has a multivariate (univariate if only one component is taken) normal distribution with the corresponding means, variances and covariances.
2. Again from the definition, if $\mathbf{A}$ is a $q \times N$ matrix then

$$\mathbf{AX} \sim N(\mathbf{A}\boldsymbol{\mu}, \mathbf{A\Sigma A}')$$

3. Uncorrelated components of X are mutually independent.

The last assertion follows from the characteristic function.

Example 2.15: A Three Dimensional Orthant Probability [6]

The two-dimensional orthant probability discussed in Section 2.4.5 can be extended to the three-dimensional case using an elementary probability formula. We write AB to denote the event A and B and similarly ABC means A and B and C. Then

$$\begin{aligned} P(ABC) &= 1 - P(A^c \text{ or } B^c \text{ or } C^c) \\ &= 1 - [P(A^c) + P(B^c) + P(C^c)] \\ &\quad + [P(A^cB^c) + P(A^cC^c) + P(B^cC^c)] \\ &\quad - P(A^cB^cC^c) \end{aligned}$$

Now let $\mathbf{X} = (X_1, X_2, X_3)'$ have a multivariate normal distribution with zero means, variances equal to 1, and correlation ρ_{ij} between X_i and X_j. This implies that every pair (X_i, X_j) has a bivariate normal distribution. By the symmetry of the multivariate normal distribution

$$P(X_1 \geq 0, X_2 \geq 0, X_3 \geq 0) = P(X_1 < 0, X_2 < 0, X_3 < 0)$$

Therefore by the above formula and Example 2.13 we have

$$\begin{aligned} 2P(X_1 \geq 0, X_2 \geq 0, X_3 \geq 0) &= 1 - \sum_j P(X_j < 0) + \sum_{i<j} P(X_i < 0, X_j < 0) \\ &= 1 - \frac{3}{2} + (\frac{3}{4} + \frac{1}{2\pi}\sin^{-1}\rho_{12} + \frac{1}{2\pi}\sin^{-1}\rho_{13} + \frac{1}{2\pi}\sin^{-1}\rho_{23}) \end{aligned}$$

and so we obtain an extension of the arcsine formula

$$P(X_1 \geq 0, X_2 \geq 0, X_3 \geq 0) = \frac{1}{8} + \frac{1}{4\pi}(\sin^{-1}\rho_{12} + \sin^{-1}\rho_{13} + \sin^{-1}\rho_{23})$$

To date no simple extension of this formula to higher dimensions is known, but there are various results concerning special cases [15].

■

2.5 MODES OF CONVERGENCE

In many statistical applications the notion of stochastic convergence of a sequence of random variables is indispensable. This is particularly so in large sample estimation problems, where the asymptotic properties of estimators are of interest. An example of this is the convergence of the zero-crossing rate observed in a time series of length N, as $N \to \infty$. Another is the convergence of the sample autocorrelation. Yet another example, albeit not as obvious, is the convergence of spectral distributions when viewed as probability distributions. Also, in many cases it is much easier to obtain the asymptotic distribution of an estimator rather than its small sample distribution. Thus, in deference to convergence, we define next the principal modes of stochastic convergence, and mention related results used in applications relative to a sequence of random variables $X_1, X_2, X_3, \ldots$, and a limiting random variables X. Stochastic convergence is treated in numerous books including [3], [4], to which the reader is referred for proofs of the following mathematical statements.

We say that the sequence X_n converges to X *almost surely* (or equivalently *with probability one*) if

$$P(\lim_{n\to\infty} X_n = X) = 1$$

and write $X_n \stackrel{\text{a.s.}}{\to} X$. At times this is replaced by "$X_n \to X$ with probability one." In the above definition a tacit reference to measure theoretic concepts has been made. An alternative, perhaps more intuitive, definition can be made by the adaptation of the fact that $X_n \stackrel{\text{a.s.}}{\to} X$ if and only if

$$\lim_{n\to\infty} P(\sup_{m\geq n} |X_m - X| > \epsilon) = 0$$

for all $\epsilon > 0$.

The sequence of random variables X_n converges to X *in probability* if for all $\epsilon > 0$,

$$\lim_{n\to\infty} P(|X_n - X| > \epsilon) = 0$$

and we write $X_n \stackrel{P}{\to} X$. Almost sure convergence implies convergence in probability, but the converse is false. The limit is unique (a.s.) in the sense that if there is also a random variable Y such that $X_n \stackrel{P}{\to} Y$, then $X = Y$ with probability one. Convergence in probability implies convergence in probability of continuous "memoryless transformations": if φ is a continuous function and $X_n \stackrel{P}{\to} X$, then also $\varphi(X_n) \stackrel{P}{\to} \varphi(X)$.

A mode of convergence that is particularly useful for time series analysis is *convergence in mean square*: X_n converges to X in mean square if

$$\lim_{n\to\infty} E(|X_n - X|^2) = 0$$

provided $E(|X_n|^2) < \infty$ for all n, and $E(|X|^2) < \infty$. This is denoted by $X_n \stackrel{\text{m.s.}}{\to} X$. Mean square convergence is also referred to as convergence in quadratic mean. This is why mean square convergence is sometimes denoted alternatively by $X_n \stackrel{\text{q.m.}}{\to} X$. Convergence in mean square implies convergence in probability. The converse also holds when $\{X_n\}_{n\geq 1}$ is a sequence of uniformly bounded random variables (a.s.). That is, if for all n and a fixed K, $P(|X_n| \leq K) = 1$, and

$X_n \xrightarrow{P} X$, then $X_n \xrightarrow{\text{q.m.}} X$. If $\{X_n\}_{n\geq 1}$ is a sequence of normal random variables converging in mean square to X, then X is also a normal random variable (see Problem 14). If $X_m \xrightarrow{\text{q.m.}} X$, and $Y_n \xrightarrow{\text{q.m.}} Y$, then $E(X_m Y_n) \to E(XY)$, and $\text{Cov}(X_m, Y_n) \to \text{Cov}(X, Y)$, as $m, n \to \infty$.

Convergence in distribution is defined in terms of distribution functions. Consider a sequence $\{F_n\}_{n\geq 1}$ of cumulative distribution functions defined on the real line. We say that F_n converges to F, another cumulative distribution function on the real line, if

$$\lim_{n\to\infty} F_n(x) = F(x)$$

for all continuity points x of F. This is known as *weak convergence*, written

$$F_n \Rightarrow F$$

If in addition $X_n \sim F_n$, and $X \sim F$, we say that X_n *converges in distribution (or in law)* to X, and we write

$$X_n \xrightarrow{\mathcal{L}} X$$

Thus, $\lim_{n\to\infty} P(X_n \leq x) = P(X \leq x)$ for all x such that $P(X = x) = 0$ if and only if $X_n \xrightarrow{\mathcal{L}} X$. Each of the three previous modes of convergence implies convergence in distribution. However, if c is a constant, then $X_n \xrightarrow{\mathcal{L}} c$ if and only if $X_n \xrightarrow{P} c$. Convergence of a sequence of characteristic functions to a limiting characteristic function is equivalent to weak convergence of the corresponding sequence of distribution functions. This useful fact is referred to as the *continuity theorem* [3, p. 359]. Another useful fact is that if $\{X_n\}_{n\geq 1}$ is a sequence of uniformly bounded random variables (a.s.), and $X_n \xrightarrow{\mathcal{L}} X$, then $E(X_n^r) \to E(X^r)$, as $n \to \infty$.

On several occasions we shall appeal to *monotone* and *dominated* convergence.

Let $0 \leq X_1 \leq X_2 \leq X_3 \leq \cdots$. If $X_n \uparrow X$ a.s., then as $n \to \infty$, $E(X_n) \to E(X)$. This fact is referred to as the *monotone convergence theorem*.

Suppose $|X_n| \leq Y$ with probability one, and $E(|Y|) < \infty$. If $X_n \xrightarrow{\text{a.s.}} X$, then $E(|X_n|) < \infty$, $E(|X|) < \infty$, and as $n \to \infty$, $E(X_n) \to E(X)$. This fact is referred to as the *dominated convergence theorem*.

Example 2.16: An Application of Dominated Convergence [4, p. 178]

As an illustration of dominated convergence, we show that almost sure convergence implies convergence in distribution.

Note that $X_n \xrightarrow{\text{a.s.}} X$ implies $e^{itX_n} \xrightarrow{\text{a.s.}} e^{itX}$. Since for any t, $|e^{itX_n}| \leq 1$, by the dominated convergence theorem we have, $E[e^{itX_n}] \xrightarrow{\text{a.s.}} E[e^{itX}]$. Thus the sequence of characteristic functions $\{E[e^{itX_n}]\}$ converges to $E[e^{itX}]$ for all real t. This last fact, however, implies that $X_n \xrightarrow{\mathcal{L}} X$.

■

2.6 PROBLEMS AND COMPLEMENTS

1. Let X be distributed as $N(0, \sigma^2)$. Find the distribution, mean, and variance of $Y = |X|$.

2. Suppose X_1, X_2, X_3, X_4 are independent random variables, each distributed as $N(0, 1)$. Define:
$$V = \frac{X_1X_2 + X_3X_4}{\sqrt{X_2^2 + X_4^2}}$$
Show that V is also distributed as $N(0, 1)$. (Hint: What happens when X_2, X_4 are fixed numbers?)

3. Let X be a random variable with mean μ, and variance σ^2. The median, for any random variable X, is defined as any number m such that $P(X \geq m) \geq 1/2$, and $P(X \leq m) \geq 1/2$. Show that,
$$|\mu - m| \leq \sigma$$
(Hint: This surprising and not so well-known inequality is a consequence of the fact that $E(|X - a|)$ is minimum for $a = m$. Start by writing $|\mu - m| = |E(X - m)| \leq E|X - m|$.)

4. Suppose X_1, X_2, X_3, X_4 have a joint (multivariate) normal distribution such that $E(X_i) = 0, i = 1, 2, 3, 4$. Show that
$$E(X_1X_2X_3) = 0$$
and
$$E(X_1X_2X_3X_4) = E(X_1X_2)E(X_3X_4) + E(X_1X_3)E(X_2X_4) + E(X_1X_4)E(X_2X_3)$$
(Hint: Use the characteristic function of (X_1, X_2, X_3, X_4).)

5. Suppose that (X, Y) have a bivariate normal distribution and that the two marginal distributions are $N(0, \sigma^2)$. Let $G(\cdot)$ be any "nice" function, and define
$$b = \frac{E(XY)}{E(Y^2)}$$
Show
$$E[XG(Y)] = bE[YG(Y)]$$
(Hint: $X - bY$ and Y are independent.)

6. Let $\mathbf{\Sigma}$ be an $n \times n$ covariance matrix, and $\mathbf{b} = (b_1, \ldots, b_n)'$ any real vector. Show that
$$\mathbf{b}'\mathbf{\Sigma}\mathbf{b} \geq 0$$
That is, $\mathbf{\Sigma}$ is nonnegative definite.

7. Let X have a mixed lognormal distribution as in Example 2.3$'$. Show that for $\tau > 0$,
$$E(X) = \beta(\tau)P(X > \tau)$$
where
$$\beta(\tau) = \frac{\exp\left(\mu + \dfrac{\sigma^2}{2}\right)}{1 - \Phi\left(\dfrac{\log \tau - \mu}{\sigma}\right)} \geq \tau$$

8. Suppose a multinomial experiment may result in one of k possible outcomes $A_1, A_2, \ldots, A_k$. Repeat the experiment n independent times, and let X_i be the number of times A_i occurs. Suppose the probability that A_i occurs (on any trial) is p_i. Clearly, $\sum_{i=1}^{k} X_i = n$. Thus the X_i must be negatively correlated; if X_i grows, it grows at the expense of the other random variables. Show that

$$E(X_i) = np_i, \qquad \text{Var}(X_i) = np_i(1 - p_i)$$

and for $i \neq j$,

$$\text{Cov}(X_i, X_j) = -np_ip_j$$

(Hint: Note that $X_i + X_j$ has a binomial distribution with parameters n, and $p_i + p_j$; find $\text{Var}(X_i + X_j)$.)

9. *A variance decomposition [11].* Consider any random variable X, and define

$$Y_c \equiv \begin{cases} 1, & \text{if } X \leq c \\ 0, & \text{if } X > c \end{cases}$$

Further, let $W(c) = E(\text{Var}[X|Y_c])$, $B(c) = \text{Var}[E(X|Y_c)]$. $W(c), B(c)$ are called *within* and *between* population variance, respectively.

(a) Show

$$\text{Var}(X) = W(c) + B(c)$$

(b) Suppose X has mixed distribution defined by $G(x) = (1-p)H(x) + pF(x)$ where $0 < p < 1$,

$$H(x) = \begin{cases} 0, & \text{if } x < 0 \\ 1, & \text{if } x \geq 0 \end{cases}$$

and $F(x)$ is a continuous distribution function. Show that

$$B(c) = -E^2(X) + \frac{E^2(XI[X > c])}{P(X > c)} + \frac{E^2(XI[X \leq c])}{P(X \leq c)}$$

where $I[A] = 1$ if the event A occurs, and is 0 otherwise.

(c) Suppose that the distribution in (b) is mixed lognormal as in Example 2.3′. Set

$$x = x(c) \equiv \frac{\log c - \mu}{\sigma}$$

Show that [22],

$$B(c) = pe^{2\mu+\sigma^2} \left\{ \frac{[1 - \Phi(x-\sigma)]^2}{1 - \Phi(x)} + p \left[\frac{\Phi^2(x-\sigma)}{1 - p[1 - \Phi(x)]} - 1 \right] \right\}$$

(Hint: Suppose $L \sim \Lambda(\mu, \sigma^2)$, and $LL \sim \Lambda(\mu + j\sigma^2, \sigma^2)$. Then

$$E(L^j I[L \in A]) = e^{j\mu + (1/2)j^2\sigma^2} P(LL \in A)$$

[1, p. 12].)

(d) Plot $B(c)$ for a mixed lognormal distribution with parameters $p = 0.1, \mu = 1, \sigma = 1$, and compute the value of the threshold c that maximizes $B(c)$. This is the optimal level that minimizes the within population variance.

(e) For a mixed random variable as in (b), show that the conditional variance $V(c) = \text{Var}[X|X > c] = E(X^2|X > c) - E^2(X|X > c)$ depends on F only. Interestingly enough, for many distributions the conditional variance increases with the level c [11],[22]. Plot $V(c)$ for a mixed lognormal distribution with parameters $p = 0.1, \mu = 1, \sigma = 1$.

10. *Houdré–Kagan Variance Inequality [8].* Let $X \sim \mathcal{N}(0,1)$, and $G(x)$ a complex-valued $2n$ times differentiable function such that

$$E|G^{(k)}(X)|^2 < \infty, \qquad k = 0, 1, \ldots, 2n, \quad n \geq 1$$

Then Houdré–Kagan inequality states that

$$\sum_{k=1}^{2n} \frac{(-1)^{k+1}}{k!} E|G^{(k)}(X)|^2 \leq \text{Var}[G(X)] \leq \sum_{k=1}^{2n-1} \frac{(-1)^{k+1}}{k!} E|G^{(k)}(X)|^2$$

The special case for a real-valued function $G(x)$,

$$\text{Var}[G(X)] \leq E[G'(X)]^2$$

is called Chernoff's inequality [5]. Another generalization of Chernoff's inequality regarding infinitely divisible distributions is given in [24]. The Houdré–Kagan inequality gives tighter upper bounds. Show that for $G(x) = x^3$, Chernoff's inequality gives 27 for an upper bound versus 15 (exact!) obtained—with sufficiently large n—from Houdré–Kagan. Also show that for $G(x) = \sin(x)$, the respective bounds are approximately 0.5676 versus 0.4323. For the standard normal cdf, $\Phi(x)$, the respective bounds are $\sqrt{3}/6\pi$ versus (approx.) $0.8915\sqrt{3}/6\pi$.

11. Let $\{X_n\}$ be a sequence of random variables such that $P(X_n = 1/n) = P(X_n = -1/n) = 1/2$. Show that $X_n \to 0$ both in mean square and almost surely.

12. *Uniqueness of the limit.* Suppose $X_n \stackrel{\text{a.s.}}{\to} X$, and also $X_n \stackrel{\text{m.s.}}{\to} X^*$. Show that $X^* = X$ with probability one.

(Hint: When $Y_n \stackrel{P}{\to} Y$, then there is a subsequence Y_{n_k} which converges to Y almost surely.)

13. *A method for proving a.s. convergence.* Let $\{X_n\}$ be a sequence of random variables, and suppose that $\sum_{n=1}^{\infty} E|X_n|^2 < \infty$. Show that $X_n \stackrel{\text{a.s.}}{\to} 0$.

(Hint: Use monotone convergence to pull the expectation in front of the sum and note that a.s. $\sum_{n=1}^{\infty} |X_n|^2 < \infty$.)

As an example, let $\{X_n\}$ be a sequence of uncorrelated random variables with mean 0 and variance 1, and consider $Y_N = (1/N)\sum_{n=1}^{N} X_n$. Then $E|Y_N|^2 = 1/N$, and

$$\sum_{N=1}^{\infty} E|Y_{N^2}|^2 = \sum_{N=1}^{\infty} \frac{1}{N^2} = \frac{\pi^2}{6}$$

so that $Y_{N^2} \stackrel{\text{a.s.}}{\to} 0$.

14. *Limit of a sequence of normal random variables [4, p. 177].* Show that if $\{X_n\}_{n\geq 1}$ is a sequence of normal random variables such that $X_n \stackrel{\text{m.s.}}{\to} X$, then X also is a normal random variable.

(Hint: Mean square convergence implies $E[X_n] \to E[X]$, and $E[X_n^2] \to E[X^2]$. Thus, show that the characteristic function of X_n converges to a normal characteristic function.)

REFERENCES

[1] Aitchison, J. and J.A.C. Brown, *The Lognormal Distribution*, Cambridge, England: Cambridge Univ. Press, 1963.

[2] Anderson, T.W., *An Introduction to Multivariate Statistical Analysis* (2nd ed.), New York: Wiley, 1984.

[3] Billingsley, P., *Probability and Measure* (2nd ed.), New York: Wiley, 1986.

[4] Brémaud, P., *An Introduction to Probabilistic Modeling*, New York: Springer-Verlag, 1988.

[5] Chernoff, H., "A note on an inequality involving the normal distribution," *Ann. Probab.*, Vol. 9, pp. 533–535, 1981.

[6] David, F.N., "A note on the evaluation of the multivariate normal integral," *Biometrika*, Vol. 40, pp. 458–459, 1953.

[7] He, S. and B. Kedem, "On the Stieltjes-Sheppard orthant probability formula," Tech. Rep. TR-89-69, Mathematics Dept., Univ. of Maryland, College Park, 1989.

[8] Houdré, C. and A. Kagan, "Variance inequalities for functions of Gaussian variables." Submitted to J. Theor. Prob., 1992.

[9] Jensen, D.R., "Multivariate distributions," in *Kotz-Johnson Encyclopedia of Statistical Sciences*, Vol. 6, New York: Wiley, pp. 43–55, 1988.

[10] Kale, B.K., "Normality of linear combinations of nonnormal random variables," *Am. Math. Mon.*, Vol. 77, pp. 992–995, 1970.

[11] Karlin, S., "Some results on optimal partitioning of variance and monotonicity with truncation level, in *Statistics and Probability: Essays in Honor of C.R. Rao*, G. Kallianpur, P.R. Krishnaiah, J.K. Ghosh, eds., Amsterdam: North-Holland, pp. 375–382, 1982.

[12] Kedem, B., L.S. Chiu, and G.R. North, "Estimation of mean rain rate: Application to satellite observations," *J. Geophys. Res.*, Vol. 95, No. D2, pp. 1965–1972, Feb. 1990.

[13] McGraw, D.K. and J.F. Wagner, "Elliptically symmetric distributions," *IEEE Trans. Inform. Theory*, Vol. IT-14, No. 1, pp. 110–120, 1968.

[14] Meyer, P.L., *Introductory Probability and Statistical Applications* (2nd ed.), Reading, Mass.: Addison-Wesley, 1970.

[15] Owen, D.B., "Orthant probabilities," in *Kotz-Johnson Encyclopedia of Statistical Sciences*, Vol. 6, New York: Wiley, pp. 521–523, 1988.

[16] Papoulis,A., *Probability, Random Variables, and Stochastic Processes* (2nd ed.), New York: McGraw-Hill, 1984.

[17] Parzen, E., *Modern Probability Theory and its Applications,* New York: Wiley, 1960.

[18] Pratt, J.W and J.D. Gibbons, *Concepts of Nonparametric Theory,* New York: Springer-Verlag, 1981.

[19] Rao, C.R., *Linear Statistical Inference and Its Applications,* New York: Wiley, 1973.

[20] Ross, S.M., *Stochastic Processes*, New York: Wiley, 1983.

[21] Sheppard, W.F., "On the application of the theory of error to the cases of normal distribution and normal correlation," *Philos. Trans. R. Soc. London*, A 192, pp. 101–167, 1899.

[22] Short, D.A., K. Shimizu, and B. Kedem, "Optimal thresholds for the estimation of area rain rate moments by the threshold method," *J. Appl. Meteorol.*, Vol. 32, No. 2, pp. 182–192, 1993.

[23] Stieltjes, T.J., "Extrait d'une lettre addressee a M. Hermite," *Bull. Sci. Math.* 2eme serie, Vol. 13, pp. 170–172, 1889.

[24] Vitale, R.A., "A differential version of the Efron-Stein inequality: Bounding the variance of a function of an infinitely divisible variable," *Statist. Prob. Lett.*, Vol. 7, pp. 105–112, Sept. 1988.

3

Elements of Stationary Processes

Stationary stochastic processes in discrete time provide a natural framework for the development of the theory and applications of higher order crossings. Hence, familiarity with some elementary notions from the theory of stationary processes is essential. This theory will be presented here briefly in its most basic level sufficient for the development in chapters to come, and will be concerned mainly with discrete time processes. All the results carry over to the continuous time case with the appropriate modifications. In this chapter we highlight the basic notions of autocovariance and autocorrelation and their relation to the spectrum. Also, we explain the notion of mixed spectrum and its manifestation in the spectral distribution function and in the process itself. We do not intend to prove much by way of formal proofs, but rather bring home certain useful facts by resorting to discussions and examples. Thus we will explain how to use the spectral representation of a wide sense stationary process, but will not prove this celebrated result. For our purposes it is important to know that the spectral representation is a useful as well as convenient operational tool in the study of spectral properties. For example, it can be used in determining the effect of a time-invariant linear operation on a process and on its spectrum.

There are many excellent books on stationary processes and their applications in various fields. Here is a scanty description of some of these books. Introductory level treatments with an inclination toward engineering applications can be found in [8], [21]. References that emphasize spectral and correlation analyses and related statistical procedures and applications are [1], [3], [4], [9], [19], [20], [22], [24], and [30], while more theoretical treatments can be found in [6] and [26].

3.1 THE NOTION OF STATIONARITY

3.1.1 Types of Oscillation

The physical sciences and the various engineering fields are abundant with examples of what appears to be oscillatory random data in steady-state measured in time. For example, such steady-state oscillation can be observed as a function of time in the velocity of a turbulent flow in a wind tunnel, in any of the components of the earth polar motion, in acoustic signals, and in electroencephalogram (EEG) records. When observing such data on an oscilloscope or a time plot, the first thing that strikes the mind is the monotonous repetitiveness or "periodicity" that the data display about a fixed level. Clearly, since there are many different types of mechanisms that generate oscillatory or "periodic" data, some "noisier" than others, from physical considerations alone it is understood that the nature of the observed oscillation cannot possibly be the same in all cases. Thus we are led to the problem of classifying or modeling oscillation observed in time, and proceed to describe a few typical cases.

Consider an infinite sequence $\{\epsilon_t\}$, $t = 1, 2, 3, \ldots$, of independent normal random variables, each with mean zero and variance σ_ϵ^2. Such a sequence is referred to as a sequence of *independently and identically distributed* random variables or simply an independently and identically distributed (IID) sequence. The probability that ϵ_1 is above 0 is $1/2$, the probability that ϵ_1, ϵ_2 are both above 0 is $1/4$, the probability that $\epsilon_1, \epsilon_2, \epsilon_3$ are all above 0 is $1/8$, and so on. Because the probability decreases exponentially fast, there is an extremely small chance that a long sequence will have all its values above 0. Similarly, there is an equally small chance that a long sequence will have all its value below 0. The infinite random sequence therefore must fluctuate around the 0 level ad infinitum in what we may call completely random oscillation, and will appear to the eye surprisingly "periodic" because of the many passes from positive to negative values and vice versa, an observation made by Slutsky [27], and illustrated in Figure 3.1(a). Observe that since $\{\epsilon_t\}$ is an IID sequence, the probability distribution of any subsequence $\epsilon_{t_1}, \epsilon_{t_2}, \ldots, \epsilon_{t_n}$ is the same as that of the shifted subsequence $\epsilon_{t_1+\tau}, \epsilon_{t_2+\tau}, \ldots, \epsilon_{t_n+\tau}$. This means that in the absence of any external interference, any section of the sequence $\{\epsilon_t\}$ observed in time should appear to the eye quite similar as a quick glance at Figure 3.1(a) shows.

In contrast, we consider next a pure random harmonic oscillation. Let A, B be two independent normal random variables, each with mean 0 and variance σ^2, and consider the infinite sequence $\{X_t\}$, $t = 1, 2, 3, \ldots$,

$$X_t = A\cos(\frac{\pi}{3}t) + B\sin(\frac{\pi}{3}t)$$

When A, B admit values from an appropriate random experiment compatible with the normal assumption, the sequence oscillates about level 0 as a pure nonrandom sinusoid with a period of 6 time units as in Figure 3.1(b). From Figure 3.1(a) and 3.1(b) it is clear that the two types of oscillation, completely random

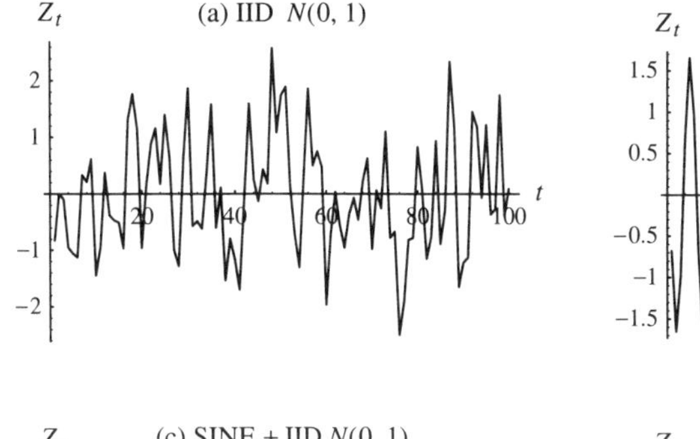

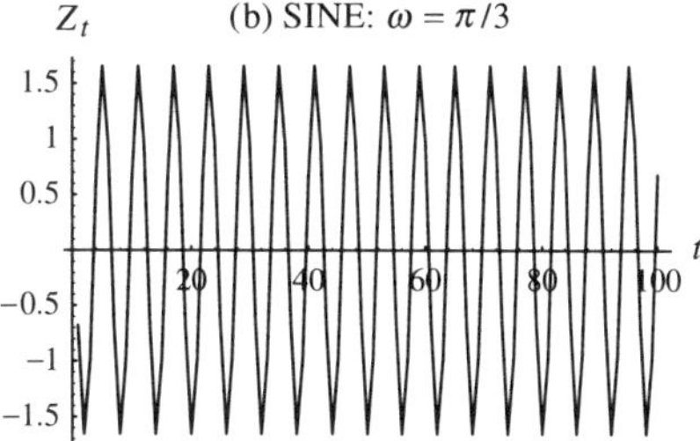

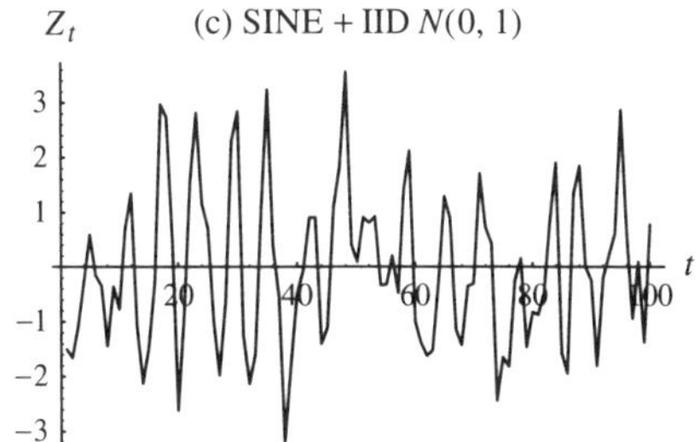

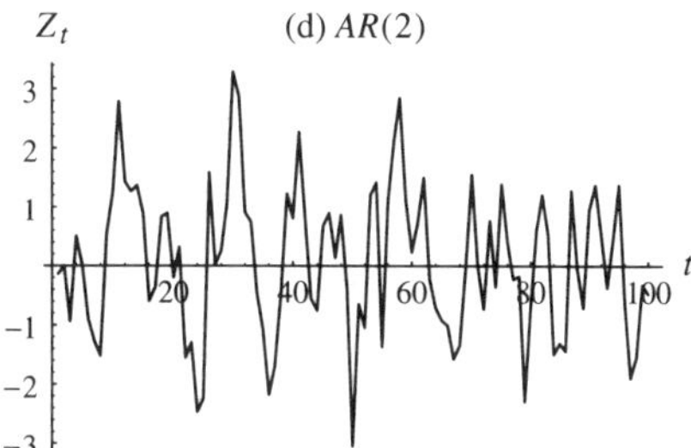

Figure 3.1: *Realizations from*: (*a*) ϵ_t, IID $N(0,1)$. (*b*) $Z_t = 0.975902\cos(\pi t/3) - 1.34624\sin(\pi t/3)$. (*c*) $Z_t = 0.975902\cos(\pi t/3) - 1.34624\sin(\pi t/3) + \epsilon_t$. (*d*) $Z_t = 0.5Z_{t-1} - 0.25Z_{t-2} + \epsilon_t$.

and sinusoidal, are profoundly different, and yet the monotonous steady-state appearance of both as random functions of time points to some common denominator after all.

We can mix the two types of oscillation to create a third. By adding to the pure harmonic sequence the IID random sequence the pure sinusoidal oscillation is altered to produce a hybrid whose graph for $t = 1, 2, \ldots, 100$ is shown in Figure 3.1(c). The new sequence of "signal plus noise,"

$$Z_t = A\cos(\frac{\pi}{3}t) + B\sin(\frac{\pi}{3}t) + \epsilon_t$$

where A, B are assumed independent of $\{\epsilon_t\}$, will appear more sinusoidal with a period of 6 time units if A, B are large by comparison with ϵ_t, and will appear quite random if the opposite holds. Yet, the monotonous overall oscillation about level 0 still persists. Processes made of the superposition of one or more sinusoids plus "noise" are sometimes called models of hidden periodicities, a term coined last century [25].

It is even possible to construct a fourth type of sequence that oscillates randomly with a *pseudo*period of 6 time units and that is neither a pure sinusoid nor an additive mixture of a sinusoid and an IID sequence. Such a sequence is given by the linear system

$$X_t - 0.5X_{t-1} + 0.25X_{t-2} = \epsilon_t$$

driven by the IID sequence $\epsilon_t, t = 1, 2, 3, \ldots$. A representative plot of $\{X_t\}, t = 1, 2, \ldots, 100$, is shown in Figure 3.1(d). This random sequence, which consists of *dependent* random variables, has an intrinsic period of 6 time units (obtained by solving the homogeneous difference equation $Y_t - 0.5Y_{t-1} + 0.25Y_{t-2} = 0$; the solution is a damped sinusoid with a period of 6 time units [29, p. 35]) and thus displays a monotonous pseudoperiodic fluctuation about level 0. The sequence is a special case of the second order stationary autoregressive process introduced by Yule [31] in 1927.

We have discussed four types of oscillation produced by random sequences where the common theme is the monotonous fluctuation about a fixed level. How is it possible to explain or describe the different types of oscillatory behavior illustrated by these examples? A comprehensive answer to this problem is given by the rich theory of stationary stochastic processes, the subject matter of this chapter. From the foregoing discussion we may guess that the displayed monotonous behavior or "sameness" of subsequences and oscillation are connected in some sense, and to a large degree this is indeed the case. In the theory of stationary processes the intuitive "sameness" of subsequences is replaced by the formal stationarity assumption that requires the invariance of joint distributions with respect to time shifts, and as we shall see, the mere notion of stationarity itself leads to an elegant classification of oscillatory phenomena as expressed by the spectral distribution function.

One of the important achievements of the theory of stationary processes is that it is able to put on the same footing different types of models, and in particular models of hidden periodicities and pseudoperiodic models. Under conditions that guarantee stationarity one theory explains them all. Thus all the four cases discussed above are special cases of stationary processes (both strict and wide sense) that differ in their spectra, and the theory of stationary processes provides the necessary tools for their analysis. The theory tells us that the completely random sequence and the second order autoregressive process possess continuous spectra (in this sense they belong to the same class), the random sinusoid has a discrete spectrum, and the hybrid of a pure random sinusoid plus an independent IID sequence is an example of a mixed spectrum stationary process.

It should be noted that not all steady-state phenomena are stationary and that fluctuations about a fixed level do not guarantee stationarity. However, in many cases it is still possible to stationarize these processes by various tricks such as random time shifts, repeated differencing, or removal of a trend component. The transformation from nonstationary to stationary behavior is useful for estimation purposes, because a great deal is known about estimation in stationary processes. See Example 3.8 and Problem 21.

3.1.2 Definition of a Stationary Process

A stochastic or random process $\{Z_t\}$ is a collection of random variables, real or complex-valued, indexed by t where t takes values in some index set T.

Examples of T are $[0, 1]$, $[0, \infty)$, $\{\ldots, -2, -1, 0, 1, 2, \ldots\}$, and $\{1, 2, 3, \ldots\}$. For our purposes it is convenient to think of t as time. When T is an interval, such as $(-\infty, \infty)$, the process is called a continuous time process, and it is called a discrete time process when T is a discrete set such as $\{\ldots, -2, -1, 0, 1, 2, \ldots\}$. In the sequel we confine ourselves almost exclusively to discrete time processes. But clearly, regardless of the type of the time parameter t, the random variables Z_t may be continuous, discrete, or of mixed type.

Consider first a real-valued process $\{Z_t\}$. As the process evolves or unfolds in time, that is, being observed in time, it produces a sequence in discrete time and a function (not necessarily continuous) in continuous time. Both are referred to as *realizations* or *sample functions* or simply as *time records*. Since for each fixed t, Z_t is a random variable, each observation of the whole process at all possible time points results, in most cases of practical interest, almost certainly in a new realization. The collection of all possible realizations is called the *ensemble*. The ensemble enables us to interpret a stochastic process as a collection of functions where randomness is interjected by a "random" experiment whose outcomes are sample functions. From this point of view, for each fixed t, the values of Z_t fall on realizations, thus creating a distribution of all possible values at time t across the ensemble. Shifting between the two interpretations—collection of random variables versus collection of sample functions—is a useful maneuver in the study of stochastic processes.

Sections of infinite realizations are also called realizations or simply *time series*. Thus, Figure 3.1 displays four discrete time finite realizations, or time series, which appear to be continuous because we always connect the resulting time values by line segments.

A remark about notation. In our notation above, $\{Z_t\}$ stands for either discrete or continuous time process. However, the more conventional notation for continuous time processes is $\{Z(t)\}$, while the symbolism $\{Z_t\}$ is reserved for discrete time processes. From now on, with only very few exceptions, we shall comply with the conventional notation.

In the remainder of the chapter we let $T = \{\ldots, -2, -1, 0, 1, 2, \ldots\}$. In the exceptional cases when reference is made to "continuous time," we have in mind $T = (-\infty, \infty)$.

For any collection of n time points $t_1 < t_2 < \cdots < t_n$, the probability distribution of the random vector $(Z_{t_1}, Z_{t_2}, \ldots, Z_{t_n})$,

$$P(Z_{t_1} \leq z_1, Z_{t_2} \leq z_2, \ldots, Z_{t_n} \leq z_n)$$

is called a *finite dimensional (probability) distribution*. In principle, we must specify the finite dimensional distributions in order to describe the stochastic process probabilistically. In practice, however, we are quite content with much less, and instead of distributions provide some moment conditions to describe the process.

From the finite dimensional distributions we can obtain probabilities and, when they exist, moments. Suppose $f_{t_1}(z)$ is the probability density function of Z_{t_1}. Then the mean (first moment) of Z_{t_1} is given by

$$E[Z_{t_1}] = \int_{-\infty}^{\infty} z f_{t_1}(z)\, dz$$

This may be interpreted as an "average across the ensemble" at time t_1. That is, the average of all the possible values which Z_{t_1} can assume across the ensemble. In general, the mean may be different at different time points, and if we define $m(t) \equiv E[Z_t]$, the result is a function of time called the *mean function*. The mean function may assume any form, but $E[Z_t - m(t)] = 0$ always.

Suppose $f_{t_1,t_2}(z_1, z_2)$ is the joint probability density function of (Z_{t_1}, Z_{t_2}). Then a second order moment is given by the expectation

$$E[Z_{t_1} Z_{t_2}] = \int_{-\infty}^{\infty} \int_{-\infty}^{\infty} z_1 z_2 f_{t_1,t_2}(z_1, z_2)\, dz_1\, dz_2$$

This may be interpreted as the average of the values which the product $Z_{t_1} Z_{t_2}$ assumes across the ensemble. Recall that the covariance $\mathrm{Cov}[Z_{t_1}, Z_{t_2}]$ of Z_{t_1} and Z_{t_2} is given by

$$\mathrm{Cov}[Z_{t_1}, Z_{t_2}] = E\{[Z_{t_1} - E(Z_{t_1})][Z_{t_2} - E(Z_{t_2})]\} = E[Z_{t_1} Z_{t_2}] - E[Z_{t_1}]E[Z_{t_2}]$$

Clearly, $\mathrm{Cov}[Z_{t_1}, Z_{t_1}] = \mathrm{Var}[Z_{t_1}]$. The function $R(s,t) \equiv \mathrm{Cov}[Z_s, Z_t]$ is called the *covariance* or *autocovariance function*.

The *correlation* or *autocorrelation function*, $\rho(s,t)$, is defined from the normalized covariance function,

$$\rho(s,t) \equiv \frac{R(s,t)}{\sqrt{R(s,s)}\sqrt{R(t,t)}}$$

and satisfies $\rho(s,t) = \rho(t,s)$, and $|\rho(s,t)| \le \rho(t,t) = 1$.

In the same way we can obtain higher order moments and moments of functions of random variables from the process. For example, let

$$f_{t_1,t_2,\ldots,t_n}(z_1, z_2, \ldots, z_n)$$

be the n-dimensional probability density of the random vector from the process $(Z_{t_1}, Z_{t_2}, \ldots, Z_{t_n})$. Then,

$$E[\varphi(Z_{t_1}, Z_{t_2}, \ldots, Z_{t_n})] =$$

$$\int_{-\infty}^{\infty} \int_{-\infty}^{\infty} \cdots \int_{-\infty}^{\infty} \varphi(z_1, z_2, \ldots, z_n) f_{t_1,t_2,\ldots,t_n}(z_1, z_2, \ldots, z_n)\, dz_1\, dz_2 \ldots dz_n$$

provided the expectation exists. Note that in all the expressions involving integrals it is assumed that the random variables are continuous (nothing to do with continuous or discrete time!). When the random variables are discrete, the integrals are replaced by summations.

An important example of a stochastic process is a *Gaussian* or *normal* process. A real-valued process $\{Z_t\}$, $t \in T$, is called Gaussian process if for all

$t_1, t_2, \ldots, t_n \in T$, the joint distribution of $(Z_{t_1}, Z_{t_2}, \ldots, Z_{t_n})$ is multivariate normal. Since a multivariate normal is completely specified by the vector of means and the covariance matrix, the finite dimensional distributions are completely determined from the mean function $m(t) = E[Z_t]$ and the covariance function $R(s,t) = \text{Cov}[Z_s, Z_t]$. Thus, for a Gaussian process we need to supply only first and second order moment information in order to obtain the finite dimensional distributions, a fact which makes Gaussian processes easier to handle than many other types of processes.

A stochastic process $\{Z_t\}$ is said to be a *strictly stationary* process if the joint distribution of $(Z_{t_1}, Z_{t_2}, \ldots, Z_{t_n})$ is the same as the joint distribution of $(Z_{t_1+\tau}, Z_{t_2+\tau}, \ldots, Z_{t_n+\tau})$ for all $t_1, t_2, \ldots, t_n$, n, and τ. In other words, as long as the relative distances between the time points are fixed, the joint distribution does not change.

Strict stationarity implies that Z_t and $Z_{t+\tau}$ have the same distribution for all t and τ. Hence, if first order moments exist, put $\tau = -t$ to see that

$$E[Z_t] = E[Z_{t+\tau}] = E[Z_0] = m \tag{3.1}$$

where m is a constant. Likewise, stationarity means that (Z_t, Z_s) and $(Z_{t+\tau}, Z_{s+\tau})$ have the same distribution for all t, s, and τ, and in particular for $\tau = -s$. Therefore, if second order moments exist,

$$E[Z_t Z_s] = E[Z_{t+\tau} Z_{s+\tau}] = E[Z_{t-s} Z_0]$$

or

$$R(t,s) = \text{Cov}[Z_t, Z_s] = R(t-s)$$

where we use the same symbol R in an obvious way. Thus, the autocovariance function of a real-valued stationary process is a function of the time *lag* τ only,

$$R(\tau) = \text{Cov}[Z_{t+\tau}, Z_t] = \text{Cov}[Z_\tau, Z_0] \tag{3.2}$$

Clearly, $\text{Var}[Z_t] = R(0)$, and $R(\tau) = R(-\tau)$. Likewise, the autocorrelation is a function of τ only,

$$\rho(\tau) = \frac{R(\tau)}{R(0)} \tag{3.3}$$

The autocorrelation $\rho(\tau)$ measures the correlation between $Z_{t+\tau}$ and Z_t as a function of the indices difference independently of t. Obviously, $\rho(\tau) = \rho(-\tau)$, and $|\rho(\tau)| \leq \rho(0) = 1$.

The notation $R(\tau), \rho(\tau)$ is usually reserved for continuous time. We shall mostly use the more suggestive notation R_k, ρ_k, $k = \ldots, -2, -1, 0, 1, 2, \ldots$, when dealing with discrete time stationary processes. With this convention, R_0 is the variance and ρ_1 is the first order autocorrelation.

In general, strict stationarity and the existence of higher order moments imply

$$E[Z_t Z_{t+t_1} Z_{t+t_2}] = E[Z_0 Z_{t_1} Z_{t_2}]$$

which is a function of t_1 and t_2 only, and

$$E[Z_t Z_{t+t_1} Z_{t+t_2} Z_{t+t_3}] = E[Z_0 Z_{t_1} Z_{t_2} Z_{t_3}]$$

which is a function of t_1,t_2, and t_3, and so on. It is worth noting that in recent years there is a growing interest in higher order moments and functions thereof called cumulants [3], [24]. For example, assuming that $E[Z_t] = 0$, the fourth order cumulant function, $\kappa(t_1, t_2, t_3)$, is given by

$$\kappa(t_1, t_2, t_3) = E[Z_0 Z_{t_1} Z_{t_2} Z_{t_3}] - \{R_{t_1} R_{t_2 - t_3} + R_{t_2} R_{t_1 - t_3} + R_{t_3} R_{t_1 - t_2}\}$$

and comes into play in the statistical estimation of the autocovariance. If the process is Gaussian, $\kappa(t_1, t_2, t_3) \equiv 0$. Hence, for a zero mean stationary Gaussian process,

$$E[Z_0 Z_{t_1} Z_{t_2} Z_{t_3}] = R_{t_1} R_{t_2 - t_3} + R_{t_2} R_{t_1 - t_3} + R_{t_3} R_{t_1 - t_2}$$

A great deal of the theory of stationary processes only requires the fulfillment of the conditions (3.1) and (3.2). In general, a process which satisfies (3.1) and (3.2) is called *weakly stationary* or *stationary in the wide sense* or sometimes is said to be *second order stationary*. A strictly stationary process need not be weakly stationary unless second order moments exist, and conversely, since weak stationarity is less restrictive it does not imply strict stationarity. *However, for the important special case of a Gaussian process, strict and wide sense stationarity coincide* (see Problem 1).

In this book, in order to avoid confusion, *stationary* is taken to mean strict stationarity together with the assumption that second order moments exist. By this convention, a stationary process is stationary in both wide and strict senses. Since the Gaussian assumption plays an important role in the coming chapters, the convention becomes quite palatable. However, the appropriate adjective "strict" or "weak" (or "wide sense") will be added explicitly whenever needed in emphasizing a particular property associated with one of the stationarity senses.

3.1.3 Complex-Valued Stationary Processes

So far we have spoken of real-valued processes; however, there is nothing special about real processes, and the theory can easily be extended to cover complex-valued processes as well. Complex processes may arise in several ways, one of which is through filters with a complex impulse response function applied to a real process. In this case, the original process is real, but the filtered process is complex. Clearly, every result concerning the complex case automatically holds for real processes as well, and in what follows we find it convenient to introduce new concepts in terms of complex-valued stationary processes.

A complex random variable X has the general form $X = U + iV$ where U, V are real random variables. We have the complex conjugate $\overline{X} = U - iV$, $|X|^2 = X\overline{X} = U^2 + V^2$, $E[X] \equiv E[U] + iE[V]$, $E[|X|^2] = E[U^2] + E[V^2]$. For two complex-valued random variables X, Y, the covariance is defined by

$$\text{Cov}(X, Y) \equiv E\{[X - E(X)][\overline{Y - E(Y)}]\} = E[X\overline{Y}] - E[X]E[\overline{Y}]$$

and $\text{Var}[X] = \text{Cov}[X, X] = E[|X - E(X)|^2]$. When $E(X) = E(Y) = 0$, then $\text{Cov}(X, Y) = E[X\overline{Y}]$, and when $\text{Cov}(X, Y) = 0$, we say that X and Y are uncorrelated.

Similarly, we define a complex-valued stochastic process $\{Z_t\}$ from two real-valued processes $\{U_t\}$,$\{V_t\}$,

$$Z_t = U_t + iV_t$$

The complex process is described probabilistically from the finite dimensional distributions of the vectors, $(U_{t_1}, U_{t_2}, \ldots, U_{t_n}, V_{t_1}, V_{t_2}, \ldots, V_{t_n})$, for all possible $t_1, t_2, \ldots, t_n$, and n, and is said to be strictly stationary if the finite dimensional distributions remain unaltered under time shifts, just as in the real case. When $\{Z_t\}$ is strictly stationary, so are the real-valued processes $\{U_t\}$ and $\{V_t\}$ (why?).

Wide or weak sense stationarity is defined as in the real case with the appropriate modifications. Thus we say that $\{Z_t\}$ is weakly stationary if the mean is a (complex) constant m, say,

$$E[Z_t] = E[U_t] + iE[V(t)] = m$$

and the autocovariance is a function of the time lag,

$$R_k = E[Z_{t+k} - m][\overline{Z_t - m}] = E[Z_{t+k}\overline{Z_t}] - |m|^2$$

Observe that the *variance* $R_0 = E[|Z_t - m|^2]$ *is always real-valued.* As in the real case, the autocorrelation function is obtained as the quotient (assuming of course $R_0 \neq 0$),

$$\rho_k = \frac{R_k}{R_0}$$

We have

$$R_{-k} = \overline{R_k}, \qquad |R_k| \leq R_0 \tag{3.4}$$

and

$$\rho_{-k} = \overline{\rho_k}, \qquad |\rho_k| \leq 1$$

Observe that for all complex numbers $a_1, a_2, \ldots, a_N$, and integers t_1, $t_2, \ldots, t_N$, with $N \geq 1$,

$$\begin{aligned} 0 \leq \text{Var}[\sum_{j=1}^{N} a_j Z_{t_j}] &= \text{Cov}[\sum_{j=1}^{N} a_j Z_{t_j}, \sum_{l=1}^{N} a_l Z_{t_l}] \\ &= \sum_{j=1}^{N}\sum_{l=1}^{N} a_j \overline{a_l} \,\text{Cov}[Z_{t_j}, Z_{t_l}] \\ &= \sum_{j=1}^{N}\sum_{l=1}^{N} a_j \overline{a_l} R_{t_j - t_l} \end{aligned} \tag{3.5}$$

That is, R_k is *nonnegative definite*. The important implication of this ostensibly simple fact is that R_k admits a Fourier or spectral representation in terms of a uniquely defined bounded monotone nondecreasing real function F such that $F(-\pi) = 0$,

$$R_k = \int_{-\pi}^{\pi} e^{ik\lambda}\, dF(\lambda), \qquad k = 0, \pm 1, \pm 2, \ldots \tag{3.6}$$

This fact is due to the German mathematician G. Herglotz (1911) [9, p. 34], and $F(\lambda)$ is called *spectral distribution function*. Conversely, it is easy to see that a complex-valued sequence R_k which admits the Fourier representation (3.6) is nonnegative definite (see Problem 7). Note that the Hermitian property (3.4) of the autocovariance implies that $F(\omega)$ is real-valued.

An alternative representation of (3.6), proved in the appendix to this chapter, is as an integral[1] with respect to a unique (see appendix to this chapter) *spectral distribution* (measure) $F(\Lambda)$ defined over (measurable) subsets Λ of $(-\pi, \pi]$,

$$R_k = \int_{-\pi}^{\pi} e^{ik\lambda} F\,(d\lambda), \qquad k = 0, \pm 1, \pm 2, \ldots \tag{3.7}$$

where the relationship between the spectral distribution function and the spectral distribution is,

$$F(\lambda) = F((-\pi, \lambda])$$

with the understanding that we use the same symbol to denote two different, yet closely related, entities one of which is a function and the other a set function. We sometimes refer to $F(\lambda)$ or $F(\Lambda)$ as "spectrum." Both give complete information as to the distribution of power over frequency, but it is a little easier to interpret $F(\Lambda)$. The spectral measure $F(\Lambda)$ may be interpreted as the amount of power or variance associated with the spectral band Λ, where Λ is a subset of $(-\pi, \pi]$. The total power then is $R_0 = F(\pi) = F(-\pi, \pi]$.

By extending the definition of $F(\lambda)$ such that $F(\lambda) = 0$ for $\lambda \leq -\pi$, and $F(\lambda) = F(\pi)$ for $\lambda \geq \pi$, $F(\lambda)$ becomes, except for a normalization by a constant, a probability distribution function. *Thus, in summary, the autocovariance of every weakly stationary process in discrete time admits a Fourier representation in terms of a uniquely defined distribution function which, up to a constant, is a probability distribution function supported on* $(-\pi, \pi]$.

As is the case with probability distributions, a spectral distribution function may have a *spectral density* $f(\lambda)$ such that

$$F(\lambda) = \int_{-\pi}^{\lambda} f(\omega)\, d\omega, \qquad f(\lambda) = F'(\lambda), \qquad -\pi \leq \lambda \leq \pi$$

---

[1] Both (3.6) and (3.7) are called Lebesgue–Stieltjes integrals. In the symbolism $F(d\lambda)$, $\xi(d\lambda)$, etc., it is helpful to think of $d\lambda$ as a very small interval containing λ.

A sufficient condition for this is the absolute summability of the autocovariance, $\sum_{k=-\infty}^{\infty} |R_k| \leq \infty$, in which case we have the inverse transform,

$$f(\lambda) = \frac{1}{2\pi} \sum_{k=-\infty}^{\infty} e^{-i\lambda k} R_k \tag{3.8}$$

When a spectral density exists, $F(d\lambda) = f(\lambda)\, d\lambda$, and we can express (3.7) as

$$R_k = \int_{-\pi}^{\pi} e^{ik\lambda} f(\lambda)\, d\lambda \tag{3.9}$$

As an immediate consequence,

$$\text{Var}[Z_t] = R_0 = \int_{-\pi}^{\pi} f(\lambda)\, d\lambda \tag{3.10}$$

Thus, the spectral density describes how the variance—that is, power—is distributed over frequency. Since $F(\lambda)$ is monotone nondecreasing, and since $f(\lambda) = F'(\lambda)$, $f(\lambda) \geq 0\,, -\pi \leq \lambda \leq \pi$.

Everything we said about discrete time extends naturally to continuous time as well. The continuous time version of (3.6) is referred to as the *Wiener–Khintchine relationship* or *Wiener–Khintchine theorem* [30, p. 93],

$$R(\tau) = \int_{-\infty}^{\infty} e^{i\tau\lambda}\, dF(\lambda) \tag{3.11}$$

or in terms of the spectral distribution,

$$R(\tau) - \int_{-\infty}^{\infty} c^{i\tau\lambda} F(d\lambda) \tag{3.12}$$

with the understanding that

$$F(\lambda) = F((-\infty, \lambda])$$

When $R(\tau)$ tends to 0 as $|\tau|$ increases such that $\int_{-\infty}^{\infty} |R(\tau)|\, d\tau < \infty$, then a spectral density $f(\lambda)$ exists and the analog of (3.9) becomes,

$$R(\tau) = \int_{-\infty}^{\infty} e^{i\tau\lambda} f(\lambda)\, d\lambda \tag{3.13}$$

It is clear that, except for a normalizing constant, the autocorrelation is related to the spectral distribution in the same way as is the autocovariance. In discrete time this takes the form,

$$\rho_k = \int_{-\pi}^{\pi} e^{ik\lambda} \bar{F}(d\lambda) \tag{3.14}$$

where $\bar{F}(d\lambda) \equiv F(d\lambda)/R_0$ is the normalized spectral distribution. In the sequel we shall refer to any of the relationships between the autocovariance or the autocorrelation and the spectral distribution somewhat loosely as the Wiener–Khintchine relationship, without reference to discrete or continuous time. Note that this is a relationship between *time domain* and *spectral domain* functions. Today we know that the relationship between the spectrum and the autocorrelation also was understood by A. Einstein early in this century. See the reference to and a summary of Einstein's 1914 work on "fluctuating observations" in Notes 13a and 41 of [30, pp. 32, 88, Vol II].

The Real Case

In the real case, $R_k = R_{-k}$, and hence (3.7) reduces to

$$R_k = \int_{-\pi}^{\pi} \cos(k\lambda)F(d\lambda)$$

and similarly in continuous time $R(\tau) = R(-\tau)$ so that (3.12) becomes

$$R(\tau) = \int_{-\infty}^{\infty} \cos(\tau\lambda)F(d\lambda)$$

When a spectral density exists, by the symmetry of the autocovariance it is symmetric, $f(\lambda) = f(-\lambda)$, and in the preceding two equations we can replace $F(d\lambda)$ by $f(\lambda)\,d\lambda$:

$$R_k = \int_{-\pi}^{\pi} \cos(k\lambda)f(\lambda)\,d\lambda$$

and similarly for continuous time,

$$R(\tau) = \int_{-\infty}^{\infty} \cos(\tau\lambda)f(\lambda)\,d\lambda$$

3.1.4 Examples of Stationary Processes

Example 3.1: An IID Sequence

Suppose $\{Z_t\}$ constitutes a sequence of independently and identically distributed complex-valued random variables. Then $\{Z_t\}$ is strictly stationary. If in addition $E[Z_t] = m$, and $\text{Var}[Z_t] = \sigma^2 < \infty$, then

$$R_k = \begin{cases} \sigma^2, & \text{for } k = 0 \\ 0, & \text{for } k = \pm 1, \pm 2, \pm 3, \ldots \end{cases}$$

and the process is also wide sense stationary. In this case we can write

$$R_k = \frac{\sigma^2}{2\pi} \int_{-\pi}^{\pi} e^{ik\lambda}\,d\lambda$$

and this defines uniquely the spectral distribution function as

$$F(\lambda) = \frac{\sigma^2}{2\pi}(\lambda + \pi)$$

which can also be expressed in terms of the spectral density $f(\lambda) = \sigma^2/2\pi$,

$$F(\lambda) = \int_{-\pi}^{\lambda} f(\omega)\, d\omega = \frac{\sigma^2}{2\pi}\int_{-\pi}^{\lambda} d\omega, \qquad -\pi \le \lambda \le \pi$$

Note that we also have,

$$f(\lambda) = \frac{1}{2\pi}\sum_{k=-\infty}^{\infty} e^{-ik\lambda} R_k = \cdots + 0 + 0 + \frac{\sigma^2}{2\pi} + 0 + 0 + \cdots$$

If we extend the definition of $F(\lambda)$ such that $F(\lambda) = 0$, $\lambda \le -\pi$, and $F(\lambda) = \sigma^2$, $\lambda \ge \pi$, then, except for a normalizing constant, $F(\lambda)$ behaves like a continuous probability distribution function with a density supported on $[-\pi, \pi]$.

We observe, in particular, that for any sequence of uncorrelated real- or complex-valued random variables $\{u_t\}$ (this is a much weaker assumption than IID) with mean 0 and variance σ^2,

$$R_u(k) \equiv E[u_t \overline{u_{t-k}}] = \begin{cases} \sigma^2, & \text{for } k = 0 \\ 0, & \text{for } k - \perp 1, \perp 2, \perp 3, \ldots \end{cases}$$

has the same autocovariance as does the preceding iid sequence. Therefore such a sequence is weakly stationary with a flat spectral density $f(\lambda) = \sigma^2/2\pi$, $-\pi \le \lambda \le \pi$. For this reason a sequence of uncorrelated random variables with constant mean and variance is termed *white noise* by analogy with white light, which consists of waves with the same power regardless of frequency. As a rule, we always take the mean of white noise to be 0.

■

Example 3.2: Pure Random Sinusoid

Let A, B be uncorrelated real-valued random variables with mean 0 and variance σ^2, and define a real-valued process by

$$Z_t = A\cos(\omega_1 t) + B\sin(\omega_1 t), \qquad t = 0, \pm 1, \pm 2, \ldots \tag{3.15}$$

where $\omega_1 \in (0, \pi]$ is a constant frequency. Then since $E[A] = E[B] = 0, E[A^2] = E[B^2] = \sigma^2$, and $E[AB] = 0$, and remembering the formula for $\cos(\alpha - \beta)$,

$$\begin{aligned} R_k &= \text{Cov}[Z_{t+k}, Z_t] = E[Z_{t+k}Z_t] \\ &= E\{[A\cos\omega_1(t+k) + B\sin\omega_1(t+k)][A\cos(\omega_1 t) + B\sin(\omega_1 t)]\} \\ &= E\{[A\cos\omega_1(t+k)][A\cos(\omega_1 t)]\} + E\{[B\sin\omega_1(t+k)][B\sin(\omega_1 t)]\} \\ &= \sigma^2[\cos\omega_1(t+k)\cos(\omega_1 t) + \sin\omega_1(t+k)\sin(\omega_1 t)] \\ &= \sigma^2\cos(\omega_1 k), \qquad k = 0, \pm 1, \pm 2, \ldots \end{aligned} \tag{3.16}$$

Since the autocovariance depends on the lag k only, the process is weakly stationary. Also,

$$\rho_k = \frac{R_k}{R_0} = \cos(\omega_1 k), \qquad k = 0, \pm 1, \pm 2, \ldots \tag{3.17}$$

Hence both the autocovariance and autocorrelation are periodic with the same period $2\pi/\omega_1$ as the process itself. However, whereas the process is random, the autocovariance and autocorrelation are nonrandom (deterministic).

Define a function $F(\lambda)$ by,

$$F(\lambda) \equiv \begin{cases} 0, & -\pi \le \lambda < -\omega_1 \\ \frac{\sigma^2}{2}, & -\omega_1 \le \lambda < \omega_1 \\ \sigma^2, & \omega_1 \le \lambda \le \pi \end{cases} \tag{3.18}$$

Then $F(\lambda)$ is a right-continuous step function with two jumps of size $\frac{1}{2}\sigma^2$ at $\pm\omega_1$, such that $F(-\pi) = 0$, $F(\pi) = \sigma^2$, and we see that

$$\int_{-\pi}^{\pi} \cos(k\lambda)\, dF(\lambda) = \frac{\sigma^2}{2}\cos(-k\omega_1) + \frac{\sigma^2}{2}\cos(k\omega_1) = \sigma^2 \cos(\omega_1 k) = R_k$$

Therefore, by uniqueness, $F(\lambda)$ must be the spectral distribution function of $\{Z_t\}$.

So far we have not made use of any distributional assumption regarding the amplitudes except for the requirement that they are uncorrelated with mean 0 and variance σ^2. Assume now in addition that A, B are (jointly) normally distributed. Then, since A, B are uncorrelated, the additional normal assumption makes them independent $\mathcal{N}(0, \sigma^2)$ random variables. Consequently, linear combinations in the Z_t are normal, and hence the finite dimensional distributions are multivariate normal, which makes the process Gaussian. Therefore, the process is also strictly stationary (see Problem 1).

Now, write $A = R\cos\Theta$, $B = R\sin\Theta$ (note the difference between R and the autocovariance). Then

$$\begin{aligned} Z_t &= R\cos\Theta\cos(\omega_1 t) + R\sin\Theta\sin(\omega_1 t) \\ &= R\cos(\omega_1 t - \Theta) \end{aligned}$$

To investigate the distribution of the phase and amplitude in the new representation under the assumption of normal amplitudes, notice that $Y \equiv R^2 = A^2 + B^2$, and $\Theta = \tan^{-1}(B/A)$, and that the Jacobian of the transformation is $J = \frac{1}{2}\cos^2\theta + \frac{1}{2}\sin^2\theta = \frac{1}{2}$. Then the joint probability density of (R^2, Θ) is

$$g(y, \theta) = \frac{1}{4\pi\sigma^2}\exp\left\{-\frac{y}{2\sigma^2}\right\}$$

Hence, Θ is uniformly distributed in $(0, 2\pi)$ independently of $Y = R^2$, which is distributed as $\sigma^2\chi^2_{(2)}$. That is, R^2 is proportional to a chi square random variable with 2 degrees of freedom. This in turn implies that the distribution of $R = \sqrt{Y}$ is Rayleigh with probability density

$$g_R(r) = \begin{cases} \frac{r}{\sigma^2}\exp\left\{-\frac{r^2}{2\sigma^2}\right\}, & r \ge 0 \\ 0, & r < 0 \end{cases}$$

From this discussion we can see that, *in a sinusoidal model, the assumption of a uniform phase independently distributed of a random amplitude is quite plausible*. In fact, in applications the use of the random phase sinusoidal model

$$Z_t = \beta \cos(\omega_1 t + \phi)$$

where ϕ is uniformly distributed in $(0, 2\pi)$, and β is a *constant*, is widespread. This process is not Gaussian, but it is still strictly stationary since a time shift only gives a new phase which is again uniformly distributed over an interval of length 2π (see Problem 4). In the present representation, the only source of randomness is the uniformly distributed phase with a constant probability density $1/2\pi$ supported over $[0, 2\pi]$, and hence for every t,

$$E[Z_t] = \frac{\beta}{2\pi} \int_0^{2\pi} \cos(\omega_1 t + \phi)\, d\phi = 0$$

and the autocovariance is given by,

$$\begin{aligned} R_k = E[Z_t Z_{t-k}] &= \frac{\beta^2}{2\pi} \int_0^{2\pi} \cos(\omega_1 t + \phi) \cos(\omega_1 t - \omega_1 k + \phi)\, d\phi \\ &= \frac{\beta^2}{4\pi} \int_0^{2\pi} [\cos(2\omega_1 t - \omega_1 k + 2\phi) + \cos(\omega_1 k)]\, d\phi \\ &= \frac{\beta^2}{2} \cos(\omega_1 k), \qquad k = 0, \pm 1, \pm 2, \ldots \end{aligned} \tag{3.19}$$

Thus the autocovariance also is a sinusoid whose amplitude depends on the fixed amplitude of the process, while in (3.16) it depends on the variance of the process, the latter having a random amplitude. On the other hand, the autocorrelation is identical to (3.17),

$$\rho_k = \frac{R_k}{R_0} = \cos(\omega_1 k), \qquad k = 0, \pm 1, \pm 2, \ldots \tag{3.20}$$

By employing the same reasoning as in the random amplitude case, we find that $F(\lambda)$ is again a step function with two jumps of size $\beta^2/4$ at $\pm\omega_1$, such that $F(-\pi) = 0$, $F(\pi) = \beta^2/2$. Thus, the spectral distribution functions in the fixed and random amplitude models are of the same type, and when $\beta = \sqrt{2}\sigma$, they are identical. This shows that *two different processes may still have the same spectral distribution*.

Notice that F is very reminiscent of a discrete probability distribution supported at $\pm\omega_1$. For F to be a genuine discrete probability distribution function it has to be extended to the whole real line, as in the previous example, and normalized such that $F(\pi) = 1$.

■

Example 3.3: Sum of Random Sinusoids

The development in Example 3.2 can be easily extended to the case of a sum of random sinusoids. First consider random amplitudes. Let

$$Z_t = \sum_{j=1}^{p} \{A_j \cos(\omega_j t) + B_j \sin(\omega_j t)\}, \qquad t = 0, \pm 1, \pm 2, \ldots \tag{3.21}$$

where $A_1, \ldots, A_p, B_1, \ldots, B_p$ are (pairwise) uncorrelated real-valued random variables with mean 0, and $\text{Var}[A_j] = \text{Var}[B_j] = \sigma_j^2$ for all j. The frequencies ω_j are ordered real constants in $(0, \pi]$, $0 < \omega_1 < \omega_2 < \cdots < \omega_p \leq \pi$. Then for all t, $E[Z_t] = 0$, and

$$\begin{aligned} R_k &= E[Z_t Z_{t-k}] = \sum_{j=1}^{p} \sigma_j^2 \cos(\omega_j k) \\ &= \sum_{j=1}^{p} \{\frac{1}{2}\sigma_j^2 \cos(\omega_j k) + \frac{1}{2}\sigma_j^2 \cos(-\omega_j k)\}, \qquad k = 0, \pm 1, \pm 2, \ldots \end{aligned}$$

Therefore the process is weakly stationary. The autocorrelation is

$$\rho_k = \frac{R_k}{R_0} = \frac{\sum_{j=1}^{p} \sigma_j^2 \cos(\omega_j k)}{\sum_{j=1}^{p} \sigma_j^2}, \qquad k = 0, \pm 1, \pm 2, \ldots \tag{3.22}$$

By inspection, the spectral distribution function $F(\omega)$ is a nondecreasing step function with jumps of size $\frac{1}{2}\sigma_j^2$ at $\pm\omega_j$, and $F(-\pi) = 0, F(\pi) = R_0 = \sum_{j=1}^{p} \sigma_j^2$. The thing to remember is that the power, or intensity, at ω_j in (3.21) is proportional to σ_j^2 rather than the size of the amplitudes A_j, B_j.

As in the previous example, when in addition the A's and B's are (jointly) normally distributed, the process is Gaussian and thus also strictly stationary. Furthermore, the process can be expressed as a sum of cosines with independent random phases uniformly distributed in $(0, 2\pi)$, and independent of the amplitudes which are independent Rayleigh random variables.

With the frequencies as above, a model of random phases ϕ_j, uniformly and independently distributed in $(0, 2\pi)$, and fixed (nonrandom) amplitudes β_j, is given by

$$Z_t = \sum_{j=1}^{p} \beta_j \cos(\omega_j t + \phi_j), \qquad t = 0, \pm 1, \pm 2, \ldots \tag{3.23}$$

The process is stationary with mean 0 and autocovariance,

$$R_k = E[Z_t Z_{t-k}] = \sum_{j=1}^{p} \frac{\beta_j^2}{2} \cos(\omega_j k), \qquad k = 0, \pm 1, \pm 2, \ldots$$

The autocorrelation function is

$$\rho_k = \frac{R_k}{R_0} = \frac{\sum_{j=1}^{p} \beta_j^2 \cos(\omega_j k)}{\sum_{j=1}^{p} \beta_j^2}, \qquad k = 0, \pm 1, \pm 2, \ldots \tag{3.24}$$

The spectral distribution function $F(\omega)$ is a nondecreasing step function with jumps of size $\frac{1}{4}\beta_j^2$ at $\pm\omega_j$, and $F(\pi) = R_0 = \sum_{j=1}^{p} \frac{1}{2}\beta_j^2$. Thus, in the model (3.23) the power at ω_j is proportional to β_j^2, the square of the amplitude.

■

Example 3.4: Sum of Random Complex Exponentials

The complex version of the models (3.21), (3.23), is the almost periodic sequence

$$X_t = \sum_{j=-\infty}^{\infty} \xi_j e^{i\omega_j t}, \qquad t = 0, \pm 1, \pm 2, \ldots \tag{3.25}$$

where the ω_j are in $(-\pi, \pi]$, and where the ξ_j are zero mean (pairwise) uncorrelated, or *orthogonal*, complex-valued random variables, $E[\xi_j \overline{\xi}_k] = 0$, $j \neq k$, and $E[|\xi_j|^2] = \sigma_j^2$. The sum (3.25) converges in mean square when $\sum_{j=-\infty}^{\infty} \sigma_j^2 < \infty$, and can be made real by appropriate conditions on the frequencies and amplitudes (see Problem 6). By orthogonality we have,

$$\begin{aligned} R_k = EX_t\overline{X}_{t-k}] &= E\left[\sum_j \xi_j e^{i\omega_j t} \overline{\sum_n \xi_n e^{i\omega_n (t-k)}}\right] \\ &= \sum_j \sum_n e^{i[\omega_j t - \omega_n (t-k)]} E[\xi_j \overline{\xi}_n] \\ &= \sum_j \sigma_j^2 e^{i\omega_j k}, \qquad k = 0, \pm 1, \pm 2, \ldots \end{aligned} \tag{3.26}$$

Therefore, the process is weakly stationary. As in the real case the spectral distribution function is a step function,

$$F(\omega) = \sum_{\omega_j \le \omega} \sigma_j^2 \tag{3.27}$$

where the summation is over all j such that $\omega_j \le \omega$. Indeed this gives

$$R_k = \int_{-\pi}^{\pi} e^{ik\omega}\, dF(\omega) = \int_{-\pi}^{\pi} e^{ik\omega} F\,(d\omega)$$

Notice that in the complex case the jump in $F(\omega)$ at ω_j is of size σ_j^2, while in the real case—due to symmetry—it is of size $\frac{1}{2}\sigma_j^2$.

■

Example 3.5: Signal Plus Noise

Let $\{X_t\}$ be a "signal" as in (3.25), and let $\{\epsilon_t\}$ be a zero-mean wide sense stationary "noise" uncorrelated with $\{X_t\}$ and with a spectrum which possesses a spectral density $f_\epsilon(\omega)$,

$$F_\epsilon(\omega) = \int_{-\pi}^{\omega} f_\epsilon(\lambda)\, d\lambda$$

Similarly, use $F_x(\omega)$ to denote (3.27). Define a signal plus noise process $\{Z_t\}$ as

$$Z_t = X_t + \epsilon_t = \sum_{j=-\infty}^{\infty} \xi_j e^{i\omega_j t} + \epsilon_t, \qquad t = 0, \pm 1, \pm 2, \ldots \tag{3.28}$$

Then with obvious notation, since the signal and noise are uncorrelated,

$$R_k = E[Z_t \overline{Z}_{t-k}] = E[X_t \overline{X}_{t-k}] + E[\epsilon_t \bar{\epsilon}_{t-k}] \equiv R_k^x + R_k^\epsilon$$

We can see that if we define

$$F(\omega) \equiv F_x(\omega) + F_\epsilon(\omega) = \sum_{\omega_j \le \omega} \sigma_j^2 + \int_{-\pi}^{\omega} f_\epsilon(\lambda)\, d\lambda \tag{3.29}$$

then $F(\omega)$ must be the spectral distribution function of $\{Z_t\}$. This is an illustration of the general fact that *the spectrum of the sum of uncorrelated stationary processes is the sum of the individual spectra*.

Apart from a constant, by extending its definition to the whole real line, we recognize $F(\omega)$ as the distribution function of a mixed probability distribution whose atoms (points of jump) are the frequencies ω_j. Indeed this is an example of a *mixed spectrum*. In the jargon of stationary processes, the spectra of $\{X_t\}$,$\{\epsilon_t\}$, and $\{Z_t\}$ are *discrete*, *continuous*, and *mixed*, respectively.

As for the autocorrelation, it can be written as a convex combination of the autocorrelations of the signal, ρ_k^x, and that of the noise, ρ_k^ϵ,

$$\rho_k = \frac{R_0^x}{R_0^x + R_0^\epsilon} \rho_k^x + \frac{R_0^\epsilon}{R_0^x + R_0^\epsilon} \rho_k^\epsilon \tag{3.30}$$

As $R_0^x \to \infty$, $\rho_k \to \rho_k^x$.

■

Example 3.6: The Stationary *AR*(1) Process

Let $\{\epsilon_t\}$, $t = 0, \pm 1, \pm 2, \ldots$, be a sequence of uncorrelated real-valued random variables with mean zero and variance σ_ϵ^2 (i.e., white noise) and define a real-valued weakly stationary process $\{Z_t\}$ by the stochastic difference equation

$$Z_t = \phi_1 Z_{t-1} + \epsilon_t, \qquad t = 0, \pm 1, \pm 2, \ldots \tag{3.31}$$

where $|\phi_1| < 1$. The process (3.31) is called a *first order autoregressive process* and is commonly denoted by $AR(1)$. By repeated substitution we obtain,

$$Z_t = \epsilon_t + \phi_1 \epsilon_{t-1} + \phi_1^2 \epsilon_{t-2} + \cdots + \phi_1^{p-1} \epsilon_{t-(p-1)} + \phi_1^p Z_{t-p}$$

Observe now that since the process is weakly stationary, $E[Z_t^2]$ is equal to a finite constant uniformly in t. This, coupled with the fact that $|\phi_1| < 1$, gives as $p \to \infty$,

$$E[Z_t - (\epsilon_t + \phi_1 \epsilon_{t-1} + \phi_1^2 \epsilon_{t-2} + \cdots + \phi_1^{p-1} \epsilon_{t-(p-1)})]^2 = \phi_1^{2p} E[Z_{t-p}^2] \to 0$$

That is,

$$Z_t = \sum_{j=0}^{\infty} \phi_1^j \epsilon_{t-j} \tag{3.32}$$

where the sum converges in mean square. In fact, with $|\phi_1| < 1$, the representation (3.32) is *the stationary solution* of the stochastic difference equation (3.31). In other words, with

$|\phi_1| < 1$, there exists a stationary solution in terms of present and past ϵ_t, and hence our stationarity assumption is not without foundation. Since limits in mean square and E commute, we find

$$E[Z_t] = E\left\{\lim_{n\to\infty}\sum_{j=0}^{n}\phi_1^j\epsilon_{t-j}\right\} = \lim_{n\to\infty} E\left\{\sum_{j=0}^{n}\phi_1^j\epsilon_{t-j}\right\} = 0$$

Thus $E[Z_t] = 0$ for all t. Similarly, by the orthogonality of the ϵ_t ($E[\epsilon_t\epsilon_s] = 0,\ s \neq t$)

$$\begin{aligned}
R_0 = E[Z_t^2] &= E\left\{\lim_{m,n\to\infty}\sum_{j=0}^{m}\phi_1^j\epsilon_{t-j}\sum_{k=0}^{n}\phi_1^k\epsilon_{t-k}\right\} \\
&= \lim_{m,n\to\infty} E\left\{\sum_{j=0}^{m}\phi_1^j\epsilon_{t-j}\sum_{k=0}^{n}\phi_1^k\epsilon_{t-k}\right\} \\
&= \lim_{m,n\to\infty}\sum_{j=0}^{m}\sum_{k=0}^{n}\phi_1^{j+k}E[\epsilon_{t-j}\epsilon_{t-k}] \\
&= \sum_{j=0}^{\infty}\phi_1^{2j}\sigma_\epsilon^2 = \frac{\sigma_\epsilon^2}{1-\phi_1^2} \qquad (3.33)
\end{aligned}$$

In the same way, using the representation (3.32) and the orthogonality of the ϵ_t, we obtain $E[\epsilon_t Z_{t-k}] = 0,\ k = 1, 2, \ldots$, so that by multiplying both sides of (3.31) by Z_{t-k} and taking expectations,

$$R_k = \phi_1 R_{k-1}$$

This gives $\rho_1 = \phi_1$, and more generally , since $R_k = R_{-k}$,

$$\rho_k = \phi_1^{|k|}, \qquad k = 0, \pm 1, \pm 2, \ldots \qquad (3.34)$$

From this and (3.33) we obtain the autocovariance as

$$R_k = \frac{\sigma_\epsilon^2\phi_1^{|k|}}{1-\phi_1^2}, \qquad k = 0, \pm 1, \pm 2, \ldots \qquad (3.35)$$

Since the autocovariance is absolutely summable, a spectral density exists and is given by

$$\begin{aligned}
f(\lambda) &= \frac{R_0}{2\pi}\sum_{k=-\infty}^{\infty}\phi_1^{|k|}\cos(k\lambda) \\
&= \frac{R_0}{2\pi}\left\{1 + \sum_{k=1}^{\infty}\phi_1^k[\exp(ik\lambda) + \exp(-ik\lambda)]\right\} \\
&= \frac{R_0}{2\pi}\left\{1 + \frac{\phi_1 e^{i\lambda}}{1-\phi_1 e^{i\lambda}} + \frac{\phi_1 e^{-i\lambda}}{1-\phi_1 e^{-i\lambda}}\right\} \\
&= \frac{R_0}{2\pi}\cdot\frac{1-\phi_1^2}{1-2\phi_1\cos(\lambda)+\phi_1^2} \\
&= \frac{\sigma_\epsilon^2}{2\pi}\cdot\frac{1}{1-2\phi_1\cos(\lambda)+\phi_1^2}, \qquad -\pi \leq \lambda \leq \pi \qquad (3.36)
\end{aligned}$$

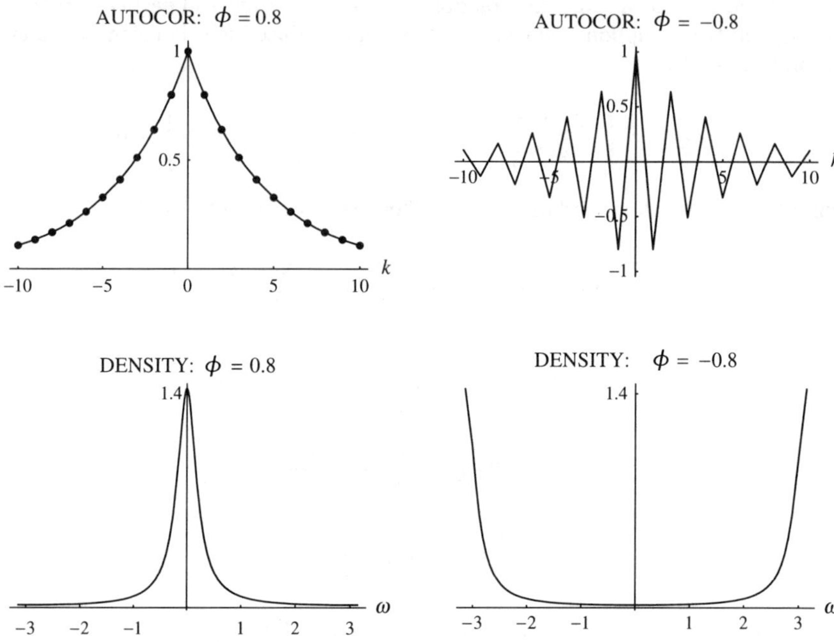

Figure 3.2: *Autocorrelation and corresponding spectral density of $AR(1)$ with the indicated parameter.*

Thus, the (weakly) stationary $AR(1)$ process has a continuous spectrum.

The $AR(1)$ process is special in that its parameter $\phi_1(=\rho_1)$ single-handedly controls both the autocorrelation and the normalized spectral density (i.e., the density divided by R_0 so that the total area under the curve is 1). The autocorrelation and the normalized spectral density for $\phi_1 = 0.8$ and $\phi_1 = -0.8$ are given in Figure 3.2. For $\phi_1 = 0.8$, neighboring observations are positively correlated and we expect dominance of low frequencies. Indeed, the spectral sensity gives most of its weight to (positive) low frequencies. For $\phi_1 = -0.8$, neighboring observations are negatively correlated, which means higher oscillation and an emphasis of (positive) high frequencies. Note that since the process is real, the spectral density is symmetric, and hence it is sufficient to consider it over the interval $[0, \pi]$.

■

Remark. In Example 3.6 we imposed the condition $|\phi_1| < 1$, and showed that the stochastic difference equation (3.31) has a stationary solution in terms of past and present ϵ_t. When $|\phi_1| > 1$, the stochastic difference equation still has a stationary solution in terms of *future* but not in terms of past and present ϵ_t (see Problem 13).

Example 3.7: The Stationary *AR(p)* Process

The stationary $AR(1)$ process (3.31) can be generalized by extending the order of the stochastic difference equation,

$$Z_t = \phi_1 Z_{t-1} + \phi_2 Z_{t-2} + \cdots + \phi_p Z_{t-p} + \epsilon_t, \qquad t = 0, \pm 1, \pm 2, \ldots \tag{3.37}$$

and $\{\epsilon_t\}$ is real-valued white noise as in the $AR(1)$ process. As such this defines a real-valued process referred to as an *autoregressive process of order p*, or simply $AR(p)$. In order to guarantee stationarity, the ϕ_j must satisfy certain restrictions expressed in terms of the roots (zeros) of the *characteristic equation*

$$\phi(z) \equiv 1 - \phi_1 z - \phi_2 z^2 - \cdots - \phi_p z^p = 0$$

Several stationary solutions exist, but this will be addressed later when we introduce the spectral representation of a weakly stationary process. In particular, we shall show then that if the roots of the characteristic polynomial are all outside the unit circle, then (3.37) has a unique (weakly) stationary solution given by the infinite mean square convergent sum,

$$Z_t = \sum_{j=0}^{\infty} h_j \epsilon_{t-j} \tag{3.38}$$

This implies that $E[Z_t] = 0$ for all t. The thing to notice is that only past and present ϵ_t are involved, and that (3.32) is only a special case with $h_j = \phi_1^j$. A representation of the form (3.38) is called an *infinite moving average*. To emphasize the fact that only past and present ϵ_t are involved, we say that the infinite moving average is *one-sided*, in which case the process is said to be *linear*.

From now on, by a (weakly) stationary $AR(p)$ process we shall mean the process (3.37) with the roots of $\phi(z)$ outside the unit circle, and a consequent infinite moving average representation (3.38). Then, $E[Z_t] = 0$, and $E[\epsilon_t Z_{t-k}] = 0$, $k = 1, 2, \ldots$.

With $k > 0$, multiply both sides of (3.37) by Z_{t-k} and take expectations. This gives a pth order difference equation,

$$R_k = \phi_1 R_{k-1} + \phi_2 R_{k-2} + \cdots + \phi_p R_{k-p}$$

or, dividing by R_0,

$$\rho_k = \phi_1 \rho_{k-1} + \phi_2 \rho_{k-2} + \cdots + \phi_p \rho_{k-p}, \qquad k > 0 \tag{3.39}$$

For $k = 1, 2, \ldots, p$, we obtain from (3.39) p linear equations called the *Yule–Walker* equations which allow us to determine the AR parameters $\phi_1, \phi_2, \ldots, \phi_p$ from $\rho_1, \rho_2, \ldots, \rho_p$, where the latter are estimated from data.

Given a time series, $Z_1, Z_2, \ldots, Z_N$, the ρ_k can be estimated by the *sample correlation* [22, p. 330],

$$\hat{\rho}_k = \frac{\sum_{t=1}^{N-k}(Z_t - \hat{m})(Z_{t+k} - \hat{m})}{\sum_{t=1}^{N-k}(Z_t - \hat{m})^2}, \qquad k = 1, 2, 3, \ldots \tag{3.40}$$

where

$$\hat{m} = \frac{1}{N}\sum_{t=1}^{N} Z_t$$

is the *sample mean* (see Problem 12). By substituting the $\hat{\rho}_k$ for the ρ_k, we obtain estimates for the ϕ_k by solving the Yule–Walker equations.

The general solution of (3.39) is [2, p. 55],

$$\rho_k = A_1 G_1^k + A_2 G_2^k + \cdots + A_p G_p^k$$

where the G_j are the reciprocal of the roots of $\phi(z) = 0$. Thus $|G_j| < 1, j = 1, \ldots, p$, which means that ρ_k falls off *exponentially fast* as $|k|$ increases. The fact that ρ_k is absolutely summable implies the existence of a spectral density which we shall derive later on.

■

3.2 REPRESENTATION OF STATIONARY PROCESSES

We have seen above that the autocovariance admits a spectral representation in terms of a spectral (measure) distribution $F(\Lambda)$,

$$R_k = \int_{-\pi}^{\pi} e^{ik\lambda} F(d\lambda), \qquad k = 0, \pm 1, \pm 2, \ldots$$

It turns out that, due to the spectral representation of the autocovariance, the process itself admits a similar representation, but in terms of a *random spectral measure* $\xi(\Lambda)$ which assigns uncorrelated (orthogonal) weights to nonoverlapping (Borel) subsets of $(-\pi, \pi]$. This, as we shall see, is a very useful fact.

3.2.1 A Stochastic Integral

In order to introduce our next result, it is helpful first to consider a new mathematical entity, $\xi(\Lambda)$, called *random spectral measure* and defined over subsets[2] Λ of $(-\pi, \pi]$. The random measure $\xi(\Lambda)$ acts very much like a nonrandom measure such as probability measure, except that the weight assigned to sets is random and complex-valued. Thus, for each fixed Λ, $\xi(\Lambda)$ is a complex-valued random variable. As with probability measures, we assume that for any two nonintersecting subsets Λ_1 and Λ_2,

$$\xi(\Lambda_1 \cup \Lambda_2) = \xi(\Lambda_1) + \xi(\Lambda_2), \qquad \Lambda_1 \cap \Lambda_2 = \emptyset \tag{3.41}$$

Since $\xi(\Lambda)$ is random, it makes sense to speak of its moments. In particular, it is required that for any subset Λ,

$$E[\xi(\Lambda)] = 0 \tag{3.42}$$

Now assume that over the subsets of $(-\pi, \pi]$ there is defined a real-valued nonrandom measure (simply think of probability) $F(\Lambda)$. As with probability, $F(\emptyset) = 0$. From the nonrandom $F(\Lambda)$, we formulate the third property of $\xi(\Lambda)$ as,

$$E[\xi(\Lambda_1)\overline{\xi(\Lambda_2)}] = F(\Lambda_1 \cap \Lambda_2) \tag{3.43}$$

---

[2]Technically, all subsets are assumed to be Borel subsets [6, p. 600].

The third property implies that

$$E|\xi(\Lambda)|^2 = F(\Lambda)$$

and, when $\Lambda_1 \cap \Lambda_2 = \emptyset$ (disjoint),

$$E[\xi(\Lambda_1)\overline{\xi(\Lambda_2)}] = F(\emptyset) = 0$$

In other words, the random variables $\xi(\Lambda_1)$ and $\xi(\Lambda_2)$ are *uncorrelated* or *orthogonal* when Λ_1 and Λ_2 are disjoint. In short, $\xi(\Lambda)$ is an orthogonal complex-valued set function.

Next, we construct a *stochastic integral* with respect to $\xi(\Lambda)$. Partition $(-\pi, \pi]$ into n disjoint intervals Λ_j, and consider the indicator function $I_{\Lambda_j}(\lambda)$, which is 1 for $\lambda \in \Lambda_j$, and is 0 otherwise. For complex a_j, let

$$g_n(\lambda) \equiv \sum_{j=1}^{n} a_j I_{\Lambda_j}(\lambda), \qquad \lambda \in (-\pi, \pi]$$

The integral of $g_n(\lambda)$ with respect to $\xi(\Lambda)$ is defined as

$$\int_{-\pi}^{\pi} g_n(\lambda)\xi\,(d\lambda) \equiv \sum_{j=1}^{n} a_j \xi(\Lambda_j)$$

Note that the partial sum defines a sequence of random variables. We are interested in limits in mean square of such sequences.

Suppose $g(\lambda)$, $\lambda \in (-\pi, \pi]$, is complex-valued (nonrandom) such that $\int_{-\pi}^{\pi} |g(\lambda)|^2 F\,(d\lambda) < \infty$. Then it can be approximated by functions $g_n(\lambda)$ with respect to F,

$$\lim_{n\to\infty} \int_{-\pi}^{\pi} |g_n(\lambda) - g(\lambda)|^2 F\,(d\lambda) = 0$$

We define

$$\int_{-\pi}^{\pi} g(\lambda)\xi\,(d\lambda) \equiv \lim_{n\to\infty} \int_{-\pi}^{\pi} g_n(\lambda)\xi\,(d\lambda) \tag{3.44}$$

where the limit is a limit in mean square. This is what we mean by the stochastic integral of $g(\lambda)$ with respect to $\xi(\Lambda)$. It can be shown that the limit exists and is independent of the particular sequence of approximating functions $g_n(\lambda)$ [6], [9].

The newly defined integral (3.44) is linear,

$$\int_{-\pi}^{\pi} [ag(\lambda) + bh(\lambda)]\xi\,(d\lambda) = a\int_{-\pi}^{\pi} g(\lambda)\xi\,(d\lambda) + b\int_{-\pi}^{\pi} h(\lambda)\xi\,(d\lambda) \tag{3.45}$$

and by the orthogonality of $\xi(\Lambda)$,

$$\begin{aligned} E\left\{\int_{-\pi}^{\pi} g(\lambda)\xi\,(d\lambda)\overline{\int_{-\pi}^{\pi} h(\omega)\xi\,(d\omega)}\right\} &= \int_{-\pi}^{\pi}\int_{-\pi}^{\pi} g(\lambda)\overline{h(\omega)}E[\xi\,(d\lambda)\overline{\xi\,(d\omega)}] \\ &= \int_{-\pi}^{\pi} g(\lambda)\overline{h(\lambda)}F\,(d\lambda) \end{aligned} \tag{3.46}$$

where we have used the operational convention

$$E[\xi(d\lambda)\overline{\xi(d\omega)}] = \begin{cases} F(d\lambda), & \text{if } \lambda = \omega \\ 0, & \text{if } \lambda \neq \omega \end{cases} \tag{3.47}$$

Here we think of $d\lambda$ and $d\omega$ intuitively as two very small disjoint intervals containing λ and ω, respectively. Since the mean is 0, (3.46) gives

$$\text{Var}\left\{\int_{-\pi}^{\pi} g(\lambda)\xi\,(d\lambda)\right\} = E\left|\int_{-\pi}^{\pi} g(\lambda)\xi\,(d\lambda)\right|^2 = \int_{-\pi}^{\pi} |g(\lambda)|^2 F\,(d\lambda) \tag{3.48}$$

3.2.2 Spectral Representation

We define a stochastic process $\{Z_t\}$ with mean zero from the integral (3.44) by replacing $g(\lambda)$ by $g(t,\lambda)$, $t = 0, \pm1, \pm2, \ldots$,

$$Z_t = \int_{-\pi}^{\pi} g(t,\lambda)\xi\,(d\lambda) \tag{3.49}$$

Then, from (3.46),

$$E[Z_s\overline{Z_t}] = \int_{-\pi}^{\pi} g(s,\lambda)\overline{g(t,\lambda)}F\,(d\lambda) \tag{3.50}$$

We derived (3.50) from (3.49). Conversely, *if (3.50) holds, then so does (3.49)* [9, p. 27].

Let $\{Z_t\}$, $t = 0, \pm1, \pm2, \ldots$, *be a zero mean complex-valued weakly stationary process. Then*

$$E[Z_s\overline{Z_t}] = R_{s-t} = \int_{-\pi}^{\pi} e^{is\lambda}e^{-it\lambda}F\,(d\lambda), \qquad s,t = 0, \pm1, \pm2, \ldots \tag{3.51}$$

Therefore, with $g(t,\lambda) = e^{it\lambda}$,

$$Z_t = \int_{-\pi}^{\pi} e^{it\lambda}\xi\,(d\lambda), \qquad t = 0, \pm1, \pm2, \ldots \tag{3.52}$$

where now the spectral distribution satisfies

$$E[\xi(d\lambda)\overline{\xi(d\omega)}] = \begin{cases} F(d\lambda), & \text{if } \lambda = \omega \\ 0, & \text{if } \lambda \neq \omega \end{cases} \tag{3.53}$$

The Fourier representation (3.52) is the celebrated *spectral representation* of a zero mean weakly stationary process in discrete time. Apparently starting with the works of A. Kolmogorov and H. Cramér in the early 1940s, the representation has been studied by quite a few investigators who devised different methods of proof, some of which are discussed in [22, Sec. 4.11]. In the above discussion we followed [9]. For an interesting historical account of the spectral representation, see Note 17 in [30, p. 33, Vol II].

The continuous time version of (3.52) requires the continuity of $R(\tau)$ at $\tau = 0$. This implies that $R(\tau)$ is continuous for all τ. The representation in the continuous time case is entirely analogous to (3.52) except that the limits of integration extend from $-\infty$ to ∞,

$$Z(t) = \int_{-\infty}^{\infty} e^{it\lambda} \xi\,(d\lambda), \qquad -\infty < t < \infty \tag{3.54}$$

Spectral Representation for Real-Valued Processes

When $\{Z_t\}$ is real-valued, of course we can still use (3.52), as we shall do on several occasions. However, it is sometimes convenient to rewrite the spectral representation in terms of "sines" and "cosines." This is done as follows. First note that real $\{Z_t\}$ implies,[3]

$$\xi(-d\lambda) = \overline{\xi(d\lambda)}$$

If we define real-valued $\xi_1(d\lambda)$ and $\xi_2(d\lambda)$ by

$$\xi_1(d\lambda) = \xi(d\lambda) + \overline{\xi(d\lambda)}$$

$$\xi_2(d\lambda) = i\{\xi(d\lambda) - \overline{\xi(d\lambda)}\}$$

then

$$\xi(d\lambda) = \tfrac{1}{2}\{\xi_1(d\lambda) - i\xi_2(d\lambda)\}$$

and $\xi_1(d\lambda)$ is even while $\xi_2(d\lambda)$ is odd,

$$\xi_1(-d\lambda) = \xi_1(d\lambda), \qquad \xi_2(-d\lambda) = -\xi_2(d\lambda)$$

Also, from the definition,

$$\xi_1(\{0\}) = 2\xi(\{0\}), \qquad \xi_2(\{0\}) = 0$$

By making the substitution $\xi(d\lambda) = \frac{1}{2}\{\xi_1(d\lambda) - i\xi_2(d\lambda)\}$ in (3.52), the cross terms vanish and the spectral representation in discrete time in the real case reduces to [19, p. 52],

$$Z_t = \frac{1}{2}\xi_1(\{0\}) + \int_{0+}^{\pi} \cos(t\lambda)\xi_1\,(d\lambda) + \int_{0+}^{\pi} \sin(t\lambda)\xi_2\,(d\lambda) \tag{3.55}$$

It is not difficult to see that $\xi_1(d\lambda)$ and $\xi_2(d\lambda)$ are orthogonal random measures, and that they are uncorrelated. We have for all $\lambda, \omega \in (0, \pi]$,

$$E[\xi_1(d\lambda)\overline{\xi_2(d\lambda)}] = 0$$

---

[3]It is helpful to think of $-d\lambda$ as a small interval containing $-\lambda$.

$$E[\xi_1(d\lambda)\overline{\xi_1(d\omega)}] = E[\xi_2(d\lambda)\overline{\xi_2(d\omega)}] = \begin{cases} 2F(d\lambda), & \text{if } \lambda = \omega \\ 0, & \text{if } \lambda \neq \omega \end{cases}$$

This indeed gives

$$R_k = F(\{0\}) + 2\int_{0+}^{\pi} \cos(k\lambda)F(d\lambda)$$

as it should.

3.3 DECOMPOSITION OF THE SPECTRAL DISTRIBUTION

In the previous section we have discussed the interdependency between the spectral representation of the autocovariance and that of the process itself as summarized in equations (3.51) and (3.52). The interdependency manifests itself again through a certain important decomposition of the spectrum and of the random spectral measure.

On several occasions we pointed out that the spectral distribution behaves very much like a probability measure except for a normalizing constant. Indeed, as is the case with probability, the spectral distribution (equivalently the spectral distribution function) can be decomposed into continuous and discrete components,[4]

$$F(d\lambda) = F_c(d\lambda) + F_d(d\lambda) \tag{3.56}$$

where the continuous part F_c can be expressed as an integral of a nonnegative spectral density $f(\lambda)$,

$$F_c(\Lambda) = \int_\Lambda f(\lambda)\,d\lambda$$

while the discrete part F_d can be expressed as a sum of a nonnegative mass function $p(\lambda)$ which assigns positive values only to discrete frequencies λ_j,

$$F_d(\Lambda) = \sum_{\lambda_j \in \Lambda} p(\lambda_j)$$

Note that $F_d(\Lambda) = 0$ if Λ contains none of the discrete λ_j. It follows that the power associated with a frequency band Λ is given by

$$F(\Lambda) = \int_\Lambda f(\lambda)\,d\lambda + \sum_{\lambda_j \in \Lambda} p(\lambda_j) \tag{3.57}$$

The mass function $p(\lambda)$, called the *spectral function*, gives the power at discrete frequencies λ_j. That is, $F(\{\lambda_j\}) = F_d(\{\lambda_j\}) = p(\lambda_j)$. Thus, whenever a "spectral line" corresponding to λ_j exists, $p(\lambda_j) > 0$; otherwise, $p(\lambda) = 0$.

---

[4]A third "singular" component is rather rare and thus assumed absent.

When the process is real-valued, by the symmetry of $F(d\lambda)$, $f(-\lambda) = f(\lambda)$ and $p(-\lambda) = p(\lambda)$.

It then follows from the Wiener–Khintchine relationship that

$$\begin{aligned} R_k &= \int_{-\pi}^{\pi} e^{ik\lambda} F_c\,(d\lambda) + \int_{-\pi}^{\pi} e^{ik\lambda} F_d\,(d\lambda) \\ &= \int_{-\pi}^{\pi} e^{ik\lambda} f(\lambda)\,d\lambda + \sum_j e^{ik\lambda_j} p(\lambda_j) \end{aligned} \tag{3.58}$$

In particular, both components contribute to the total power,

$$R_0 = \int_{-\pi}^{\pi} f(\lambda)\,d\lambda + \sum_k p(\lambda_k)$$

From this and (3.58) we obtain the general autocorrelation presentation

$$\rho_k = \frac{\int_{-\pi}^{\pi} e^{ik\lambda} f(\lambda)\,d\lambda + \sum_j e^{ik\lambda_j} p(\lambda_j)}{\int_{-\pi}^{\pi} f(\lambda)\,d\lambda + \sum_j p(\lambda_j)} \tag{3.59}$$

When the process is real-valued and $p(0) = 0$, this becomes by symmetry,

$$\rho_k = \frac{\int_0^{\pi} \cos(k\lambda) f(\lambda)\,d\lambda + \sum_{\lambda_j>0} \cos(k\lambda_j) p(\lambda_j)}{\int_0^{\pi} f(\lambda)\,d\lambda + \sum_{\lambda_j>0} p(\lambda_j)} \tag{3.60}$$

a relationship that we shall exploit when dealing with higher order crossings.

The decomposition (3.56) of the spectral distribution also implies a decomposition of the process itself corresponding to (3.58) [6, p. 488],

$$Z_t = \int_{-\pi}^{\pi} e^{it\lambda} \xi_c\,(d\lambda) + \sum_j e^{ik\lambda_j} \xi_j, \qquad t = 0, \pm 1, \pm 2, \ldots \tag{3.61}$$

where

$$E[\xi_c(d\lambda)\overline{\xi_c(d\omega)}] = \begin{cases} F_c(d\lambda) = f(\lambda)d\lambda, & \text{if } \lambda = \omega \\ 0, & \text{if } \lambda \neq \omega \end{cases} \tag{3.62}$$

and $\xi_j = \xi(\{\lambda_j\})$. It follows that the ξ_j are uncorrelated with $\xi_c(d\lambda)$ and

$$E[\xi_j\overline{\xi_k}] = \begin{cases} F_d(\{\lambda_j\}) = p(\lambda_j), & \text{if } j = k \\ 0, & \text{if } j \neq k \end{cases} \tag{3.63}$$

Thus, the the presence of a discrete spectral component $F_d(d\lambda)$ implies the presence of complex exponentials whose amplitudes ξ_j are orthogonal random variables.

From (3.59) we observe that the autocorrelation is a sum of two components corresponding to the continuous and discrete components of $F(d\lambda)$. With obvious notation, rewrite (3.59) as $\rho_k = \rho_c(k) + \rho_d(k)$. It can be shown that the contribution to the autocorrelation from the continuous component falls off to 0 as $|k|$ increases, while the contribution from the discrete part never tends to 0. Therefore, in light of the process decomposition (3.61), the presence of sinusoidal components prevents the autocorrelation from decaying to 0. Moreover, the fact that ρ_k eventually—as k increases—behaves like $\rho_d(k)$ can serve as the basis for an approach to mixed spectrum analysis [22, p. 626].

In the basic decomposition (3.56), one of the components may be absent. When only $F_c(\Lambda)$ is present, the process is called *continuous spectrum process*. The $AR(1)$ process discussed in Example 3.6 is a continuous spectrum process. The process is called *discrete spectrum process* when only $F_d(\Lambda)$ is present. Such a process was encountered in Example 3.4. When both components are present, as in the signal plus noise process discussed in Example 3.5, the process is called *mixed spectrum process*. Thus (3.59) is the autocorrelation of a mixed spectrum zero-mean wide sense stationary process in discrete time. Mixed spectrum processes do arise in nature and also in man-made systems. For example, a diurnal cycle of 24 hours can be observed in many "noisy" meteorological and climatological time series. A specific case of this is discussed in Chapter 7. Another example from geophysics—discussed in Chapter 7—is the Chandler wobble secular motion which scientists believe contains at least one spectral line corresponding, roughly, to 14 months. Also, a vibrating machinery may release signatures containing spectral lines associated with rotating parts.

Similar remarks apply to the spectral distribution function $F(\lambda)$. In terms of the spectral distribution function, the decomposition takes the form

$$F(\lambda) = F_c(\lambda) + F_d(\lambda) \tag{3.64}$$

where

$$F_c(\lambda) = \int_{-\pi}^{\lambda} f(\omega)\, d\omega$$

and

$$F_d(\lambda) = \sum_{\lambda_j \leq \lambda} p(\lambda_j)$$

Thus, $F_c(\lambda)$ is absolutely continuous—integral of its derivative—with a nonnegative density f, while $F_d(\lambda)$ is a step function with jumps of size $p(\lambda_j)$ at λ_j. This is precisely analogous to the probability distribution function discussed in Chapter 2. From this point of view, the autocovariance plays the role of a characteristic function except for normalization.

Convenient Formalism

It is sometimes convenient to introduce the notion of a density $g(\lambda)$ that puts together the spectral density $f(\lambda)$ and spectral function $p(\lambda)$. This is done by resorting to the Dirac delta function $\delta(\lambda)$, which satisfies, for $h(\lambda)$ continuous at $\lambda_0 \in (-\pi, \pi]$, the integral relationship

$$\int_{-\pi}^{\pi} h(\lambda)\delta(\lambda - \lambda_0)\, d\lambda = h(\lambda_0)$$

Then, if we let

$$g(\lambda) = f(\lambda) + \sum_j p(\lambda_j)\delta(\lambda - \lambda_j)$$

we obtain for $h(\lambda)$ continuous at the λ_j,

$$\int_{-\pi}^{\pi} h(\lambda)g(\lambda)\, d\lambda = \int_{-\pi}^{\pi} h(\lambda)f(\lambda)\, d\lambda + \sum_j p(\lambda_j)h(\lambda_j)$$

In particular,

$$F(\Lambda) = \int_\Lambda g(\lambda)\, d\lambda = \int_\Lambda f(\lambda)\, d\lambda + \sum_{\lambda_j \in \Lambda} p(\lambda_j)$$

and

$$R_k = \int_{-\pi}^{\pi} e^{ik\lambda} g(\lambda)\, d\lambda = \int_{-\pi}^{\pi} e^{ik\lambda} f(\lambda)\, d\lambda + \sum_j p(\lambda_j)e^{ik\lambda_j}$$

Thus, $g(\lambda)$ may serve as a formal spectral density. As an illustration, consider the random real-valued stationary sinusoid in Example 3.2. Then,

$$g(\lambda) = \frac{\sigma^2}{2}[\delta(\omega + \omega_1) + \delta(\omega - \omega_1)]$$

3.4 APPLICATIONS OF THE SPECTRAL REPRESENTATION

General spectral properties of weakly stationary processes can be elicited most conveniently through the spectral representation (3.52). This will be illustrated in this section in studying the effect of linear operations on weakly stationary processes.

3.4.1 Stationary Solutions of Stochastic Difference Equations

As a first application of the spectral representation, we show how to construct stationary autoregressive processes.

Consider the stochastic difference equation of order p,

$$X_t + a_1 X_{t-1} + \cdots + a_p X_{t-p} = u_t, \qquad t = 0, \pm 1, \pm 2, \ldots \tag{3.65}$$

where the a_j are real constants such that $a_p \neq 0$, and $\{u_t\}$ is real-valued white noise with mean 0 and variance σ_u^2. We are interested in conditions which yield real-valued weakly stationary solutions of (3.65) [9, pp. 37–38].

Clearly, $\{u_t\}$ is weakly stationary so that

$$u_t = \int_{-\pi}^{\pi} e^{it\lambda} \xi_u \, (d\lambda)$$

Let $\{X_t\}$ be a real-valued stationary solution. Then again,

$$X_t = \int_{-\pi}^{\pi} e^{it\lambda} \xi_x \, (d\lambda)$$

Then we can write

$$u_t = \int_{-\pi}^{\pi} e^{it\lambda} (1 + a_1 e^{-i\lambda} + \cdots + a_p e^{-ip\lambda}) \xi_x \, (d\lambda)$$

It is convenient to introduce the polynomial,

$$\phi(z) = 1 + a_1 z + a_2 z^2 + \cdots + a_p z^p$$

Then, for $t = 0, \pm 1, \pm 2, \ldots,$

$$\int_{-\pi}^{\pi} e^{it\lambda} \phi(e^{-i\lambda}) \xi_x \, (d\lambda) = \int_{-\pi}^{\pi} e^{it\lambda} \xi_u \, (d\lambda)$$

As with Fourier transforms, we can equate the integrands,

$$\phi(e^{-i\lambda}) \xi_x(d\lambda) = \xi_u(d\lambda) \tag{3.66}$$

Therefore, we obtain a useful relationship between the respective spectral distributions $F_x(d\lambda), F_u(d\lambda)$,

$$|\phi(e^{-i\lambda})|^2 F_x(d\lambda) = F_u(d\lambda) = \frac{\sigma_u^2}{2\pi} d\lambda \tag{3.67}$$

If we let $f_x(\lambda)$ be the spectral density of $\{X_t\}$, then (3.67) implies that

$$f_x(\lambda) = \frac{\sigma_u^2}{2\pi} \cdot \frac{1}{|\phi(e^{-i\lambda})|^2}$$

But since weak stationarity implies finite variances,

$$\int_{-\pi}^{\pi} f_x(\lambda) \, d\lambda = \text{Var}[X_t] < \infty$$

it follows that $\phi(e^{-i\lambda}) \neq 0$, and we obtain from (3.66)

$$\xi_x(d\lambda) = \frac{1}{\phi(e^{-i\lambda})} \xi_u(d\lambda)$$

or, by integrating both sides,

$$X_t = \int_{-\pi}^{\pi} e^{it\lambda} \frac{1}{\phi(e^{-i\lambda})} \xi_u\,(d\lambda) \tag{3.68}$$

This is the unique general stationary solution of the difference equation (3.65). Evidently, the solution has a continuous spectrum with spectral density

$$f_x(\lambda) = \frac{\sigma_u^2}{2\pi} \cdot \frac{1}{|\phi(e^{-i\lambda})|^2} = \frac{\sigma_u^2}{2\pi} \cdot \frac{1}{|1 + a_1 e^{-i\lambda} + \cdots + a_p e^{-ip\lambda})|^2} \tag{3.69}$$

By putting conditions on $\phi(e^{-i\lambda})$ we can deduce from (3.68) some particular forms for X_t.

We saw that under stationarity, $\phi(z)$ has no zeros (roots) on the unit circle $|z| = 1$. But $\phi(z)$, being a polynomial of degree p, has p zeros, some of which may lie inside and some outside the unit circle. We shall see momentarily that the case of interest to us is when all the zeros are outside the unit circle.

Denote the roots of $\phi(z) = 0$ by $z_1, z_2, \ldots, z_p$, and suppose all the roots are outside the unit circle. That is, $|z_j| > 1$ for all j. Then we can expand $1/\phi(z)$ in partial fractions for $|z| = 1$. Noting that for $|z| = 1$, $|z/z_j| < 1$, we have for some constants A_j,

$$\frac{1}{\phi(z)} = \sum_{j=1}^{p} \frac{A_j}{1 - \frac{z}{z_j}} = \sum_{j=1}^{p} A_j \sum_{r=0}^{\infty} \left(\frac{z}{z_j}\right)^r$$

and this converges geometrically. We thus have a representation for $1/\phi(e^{-i\lambda})$ which we substitute in the general solution (3.68). By switching the order of summation we have

$$X_t = \int_{-\pi}^{\pi} e^{it\lambda} \sum_{r=0}^{\infty} e^{-ir\lambda} \left[\sum_{j=1}^{p} \frac{A_j}{z_j^r}\right] \xi_u\,(d\lambda) \tag{3.70}$$

This, however, from the spectral representation of u_t, has the form

$$\sum_{r=-\infty}^{\infty} h_r u_{t-r}$$

where

$$h_r = \begin{cases} \sum_{j=1}^{p} \frac{A_j}{z_j^r}, & \text{for } r \geq 0 \\ 0, & \text{for } r < 0 \end{cases}$$

Therefore, X_t is a one-sided infinite moving average in terms of past and present u_t,

$$X_t = \sum_{r=0}^{\infty} h_r u_{t-r}$$

In summary, *we have shown that if the roots of* $\phi(z) = 0$ *all lie outside the unit circle, the stationary solution of* (3.65) *is a one-sided infinite moving average in terms of* $u_t, u_{t-1}, u_{t-2}, \ldots$.

In the same way we can show that if all the roots are *inside* the unit circle the stationary solution is an infinite moving average in terms of $u_{t+1}, u_{t+2}, \ldots$. If some roots are inside and some outside the unit circle, the stationary solution is a two-sided infinite moving average,

$$X_t = \sum_{r=-\infty}^{\infty} h_r u_{t-r}$$

Going back to the stationary $AR(p)$ process (3.37), we conclude that if the roots of the characteristic equation

$$\phi(z) \equiv 1 - \phi_1 z - \phi_2 z^2 - \cdots - \phi_p z^p = 0$$

all lie outside the unit circle, the process can be represented as an infinite moving average in terms of past and present ϵ_t as in (3.38).

Example 3.8: Stationarizing by Differencing

Suppose $|a| < 1$, and consider a process $\{X_t\}$, $t = 0, \pm 1, \pm 2, \ldots$, defined by a special case of the stochastic difference equation (3.65),

$$X_t - X_{t-1} - aX_{t-1} + aX_{t-2} = u_t$$

Then

$$\phi(z) = 1 - z - az + az^2 = (1 - az)(1 - z)$$

and we see that the root $z_1 = 1/a$ is outside the unit circle while the second root, $z_2 = 1$, is on the unit circle. Thus, $\{X_t\}$ cannot be stationary. However, by defining $Y_t \equiv X_t - X_{t-1}$, the difference equation can be expressed as

$$Y_t = aY_{t-1} + u_t$$

and this has a stationary solution, namely $\sum_{j=0}^{\infty} a^j u_{t-j}$. Thus, in the very special case of a unit root 1, a wide sense stationary process can be obtained by differencing a nonstationary one. The same procedure applies when there are d unit roots all equal to 1, and the rest are outside the unit circle, except that now the process is differenced repeatedly d times. There is practical evidence showing that as little as one or two differences of a nonstationary process may render it stationary [2, ch. 4].

■

3.4.2 Effect of a Linear Filter

By a *time-invariant linear filter* applied to a sequence $\{x_t\}$, $t = 0, \pm 1, \pm 2, \ldots$, we mean the linear operation or convolution,

$$\mathcal{L}(\{x_t\}) = \sum_{j=-\infty}^{\infty} h_j x_{t-j} \tag{3.71}$$

where $\mathcal{L}(\{x_t\})$ provides the value at time t of the operation, and the h_j are complex constants. The sequence $\{h_j\}$ is called the *impulse response* of the filter and is assumed to fall off to 0 sufficiently fast for the operation to converge in some sense. For example, we may require absolute summability. The time invariance refers to the fact that the impulse response does not depend on t (see Problem 22). By linear filter we always mean time-invariant linear filter. When $h_j = 0$ for $j < 0$, the filter is called *physically realizable* or *causal*.

Combinations of linear filters are defined in the most straightforward manner. If $\mathcal{L}_1$ and $\mathcal{L}_2$ are two linear filters, we define for complex numbers a_1, a_2,

$$(a_1\mathcal{L}_1 + a_2\mathcal{L}_2)(\{x_t\}) \equiv a_1\mathcal{L}_1(\{x_t\}) + a_2\mathcal{L}_2(\{x_t\}) \tag{3.72}$$

and

$$\mathcal{L}_2\mathcal{L}_1(\{x_t\}) \equiv \mathcal{L}_2(\mathcal{L}_1(\{x_t\})) \tag{3.73}$$

The application of a filter to the outcome of another as in (3.73) is known as *cascaded* or *sequential* filtering.

A convenient representation of $\mathcal{L}$ in (3.71) is obtained through the *backward shift operator* $\mathcal{B}$,

$$\mathcal{B}x_t \equiv \mathcal{B}(x_t) = x_{t-1}, \qquad \mathcal{B}^{-1}x_t \equiv \mathcal{B}^{-1}(x_t) = x_{t+1}$$

That is, $\mathcal{B}$ is itself a linear filter whose value at time t when applied to the sequence $\{x_t\}$ is x_{t-1}. Repeated application of $\mathcal{B}$ gives

$$\mathcal{B}^j x_t \equiv \mathcal{B}^j(x_t) \equiv \mathcal{B}(\mathcal{B}^{j-1}(x_t)) = x_{t-j}$$

Thus we can write

$$\mathcal{L} \equiv \mathcal{L}(\mathcal{B}) = \sum_{j=-\infty}^{\infty} h_j \mathcal{B}^j \tag{3.74}$$

where $\mathcal{B}^0 \equiv 1$ is the identity operator, $\mathcal{B}^0 x_t \equiv \mathcal{B}^0(x_t) = x_t$. A specific example of (3.74) is the $AR(1)$ or "alpha" filter defined for $|\alpha| < 1$,

$$\mathcal{L} \equiv \mathcal{L}(\mathcal{B}) = \frac{1}{1 - \alpha\mathcal{B}} = 1 + \alpha\mathcal{B} + \alpha^2\mathcal{B}^2 + \alpha^3\mathcal{B}^3 + \cdots \tag{3.75}$$

Note that if $h_j = 0$ for all j except for, say, $j = 0, 1, 2, \ldots, p$, (3.74) reduces to a polynomial in $\mathcal{B}$ of degree p. In this case

$$(1 + h_1\mathcal{B} + h_2\mathcal{B}^2 + \cdots + h_p\mathcal{B}^p)Z_t = Z_t + h_1 Z_{t-1} + h_2 Z_{t-2} + \cdots + h_p Z_{t-p}$$

In general, polynomials in $\mathcal{B}$ are well defined and the usual operations of addition, multiplication, and division apply, provided we put restrictions on the coefficients when the result is an infinite polynomial.

A new sequence $\{y_t\}$ can be constructed by the linear operation

$$y_t = \mathcal{L}(\{x_t\}) = \sum_{j=-\infty}^{\infty} h_j \mathcal{B}^j(\{x_t\}) = \sum_{j=-\infty}^{\infty} h_j x_{t-j} \tag{3.76}$$

In (3.76), following the commonplace abuse of notation, y_t stands for both the value of $\mathcal{L}(\{x_t\})$ at time t and the new sequence $\{y_t\}$. Thus (3.76) defines an input–output linear relationship with $\{x_t\}$ being the input and $\{y_t\}$ the output. By a proper choice or *design* of the impulse response sequence, outputs with certain desired properties are obtained.

The first step in understanding linear filters is to realize the effect of the filter on the complex exponential $x_t = e^{it\lambda}$. For $0 < \lambda \leq \pi$,

$$\mathcal{L}(\{e^{it\lambda}\}) = e^{it\lambda} \sum_{j=-\infty}^{\infty} h_j e^{-ij\lambda}$$

Thus, when the complex exponential $x_t = e^{it\lambda}$ passes through a linear filter the output is the same complex exponential, except that now it is multiplied by $H(\lambda)$ defined by

$$H(\lambda) \equiv \sum_{j=-\infty}^{\infty} h_j e^{-ij\lambda}, \qquad 0 < \lambda \leq \pi$$

The new function $H(\lambda)$, called the *transfer function* or *frequency response* of the filter, describes the frequency domain characteristics of the filter. Clearly, different impulse responses lead to different transfer functions. A closer look reveals that the complex exponential is an eigenfunction of any linear filter with the transfer function being the corresponding eigenvalue.

Under suitable conditions the impulse response can be recovered from the transfer function. For example, if h_j is absolutely summable or merely

$$\sum_{j=-\infty}^{\infty} |h_j|^2 < \infty$$

then

$$h_j = \frac{1}{2\pi} \int_{-\pi}^{\pi} e^{ij\lambda} H(\lambda) d\lambda, \qquad j = 0, \pm 1 \pm 2, \ldots$$

and from Parseval's relation,

$$\sum_{j=-\infty}^{\infty} |h_j|^2 = \frac{1}{2\pi} \int_{-\pi}^{\pi} |H(\lambda)|^2 \, d\lambda$$

The function $|H(\lambda)|$ is called the *gain* of the linear filter. In general

$$H(\lambda) = |H(\lambda)| e^{i\theta(\lambda)}$$

and $\theta(\lambda)$ is called the *phase function* of the filter. Observe that for real impulse response,

$$H(-\lambda) = \overline{H(\lambda)}, \qquad |H(-\lambda)| = |H(\lambda)|$$

The effect of combinations of filters can be studied again by inputting the complex exponential $x_t = e^{it\lambda}$. Thus if H_j is the transfer function of $\mathcal{L}_j, j = 1, 2,$

$$\begin{aligned}(a_1\mathcal{L}_1 + a_2\mathcal{L}_2)(\{e^{it\lambda}\}) &= a_1\mathcal{L}_1(\{e^{it\lambda}\}) + a_2\mathcal{L}_2(\{e^{it\lambda}\}) \\ &= e^{it\lambda}[a_1H_1(\lambda) + a_2H_2(\lambda)]\end{aligned} \tag{3.77}$$

We find that $a_1\mathcal{L}_1 + a_2\mathcal{L}_2$ has transfer function

$$a_1H_1(\lambda) + a_2H_2(\lambda)$$

Similarly,

$$\mathcal{L}_2\mathcal{L}_1(\{e^{it\lambda}\}) = \mathcal{L}_2(\mathcal{L}_1(\{e^{it\lambda}\})) = e^{it\lambda}H_2(\lambda)H_1(\lambda) \tag{3.78}$$

Thus, the transfer function of the sequential filter $\mathcal{L}_2\mathcal{L}_1$ is

$$H_2(\lambda)H_1(\lambda)$$

with gain $|H_2(\lambda)||H_1(\lambda)|$, and the transfer function of $\mathcal{L}_3\mathcal{L}_2\mathcal{L}_1$ is

$$H_3(\lambda)H_2(\lambda)H_1(\lambda)$$

with gain $|H_3(\lambda)||H_2(\lambda)||H_1(\lambda)|$, etc. This translates into *convolution* for the corresponding impulse response sequences. Thus if $h_{1,j}$ and $h_{2,j}$ are the impulse response sequences corresponding to $\mathcal{L}_1$ and $\mathcal{L}_2$, respectively, then the impulse response of $\mathcal{L}_2\mathcal{L}_1$ is convolution

$$h_j = \sum_{k=-\infty}^{\infty} h_{2,k}h_{1,j-k}$$

provided the sum converges.

Time-invariant linear filters are used routinely in a wide range of engineering and scientific applications whenever a manipulation of the spectrum is sought. That is, whenever we wish to derive new sequences which possess certain spectral properties. The fact that this can be done is a consequence of the following important result.

Let $\{Z_t\}, t = 0, \pm1, \pm2, \ldots,$ be a zero mean weakly stationary process and consider the filter (3.71) with $\sum_j |h_j| < \infty$. This guarantees a weakly stationary output with mean zero and finite variance (see Problem 26). Define a new process by

$$Y_t = \mathcal{L}(\{Z_t\})$$

Then $\{Y_t\}$ admits the spectral representation

$$Y_t = \int_{-\pi}^{\pi} e^{it\lambda}\xi_y(d\lambda) = \int_{-\pi}^{\pi} e^{it\lambda}H(\lambda)\xi_z(d\lambda) \tag{3.79}$$

Therefore, as with Fourier transforms,

$$\xi_y(d\lambda) = H(\lambda)\xi_z(d\lambda) \tag{3.80}$$

and we arrive at an important relationship between the input and output spectra,

$$F_y(d\lambda) = E[|\xi_y(d\lambda)|^2] = |H(\lambda)|^2 F_z(d\lambda) \tag{3.81}$$

From this, the spectral densities and spectral functions obey the relationships

$$f_y(\lambda) \quad = \quad |H(\lambda)|^2 f_z(\lambda) \tag{3.82}$$

and

$$p_y(\lambda) \quad = \quad |H(\lambda)|^2 p_z(\lambda) \tag{3.83}$$

Thus, by controlling $|H(\lambda)|^2$ we can control the output spectral characteristics. We note, however, that, from (3.81), if for some Λ, $F_z(\Lambda) = 0$, then also $F_y(\Lambda) = 0$, and filtering cannot change this. In particular, a discrete spectrum cannot be rendered continuous.

From (3.81), by shaping $|H(\lambda)|$ we can let certain frequency bands "pass" while stopping others. When high positive frequencies pass while low positive frequencies are attenuated or cut out completely, the filter is termed *high-pass*. It is termed *low-pass* when low frequencies are enhanced and high ones are attenuated. A filter is called *band-pass* if it passes only a certain band of frequencies. By combining filters sequentially and in parallel we can construct different types of desired filters with special characteristics (see Problem 28).

As alluded to earlier, (3.81) together with $\sum_j |h_j| < \infty$, imply

$$\text{Var}[Y_t] = \int_{-\pi}^{\pi} |H(\lambda)|^2 F_z\,(d\lambda) < \infty \tag{3.84}$$

This is the so-called *matching condition* which we require if the output is to have a finite power [19].

For a better understanding of the meaning of the gain and phase functions of a linear filter, suppose the input $\{Z_t\}$ is a random complex exponential with frequency λ_1,

$$Z_t = \int_{-\pi}^{\pi} e^{it\lambda}\xi_z\,(d\lambda) = \xi_z(\{\lambda_1\})e^{it\lambda_1}$$

With

$$\xi_z(\{\lambda_1\}) = |\xi_z(\{\lambda_1\})|e^{i\psi(\lambda_1)}$$

we have,

$$Z_t = |\xi_z(\{\lambda_1\})|e^{i(\lambda_1 t+\psi(\lambda_1))}$$

Then the output becomes,

$$\begin{aligned} Y_t \quad &= \quad \int_{-\pi}^{\pi} e^{it\lambda}H(\lambda)\xi_z\,(d\lambda) = H(\lambda_1)\xi_z(\{\lambda_1\})e^{it\lambda_1} \\ &= \quad |H(\lambda_1)||\xi_z(\{\lambda_1\})|e^{i(\lambda_1 t+\psi(\lambda_1)+\theta(\lambda_1))} \end{aligned}$$

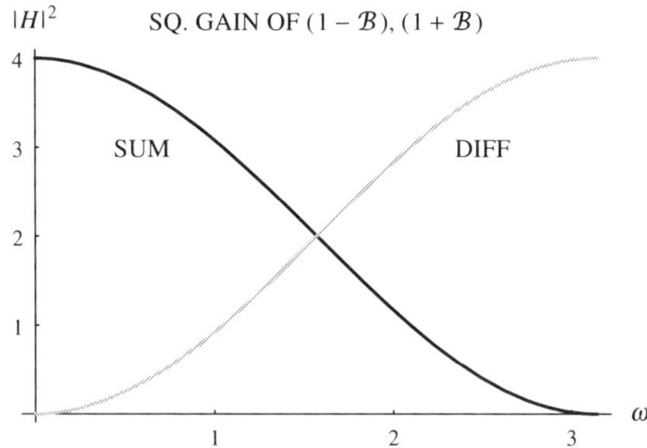

Figure 3.3: *The squared gains of the difference* $1 - \mathcal{B}$ *and sum* $1 + \mathcal{B}$ *filters.*

where we have used $H(\lambda_1) = |H(\lambda_1)|e^{i\theta(\lambda_1)}$. *Thus, on output, the modulus of the input is multiplied by* $|H(\lambda_1)|$ *and the phase is incremented by* $\theta(\lambda_1)$. When the input consists of a sum of random complex exponentials, the same holds for each term in the sum. Similarly, if the input to a filter with a real impulse response is a real-valued sinusoid with frequency λ_1 and certain (positive) amplitude and phase, the output is the same sinusoid (i.e., with the same frequency) multiplied by $|H(\lambda_1)|$ and with a phase altered additively by an amount $\theta(\lambda_1)$.

In what follows we discuss some specific filters. Many more useful examples can be found in [10], [19], [20], [30].

The Backward Shift Operator

As noted earlier, the shift $\mathcal{B}Z_t = Z_{t-1}$ is a linear filter. Since $\mathcal{B}e^{it\lambda} = e^{it\lambda}e^{-i\lambda}$, the transfer function is $e^{-i\lambda}$ and the gain is equal to 1.

The Difference Operator

With 1 being the identity operator, the difference operator is

$$\nabla Z_t \equiv (1 - \mathcal{B})Z_t = Z_t - Z_{t-1}$$

This is a linear filter with $h_0 = 1, h_1 = -1$, and $h_j = 0$ otherwise. Since $(1 - \mathcal{B})e^{it\lambda} = e^{it\lambda}(1 - e^{-i\lambda})$, the transfer function is

$$H(\lambda) = 1 - e^{-i\lambda}$$

and the squared gain is

$$|H(\lambda)|^2 = |1 - e^{-i\lambda}|^2 = 2(1 - \cos\lambda)$$

In $[0, \pi]$ the gain is monotone increasing and hence this is a high-pass filter. See Figure 3.3.

The squared gain of the second difference $(1 - \mathcal{B})^2$ is $4(1 - \cos\lambda)^2$, and hence this is a more pronounced high-pass filter. Repeated or sequential differ-

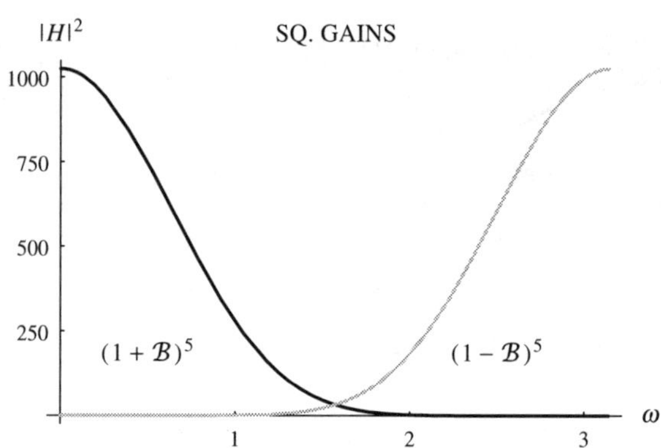

Figure 3.4: *The squared gains of the repeated difference* $(1 - \mathcal{B})^5$ *and the repeated sum* $(1 + \mathcal{B})^5$.

encing (two or more differences) is a simple way to obtain high-pass filters. See Figure 3.4.

The Summation Operator

The sum $(1 + \mathcal{B})Z_t = Z_t + Z_{t-1}$ has squared gain $2(1 + \cos\lambda)$. It is a low-pass filter as illustrated in Figure 3.3. Repeated summation produces more pronounced low-pass filters. Hence $(1+\mathcal{B})^5$ with squared gain $[2(1+\cos\lambda)]^5$ is a much more pronounced low-pass filter than a single sum $(1 + \mathcal{B})$. See Figure 3.4.

A Complex Filter

Consider the filter [11],

$$\mathcal{L}(\mathcal{B}) = \left(1 + e^{i\theta}\mathcal{B}\right)^n \tag{3.85}$$

where n is a nonnegative integer and $\theta \in [0, \pi]$. The squared gain is

$$|H(\lambda)|^2 = 4^n \cos^{2n}\left(\frac{\theta - \lambda}{2}\right) \tag{3.86}$$

For sufficiently large n the filter behaves as a band-pass filter ("centered" at θ), and is used as such in Section 7.3.4. In fact, n controls the bandwidth to such an extent that the sequence of probability densities

$$f_n(\lambda;\theta) = \frac{\cos^{2n}\left((\theta-\lambda)/2\right)}{\int_{-\pi}^{\pi} \cos^{2n}\left((\theta-\omega)/2\right)\,d\omega}, \qquad -\pi < \lambda \le \pi$$

approaches a Dirac delta as $n \to \infty$,

$$\int_{-\pi}^{\pi} f_n(\lambda;\theta) g(\lambda)\,d\lambda \to g(\theta)$$

for g continuous at θ.

Recursive Filters

The class of recursive filters is of great practical interest, as it enables the design of useful transfer functions, in addition to fast implementation.

Consider the input–output relationship,

$$Y_t = -a_1 Y_{t-1} - \cdots - a_p Y_{t-p} + b_0 X_t + b_1 X_{t-1} + \cdots + b_q X_{t-q} \qquad (3.87)$$

This defines a *recursive* filter with input $\{X_t\}$ and output $\{Y_t\}$. Define the polynomials in $\mathcal{B}$,

$$\mathcal{P}(\mathcal{B}) = 1 + a_1\mathcal{B} + \cdots + a_p\mathcal{B}^p$$

and

$$\mathcal{Q}(\mathcal{B}) = b_0 + b_1\mathcal{B} + \cdots + b_q\mathcal{B}^q$$

Then,

$$\mathcal{P}(\mathcal{B})Y_t = \mathcal{Q}(\mathcal{B})X_t$$

This gives the *formal* equation,

$$Y_t = \frac{\mathcal{Q}(\mathcal{B})}{\mathcal{P}(\mathcal{B})} X_t$$

For this to make sense, we must be able to expand $\mathcal{Q}(\mathcal{B})/\mathcal{P}(\mathcal{B})$ in a power series as in (3.74) such that h_j vanishes fast enough. Now, if the roots z_j of $\mathcal{P}(z) = a_p \prod_{j=1}^{p}(z - z_j) = 0$ all lie *outside* the unit circle, $|z_j| > 1$ for all j, then by partial fraction expansion, $1/\mathcal{P}(z)$ admits a power series representation,

$$\frac{1}{\mathcal{P}(z)} = \sum_{j=0}^{\infty} \alpha_j z^j$$

which converges for $|z| \leq 1$. It follows that

$$\mathcal{L}(\mathcal{B}) \equiv \frac{\mathcal{Q}(\mathcal{B})}{\mathcal{P}(\mathcal{B})} = \sum_{j=0}^{\infty} h_j \mathcal{B}^j$$

is a well-defined power series expansion in $\mathcal{B}$ (filter) where the coefficients are absolutely summable. This gives a *one-sided* representation,

$$Y_t = \sum_{j=0}^{\infty} h_j X_{t-j}$$

where $\sum_{j=0}^{\infty} |h_j| < \infty$. By plugging $e^{-i\lambda}$ instead of $\mathcal{B}$ we obtain the transfer function of $\mathcal{L}(\mathcal{B})$ as

$$H(\lambda) = \frac{\mathcal{Q}(e^{-i\lambda})}{\mathcal{P}(e^{-i\lambda})} \qquad (3.88)$$

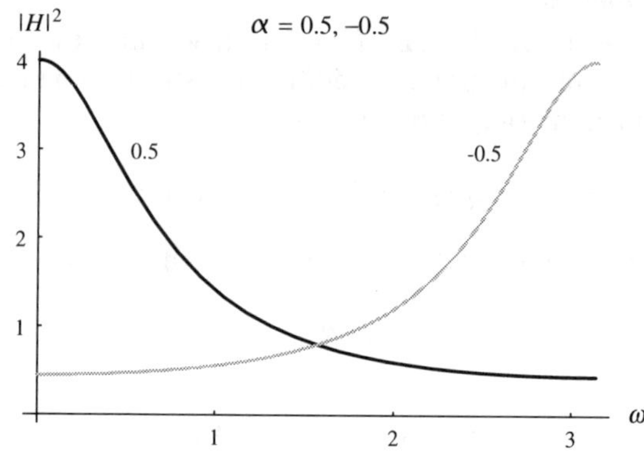

Figure 3.5: *The squared gain of the* $AR(1)$ *filter with* $\alpha = 0.5, -0.5$.

A filter is said to be *stable* when it gives a bounded output whenever the input itself is bounded. We find that for a recursive filter to be both stable and realizable it is sufficient that all the zeros of $\mathcal{P}(z)$ lie outside the unit circle.

Clearly, we can reverse the roles of X_t and Y_t. Thus, if the roots of $\mathcal{Q}(z)$ are outside the unit circle, we can express X_t in terms of a linear filter with $\{Y_t\}$ being the input.

Example 3.9: The *AR*(1) Filter

The $AR(1)$ (or α) filter is the recursive filter

$$Y_t = \alpha Y_{t-1} + X_t$$

where $|\alpha| < 1$. The transfer function is

$$H(\lambda) = \frac{1}{1 - \alpha e^{-i\lambda}}$$

Since $|\alpha| < 1$,

$$\mathcal{L}(\mathcal{B}) = \sum_{j=0}^{\infty} \alpha^j \mathcal{B}^j$$

is well defined and

$$Y_t = \sum_{j=0}^{\infty} \alpha^j X_{t-j}$$

From Figure 3.5, for $\alpha > 0$ the filter is a low-pass filter, and a high-pass for $\alpha < 0$. We sometimes refer to $Y_t = \alpha Y_{t-1} + X_t$ and the related filter $Y_t = \alpha Y_{t-1} + (1 - \alpha)X_t$ as *exponential smoothing*. In the case of $Y_t = \alpha Y_{t-1} + (1 - \alpha)X_t$, $|H(0)|^2 = 1$ for all $|\alpha| < 1$.

■

Example 3.10: *ARMA* Processes

An *autoregressive moving average* (*ARMA*) process is defined by a recursive filter applied to white noise. Let $\{\epsilon_t\}$ be (real-valued) white noise with mean 0 and variance σ_ϵ^2, and consider the process $\{Z_t\}$ defined as the output of the recursive filter,

$$Z_t = \phi_1 Z_{t-1} + \cdots + \phi_p Z_{t-p} + \epsilon_t - \theta_1 \epsilon_{t-1} - \cdots - \theta_q \epsilon_{t-q} \tag{3.89}$$

The process $\{Z_t\}$, $t = 0, \pm 1, \pm 2, \ldots$, is called *autoregressive moving average of order* (*p,q*) abbreviated to *ARMA*(p, q) or *mixed ARMA(p,q)* [2]. With

$$\mathcal{P}(z) = 1 - \phi_1 z - \phi_2 z^2 - \cdots - \phi_p z^p$$

$$\mathcal{Q}(z) = 1 - \theta_1 z - \theta_2 z^2 - \cdots - \theta_q z^q$$

(3.89) becomes,

$$\mathcal{P}(\mathcal{B}) Z_t = \mathcal{Q}(\mathcal{B}) \epsilon_t, \qquad t = 0, \pm 1, \pm 2, \ldots \tag{3.90}$$

A sufficient condition for (wide sense) stationarity and realizability is that the roots of $\mathcal{P}(z)$ lie outside the unit circle. In this case, since the spectral density of $\{\epsilon_t\}$ is $\sigma_\epsilon^2/2\pi$, (3.82) and (3.88) imply that the spectral density of the output $\{Z_t\}$ is

$$f_z(\lambda) = \frac{|\mathcal{Q}(e^{-i\lambda})|^2}{|\mathcal{P}(e^{-i\lambda})|^2} f_\epsilon(\lambda) = \frac{\sigma_\epsilon^2}{2\pi} \cdot \frac{|\mathcal{Q}(e^{-i\lambda})|^2}{|\mathcal{P}(e^{-i\lambda})|^2}, \qquad -\pi \le \lambda \le \pi \tag{3.91}$$

This clearly is a *rational* function of $e^{-i\lambda}$. When also the roots of $\mathcal{Q}(z)$ lie outside the unit circle, the process is *invertible* meaning that $\epsilon_t = [\mathcal{P}(\mathcal{B})/\mathcal{Q}(\mathcal{B})] Z_t$, which is seen to be an infinite autoregression.

■

Example 3.11: Butterworth Sine Filter

A useful recursive filter is the sine-Butterworth low-pass filter

$$Y_t = \alpha_1 Y_{t-1} + \cdots + \alpha_k Y_{t-k} + \beta_0 X_t$$

whose weights are determined from the filter squared gain,

$$|H(\omega)|^2 = \frac{1}{1 + \{[\sin(\omega/2)]/[\sin(\omega_b/2)]\}^{2k}}, \qquad \omega, \omega_b \in [0, \pi] \tag{3.92}$$

Here ω_b serves as an ideal cutoff frequency. The method of obtaining the weights is described in [20, Chs. 4 and 5]. A particular case of (3.92) with $k = 6$ and $\omega_b = 1.2$ is shown in Figure 3.6.

■

3.4.3 Mean Square Ergodicity

Let $\{Y_t\}$, $t = 0, \pm 1, \pm 2, \ldots$, be a weakly stationary process with $E[Y_t] = m$, and spectral distribution $F(d\lambda)$. If $Y_1, Y_2, \ldots, Y_N$ is a time series from the

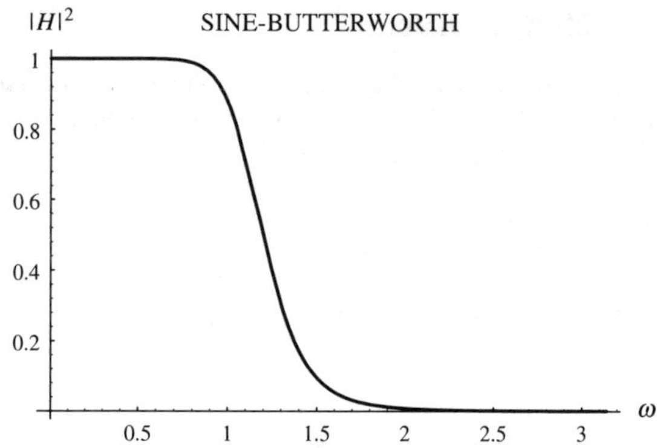

Figure 3.6: *The squared gain of the Butterworth sine filter with $k = 6$ and $\omega_b = 1.2$.*

process, the sample average or sample mean is

$$\bar{Y} = \frac{1}{N}\sum_{t=1}^{N} Y_t$$

We shall study the mean square convergence of $\bar{Y}$.

It is convenient to introduce the centered process

$$Z_t = Y_t - m, \qquad t = 0, \pm 1, \pm 2, \ldots$$

Then $\{Z_t\}$ is a zero mean weakly stationary process, with the same $F(d\lambda)$, possessing the spectral representation (3.52). The spectral representation of $\{Z_t\}$ provides a very suggestive representation for the sample average $\bar{Z}$,

$$\bar{Z} = \frac{1}{N}\sum_{t=1}^{N} Z_t = \int_{-\pi}^{\pi} \frac{1}{N}\sum_{t=1}^{N} e^{it\lambda}\xi\,(d\lambda) = \int_{-\pi}^{\pi} \varphi_N(\lambda)\xi\,(d\lambda)$$

where

$$\varphi_N(\lambda) = \frac{1}{N}\sum_{t=1}^{N} e^{it\lambda} = \begin{cases} e^{i\lambda(N+1)/2} \cdot \dfrac{\sin\frac{1}{2}\lambda N}{N\sin\frac{1}{2}\lambda}, & \lambda \neq 0 \\ 1, & \lambda = 0 \end{cases}$$

Clearly, on $(-\pi, \pi]$, as $N \to \infty$, $\varphi_N(\lambda)$ converges to 0 except for $\lambda = 0$, $\varphi_N(0) = 1$ for all N. We therefore are tempted to guess that $\bar{Z}$ converges as $N \to \infty$ to the dc component $\xi(\{0\})$.

To show that this intuition is in fact correct, we appeal to the inequality,

$$\frac{\sin x}{x} \geq \frac{2}{\pi}, \qquad \frac{-\pi}{2} \leq x \leq \frac{\pi}{2}$$

from which $|\varphi_N(\lambda)| \leq \pi/2$. Then, (3.48) and dominated convergence give

$$\lim_{N\to\infty} E\left|\int_{-\pi}^{\pi} \varphi_N(\lambda)\xi(d\lambda)\right|^2 = F(\{0\})$$

From this and the orthogonality of $\xi(d\lambda)$ we find that

$$\lim_{N\to\infty} E|\bar{Z} - \xi(\{0\})|^2 = 0$$

and therefore, $\bar{Z} \overset{m.s.}{\to} \xi(\{0\})$. Thus, switching back to Y_t,

$$\bar{Y} = \bar{Z} + m \overset{m.s.}{\to} m + \xi(\{0\}), \qquad N \to \infty \tag{3.93}$$

To get rid of the dc component $\xi(\{0\})$, it is necessary and sufficient that $F(\{0\}) = 0$. That is, that $F(\lambda)$, the spectral distribution function, be continuous at $\lambda = 0$. *We have thus shown that a necessary and sufficient condition for the mean square convergence of the sample mean to the true mean m, is that $F(\lambda)$ has no jump at* $\lambda = 0$. This statement is called *the mean square ergodic theorem* [26, p. 410].

An useful consequence of (3.93) is that

$$\text{Var}[\bar{Y}] \to \text{Var}[\xi(\{0\})] = F(\{0\}), \qquad N \to \infty \tag{3.94}$$

and thus, any approximation to $F(\{0\})$ also estimates Var$[\bar{Y}]$ and vice versa. Clearly, $E[\bar{Y}] = m$ for all N.

The convergence result (3.93) is a mean square type of the law of large numbers for weakly stationary processes. Under some conditions on the spectral measure $\xi(d\lambda)$, the result also holds with probability one. Basically what is needed is that $\xi(d\lambda)$ assigns no weight to small neighborhoods of the origin excluding 0. That is, neighborhoods of the form $(2^{-n}, 2^n) \setminus \{0\}$ for large n [7], [15].

3.5 APPENDIX TO CHAPTER 3

Proof of Herglotz's Theorem. First note that for any function g,

$$\sum_{j=1}^{N}\sum_{l=1}^{N} g(j-l) = \sum_{k=-(N-1)}^{N-1} (N - |k|)g(k) \tag{3.95}$$

Now, in (3.5) take $a_j = e^{-ij\lambda}$ and $t_j = j$. Then for $N \geq 1$ and $\lambda \in [-\pi, \pi]$, the function

$$f_N(\lambda) \equiv \frac{1}{2\pi N}\sum_{j=1}^{N}\sum_{l=1}^{N} e^{-ij\lambda}e^{il\lambda}R_{j-l} \geq 0$$

is nonnegative. Rearranging terms and using (3.95), we have

$$f_N(\lambda) = \frac{1}{2\pi}\sum_{k=-(N-1)}^{N-1}\left(1 - \frac{|k|}{N}\right)e^{-ik\lambda}R_k \geq 0, \qquad \lambda \in [-\pi, \pi]$$

Define a measure $F_N(\Lambda)$ by the integral

$$F_N(\Lambda) \equiv \int_\Lambda f_N(\lambda)\, d\lambda$$

where Λ is a (measurable) subset of $[-\pi, \pi]$. Then

$$\int_{-\pi}^{\pi} e^{in\lambda} F_N(d\lambda) = \int_{-\pi}^{\pi} e^{in\lambda} f_N(\lambda)\, d\lambda = \begin{cases} \left(1 - \frac{|n|}{N}\right) R_n, & |n| < N \\ 0, & |n| \geq N \end{cases} \tag{3.96}$$

Since the measures F_N are supported on $[-\pi, \pi]$, there exists (because no spectral mass "escapes" as N grows; this is called *tightness* [26, p. 394]) a subsequence F_{N_k} which converges (weakly) to a limit measure F, and hence from (3.96),

$$R_n = \lim_{k\to\infty} \int_{-\pi}^{\pi} e^{in\lambda} F_{N_k}(d\lambda) = \int_{-\pi}^{\pi} e^{in\lambda} F(d\lambda) \tag{3.97}$$

We can interpret the integration as one over a unit circle, in which case we identify the points $-\pi$ and π and transfer the mass or power $F(\{-\pi\})$ at the frequency $-\pi$ to π. This leaves the representation (3.97) unchanged except that now we think of F in (3.97) (really a new F) as supported on $(-\pi, \pi]$.

□

Uniqueness of the Spectral Measure. The autocovariance defines the spectral measure F uniquely. To see that, suppose F_1 and F_2 are spectral measures supported on $(-\pi, \pi]$ such that for all integers k,

$$R_k = \int_{-\pi}^{\pi} e^{ik\lambda} F_1(d\lambda) = \int_{-\pi}^{\pi} e^{ik\lambda} F_2(d\lambda)$$

Let $g(\lambda)$ be any continuous bounded function on $(-\pi, \pi]$. Then it can be approximated uniformly on $(-\pi, \pi]$ by trigonometric polynomials of the form [26, p. 395],

$$h_n(\lambda) = \sum_{|k|\leq n} a_k \exp(ik\lambda)$$

where the sum is finite. It follows that

$$\int_{-\pi}^{\pi} g(\lambda) F_1(d\lambda) = \int_{-\pi}^{\pi} g(\lambda) F_2(d\lambda)$$

and this holds in particular for indicators. Therefore, $F_1(\Lambda) = F_2(\Lambda)$ for all (measurable) subsets $\Lambda \in (-\pi, \pi]$.

□

3.6 PROBLEMS AND COMPLEMENTS

1. Explain why for a Gaussian process the notions of strict and wide sense stationarity are equivalent.

   (Hint: For a Gaussian process, first and second order moments exist.)

2. *Functions of stationary processes* [30, p. 51]. Let $\{Z_t\}$ be a stationary Gaussian process, and define a new process $\{Y_t\}$ by the transformation $Y_t = g(Z_t)$. Show that $\{Y_t\}$ is strictly stationary, but that it is not necessarily Gaussian. (Hint: Take $Y_t = 1$ if $Z_t \geq 0$ and $Y_t = 0$ otherwise.)

3. Argue that when R_0 is finite, $F(\lambda)$ must be bounded.

4. *Random phase models* [30, p. 53]. Consider the process $Z_t = A\cos(\omega_1 t + \phi)$ where ϕ is uniformly distributed in $(0, 2\pi)$, independently of the random variable A. Show that $\{Z_t\}$ is strictly stationary. Argue that if ϕ is not uniformly distributed in $(0, 2\pi)$, then $\{Z_t\}$ is not strictly stationary. (Hint: Take $\phi =$ constant $\in (0, 2\pi)$.)

5. *Random frequency* [1, p. 379]. Let Ω be a random variable uniformly distributed in $(0, 2\pi)$, and define for $t = 1, 2, 3, \ldots$, $Z_t = \cos(\Omega t)$. Show that $\{Z_t\}$ is stationary in the wide sense but not in the strict sense.

6. Show that when $\omega_{-j} = -\omega_j$, and $\xi_{-j} = \overline{\xi}_j$, the process (3.25) is real-valued.

7. Show that a complex-valued sequence R_k^* which admits the representation (3.6) in terms of a nondecreasing distribution function $F(\lambda)$ is nonnegative definite. That is, show that for all complex numbers $a_1, a_2, \ldots, a_N$, and integers $t_1, t_2, \ldots, t_N$, with $N \geq 1$,

$$\sum_{j=1}^{N}\sum_{l=1}^{N} a_j \overline{a_l} R_{t_j - t_l}^* \geq 0$$

8. Let X be a random variable with characteristic function

$$\varphi(t) = E[e^{itX}] = E[\cos tX] + iE[\sin tX], \qquad -\infty < t < \infty$$

   Show that $\varphi(t)$ is nonnegative definite.

9. Let $\{Z(t)\}$ be a real-valued wide sense stationary process with mean 0 and autocorrelation $R(\tau) = e^{-|\tau|}$. Find the variance of $\sum_{t=1}^{10} Z_t$.

10. Let $\{Z_t\}$ be a real-valued wide sense stationary process with mean 0 and variance 1, and assume $\sum_k |\rho_k| < \infty$. Show that

$$F(\omega) = \frac{\omega + \pi}{2\pi} + \frac{1}{\pi}\sum_{k=1}^{\infty} \rho_k \frac{\sin(k\omega)}{k}, \qquad \omega \in [-\pi, \pi]$$

   (Hint: Use (3.8).)

11. Consider the sum of sinusoids (3.21) with $p = 3$. Express the spectral distribution $F(\omega)$ explicitly as in (3.18). Argue that $F(\omega)$ is continuous from the right.

    (Partial Answer: For $\omega_1 \leq \omega < \omega_2$, $F(\omega) = \frac{1}{2}(\sigma_3^2 + \sigma_2^2 + 2\sigma_1^2)$.)

12. *Estimation of the mean* [1, p. 459]. Let $Z_1, Z_2, \ldots, Z_N$ be a time series from a weakly stationary process with mean m and autocovariance R_k. Define the sample mean as $\hat{m} = 1/N \sum_{t=1}^{N} Z_t$.

(a) Show that $E[\hat{m}] = m$.

(b) Show that

$$N\ \mathrm{Var}[\hat{m}] = \sum_{k=-(N-1)}^{N-1} \left(1 - \frac{|k|}{N}\right) R_k$$

(c) Argue that when the sum converges, then $\mathrm{Var}[\hat{m}]$ decreases as $1/N$, in which case $\hat{m}$ converges in mean square to m. An estimator which converges to a parameter is said to be *consistent*.

(d) Show that if the spectrum is continuous with spectral density $f(\lambda)$, then from the Wiener–Khintchine relationship,

$$N\ \mathrm{Var}[\hat{m}] = \int_{-\pi}^{\pi} \frac{\sin^2 \frac{1}{2}\lambda N}{N \sin^2 \frac{1}{2}\lambda} f(\lambda)\, d\lambda$$

(e) Show that if in addition $f(\lambda)$ is continuous at $\lambda = 0$, then as $N \to \infty$,

$$N\ \mathrm{Var}[\hat{m}] \to 2\pi f(0)$$

(Hint: The kernel tends to a Dirac delta.)

13. *Solutions of the first order stochastic difference equation.* Consider the $AR(1)$ process (3.31) with $|\phi_1| > 1$. Show that there exists a stationary solution in terms of $\epsilon_{t+1}, \epsilon_{t+2}, \epsilon_{t+3}, \ldots$, weighted by powers of $1/\phi_1$. Argue, however, that with $|\phi_1| > 1$ the representation (3.32) leads to explosion. Also, show that if $|\phi_1| = 1$, and $\sigma_\epsilon^2 > 0$, the process is nonstationary. That is, (3.31) has no stationary solutions. (Hint: Take $Z_0 = 0$ and compute the variance of Z_t [22, p. 118].)

14. Consider the weakly stationary $AR(1)$ process (3.31) with $|\phi_1| < 1$, and assume that $\{\epsilon_t\}$ is a sequence of IID $\mathcal{N}(0, \sigma_\epsilon^2)$. Argue that $\{Z_t\}$ is Gaussian and hence strictly stationary. In particular, by determining the characteristic function of Z_t, $\varphi(\zeta) = E[e^{\{i\zeta Z_t\}}]$, show that for every t,

$$Z_t \sim \mathcal{N}\left(0, \frac{\sigma_\epsilon^2}{1 - \phi_1^2}\right)$$

(Hint: Use the fact that if a sequence of normal random variables $\{X_n\}$ converges in mean square to a random variable X, then X also is normal. See Problem 14 in Chapter 2.)

15. *The Galton-Watson Process with Immigration Admits an Asymptotic $AR(1)$ Representation* [12],[28]. We define a simple branching process with immigration as follows. Let (offspring) $\{Y_{n,i}\}$, $i, n = 1, 2, 3, \ldots$, be a family of IID zero-one random variables such that $Y_{n,i} = 1$ with probability m (i.e., independent Bernoulli(m) random variables). Let (immigration) $\{I_n\}$, $n = 1, 2, 3, \ldots$, be a sequence of IID Poisson(λ) random variables, independent of the $Y_{n,i}$. Recall that the possible values of I_n are $0, 1, 2, \ldots$, and that $E[I_n] = \lambda$ for all n. Put $X_0 = 0$, and $\sum_{i=1}^{0} \equiv 0$. The Galton–Watson Process with immigration $\{X_n\}$ is defined by the equation,

$$X_n = \sum_{i=1}^{X_{n-1}} Y_{n,i} + I_n, \qquad n = 1, 2, 3, \ldots \tag{3.98}$$

Evidently, this is a *Markov chain* on the nonnegative integers. Let $\mathcal{F}_n$ be the σ-field generated by $X_0, X_1, X_2, \ldots, X_n$. Think of $\mathcal{F}_{n-1}$ as representing "past information" relative to time n.

(a) Show that $E[X_n|\mathcal{F}_{n-1}] = mX_{n-1} + \lambda$.

(b) With $\epsilon_n \equiv X_n - E[X_n|\mathcal{F}_{n-1}]$, show that $\{X_n\}$ satisfies the *stochastic regression*,

$$X_n = mX_{n-1} + \lambda + \epsilon_n, \qquad n = 1, 2, 3, \ldots$$

where $E[\epsilon_n|\mathcal{F}_{n-1}] = 0$ (i.e., $\{\epsilon_n\}$ is a *martingale difference*), and $E[\epsilon_n \epsilon_k] = 0, n \neq k$.

(c) Since $m < 1$, as $n \to \infty$, $\{X_n\}$ approaches a stationary regime. Suppose $\{X_n\}$ is in its stationary regime. Show that its mean is

$$E[X_n] \equiv \mu = \frac{\lambda}{1-m}$$

and that $Y_n \equiv X_n - \mu$, satisfies the $AR(1)$ form,

$$Y_n = mY_{n-1} + \epsilon_n, \qquad n = 1, 2, 3, \ldots$$

where $E[\epsilon_n Y_{n-k}] = 0$, $k \geq 1$. What are the possible values of Y_n?

(d) Show that $\rho_k = m^k$, identical with that of an $AR(1)$ process whose values are continuous.
(Hint: For $k \geq 1$, $E[\epsilon_n \epsilon_{n-k}] = E\{\epsilon_{n-k} E[\epsilon_n|\mathcal{F}_{n-1}]\}$.)

(e) Fix $m = 0.99$, $\lambda = 0.05$. By a computer simulation of the process (3.98), illustrate the fact that the distribution of Y_n is markedly skewed, and hence far from normal (see also [17]).

16. *A Solar Energy Storage Model* [14]. Consider the process defined as

$$X_n = \max(\beta X_{n-1}, U_n), \qquad n = 1, 2, 3, \ldots$$

where $X(0) = 0$, $0 < \beta < 1$, and $\{U_n\}$ are IID random variables uniformly distributed in $[0, 1]$.

(a) Simulate a time series from the process.

(b) Give an interpretation for β.

(c) Show that with a sufficiently long time series, the parameter β can be determined *exactly* with no error.

(d) Argue that $\{X_n\}$ is asymptotically stationary.

17. Consider the $AR(p)$ process (3.37). Show that

$$\text{Var}[Z_t] = \frac{\text{Var}[\epsilon_t]}{1 - \rho_1\phi_1 - \rho_2\phi_2 - \cdots - \rho_p\phi_p}$$

18. *The $MA(q)$ Process.* Let $\{\epsilon_t\}$, $t = 0, \pm 1, \pm 2, \ldots$, be real white noise with mean zero and variance σ_ϵ^2, and define a real-valued weakly stationary process $\{Z_t\}$ by

$$Z_t = \epsilon_t - \theta_1\epsilon_{t-1} - \theta_2\epsilon_{t-2} - \cdots - \theta_q\epsilon_{t-q}, \qquad t = 0, \pm 1, \pm 2, \ldots \tag{3.99}$$

The process is called a *moving average of order* q abbreviated to $MA(q)$ [2, p. 67]. Show that the autocorrelation is given by

$$\rho_k = \begin{cases} \dfrac{-\theta_k + \theta_1\theta_{k+1} + \cdots + \theta_{q-k}\theta_q}{1 + \theta_1^2 + \cdots + \theta_q^2}, & \text{for } k = 1, 2, \ldots, q \\ 0, & \text{for } k > q \end{cases} \tag{3.100}$$

19. *Infinite Moving Average*. With $\{\epsilon_t\}$ as in the previous problem, let

$$Z_t = \sum_{j=-\infty}^{\infty} h_j \epsilon_{t-j}, \qquad t = 0, \pm 1, \pm 2, \ldots \tag{3.101}$$

where $\sum_{j=-\infty}^{\infty} |h_j|^2 < \infty$. This is necessary and sufficient for mean square convergence, and is weaker than $\sum_{j=-\infty}^{\infty} |h_j| < \infty$.

(a) Show that $\{Z_t\}$ so defined is a weakly stationary process with autocovariance,

$$R_k = \sigma_\epsilon^2 \sum_{j=-\infty}^{\infty} h_j h_{j+k}, \qquad k = 0, 1, 2, \ldots \tag{3.102}$$

and spectral density

$$f(\lambda) = \frac{\sigma_\epsilon^2}{2\pi} \left| \sum_{j=-\infty}^{\infty} h_j e^{-i\lambda j} \right|^2 \tag{3.103}$$

(b) Use the spectral representation to show that *every zero-mean continuous spectrum weakly stationary process with a positive spectral density $f(\lambda)$ has an infinite moving average representation* [19, p. 214].

20. Argue that (3.61) can be rewritten as

$$Z_t = \int_{-\pi}^{\pi} e^{it\lambda} \xi_c(d\lambda) + \int_{-\pi}^{\pi} e^{it\lambda} \xi_d(d\lambda), \qquad t = 0, \pm 1, \pm 2, \ldots \tag{3.104}$$

where

$$\xi_d(\Lambda) = \sum_{\lambda_j \in \Lambda} \xi_j$$

Show that $E[\xi_c(d\lambda)\overline{\xi_d(d\omega)}] = 0$ for all $\lambda, \omega \in (-\pi, \pi]$, and that

$$F_d(\Lambda) = E|\xi_d(\Lambda)|^2 = \sum_{\lambda_j \in \Lambda} p(\lambda_j)$$

21. *Stationarizing PC Processes.* Periodically correlated processes (PC) were defined in Chapter 1, Problem 5. These processes are encountered in such diverse fields as radiophysics and economics [30, p. 469]. Let $\{Z_t\}$, $t = 0, \pm 1, \pm 2, \ldots$, be a PC process with mean zero and minimal period T. In general PC processes are not stationary, and $E[Z_u \overline{Z_v}] = R(u, v) = R(u+T, v+T)$ admits a spectral representation in terms of a two-dimensional spectrum whose mass is concentrated on a set S of $2T-1$ parallel diagonal lines in $C \times C$, $C \equiv (-\pi, \pi]$,

$$S \equiv \{(\lambda_1, \lambda_2) \in C \times C : \lambda_2 = \lambda_1 - 2\pi s/T, s = 0, \pm 1, \pm 2, \ldots, \pm(T-1)\}$$

The result is due to E.G. Gladyshev (1961). The continuous time version is due to H.L. Hurd (1989) [16].

(a) Argue that any zero-mean wide sense stationary process is PC with $T = 1$ and a spectrum concentrated on the main diagonal.

(b) Show that $Z_t^\circ \equiv Z_{t+\theta}$, where θ is a random shift uniformly distributed over the integers $0, 1, 2, \ldots, T-1$, is weakly stationary.

(Hint: PC sequences are *harmonizable*. That is, they admit a spectral representation in terms of a random measure $\xi(d\lambda)$, not necessarily orthogonal, such that $E[\xi(d\lambda)\overline{\xi(d\omega)}] = r_\xi(d\lambda, d\omega)$ [30, p. 476].)

22. *Linear but not time-invariant.* A general linear filter with input $\{x_t\}$ and output $\{y_t\}$ is defined by,

$$y_t = \sum_{k=-\infty}^{\infty} c_{t,k} x_k, \qquad t = 0, \pm 1, \pm 2, \ldots$$

The filter becomes time-invariant if $c_{t,k} = h_{t-k}$. Define a filter by

$$y_t = a x_{-t}, \qquad t = 0, \pm 1, \pm 2, \ldots$$

Show that the filter is linear but not time-invariant [4, p. 153].

23. *A nonlinear representation.* In recent years, due to earlier work by K. Itô, V. Volterra, and N. Wiener, there is a growing interest in nonlinear representations of the form

$$\begin{aligned} X_t = &\sum_j h_j^{(1)} u_{t-j} + \sum_j \sum_k h_{jk}^{(2)} u_{t-j} u_{t-k} \\ &+ \sum_j \sum_k \sum_l h_{jkl}^{(3)} u_{t-j} u_{t-k} u_{t-l} + \cdots \end{aligned} \tag{3.105}$$

where $t = 0, \pm 1, \pm 2, \ldots$. When $\{u_t\}$ is wide sense stationary with mean zero, we also have

$$\begin{aligned} X_t = &\int_{-\pi}^{\pi} e^{it\omega_1} H_1(\omega_1) \xi_u(d\omega_1) \\ &+ \int_{-\pi}^{\pi} \int_{-\pi}^{\pi} e^{it(\omega_1+\omega_2)} H_2(\omega_1, \omega_2) \xi_u(d\omega_1) \xi_u(d\omega_2) \\ &+ \int_{-\pi}^{\pi} \int_{-\pi}^{\pi} \int_{-\pi}^{\pi} e^{it(\omega_1+\omega_2+\omega_3)} H_3(\omega_1, \omega_2, \omega_3) \xi_u(d\omega_1) \xi_u(d\omega_2) \xi_u(d\omega_3) \\ &+ \cdots \end{aligned} \tag{3.106}$$

where the kernels H_k are Fourier transforms of the $h^{(k)}$. The expansion (3.105) is called a *Volterra series expansion*. In the form (3.106) the integrals are called *Wiener-Itô functionals* or *integrals*. In continuous time the limits of integration extend from $-\infty$ to ∞.

A problem of interest in the time series and econometrics literature is the estimation of the $h^{(k)}$, or equivalently the H_k, and testing whether the infinite expansion can be truncated after a few terms, or whether only the linear component $\sum_j h_j^{(1)} u_{t-j}$ provides an adequate representation, perhaps plus some additive noise. See for example [13], [18], [23], and the many additional references therein. In probability theory, the Wiener–Itô functionals have been shown to be a powerful tool in studying ergodic and central limit theorems of nonlinear functionals of Gaussian processes [5]. See also Chapter 6.

(a) In (3.105), determine the output X_t if $u_t = Ae^{it\omega}$ and if $u_t = A_1e^{it\omega_1} + A_2e^{it\omega_2}$. Compare with the answers when only the linear component $\sum_j h_j^{(1)} u_{t-j}$ is present [23, p. 29].

(b) If the u_t are IID $\mathcal{N}(0, 1)$, show that $\{X_t\}$ is stationary.

(c) Suppose the u_t are IID $\mathcal{N}(0, 1)$. Determine the spectrum of $\{X_t\}$ when (3.105) contains the linear and quadratic components only.

(d) With u_t IID $\mathcal{N}(0, 1)$, simulate a time series from (3.105) once when only the linear component is present, and once when the linear and quadratic components are present.

24. *Estimation of lags* [18]. Let $\{X_t\}$, $t = 0, \pm 1, \pm 2, \ldots$ be a zero-mean real-valued stationary process with autocovariance $R_{xx}(k)$, and spectral density $f_{xx}(\lambda)$, and assume all the relevant moments exist. Define a zero-mean *lag process* with lag v,

$$X_v(t) \equiv X_t X_{t-v} - R_{xx}(v), \qquad t = 0, \pm 1, \pm 2, \ldots$$

For a fixed but *unknown* lag v, consider the model,

$$Y_t = \sum_{k=-\infty}^{\infty} l_k X_{t-k} + \sum_{k=-\infty}^{\infty} b_k X_v(t-k) + \epsilon_t, \qquad t = 0, \pm 1, \ldots \tag{3.107}$$

where $\{\epsilon_t\}$ is stationary noise with mean zero independent of $\{X_t\}$, and where we assume that the infinite sums converge in mean square. Define, $B(\lambda) = \sum_{k=-\infty}^{\infty} e^{-ik\lambda} b_k$, $-\pi < \lambda \leq \pi$. The *cross covariance* of $\{X_t\}$ and $\{Y_t\}$ (with 0 means) is defined by

$$R_{xy}(k) = E[X_t Y_{t-k}], \qquad k = 0, \pm 1, \pm 2, \ldots$$

The *cross spectral density* is given by

$$f_{xy}(\lambda) = \frac{1}{2\pi} \sum_{k=-\infty}^{\infty} e^{-i\lambda k} R_{xy}(k)$$

provided the sum is finite. The quantity,

$$S_1(\lambda) \equiv \frac{|f_{xy}(\lambda)|^2}{f_{xx}(\lambda) f_{yy}(\lambda)}, \qquad -\pi < \lambda \leq \pi$$

is called the (squared) *coherence* between $\{X_t\}$ and $\{Y_t\}$ at frequency λ. It acts like a squared correlation in the frequency domain. A concrete application of the measure of coherence can be found in [3, pp. 274–275]. A quantity analogous to $S_1(\lambda)$ is defined by

$$S_2(\lambda; v) \equiv S_1(\lambda) + \frac{|B(\lambda)|^2}{f_{yy}(\lambda)} \left[f_{x_v x_v}(\lambda) - \frac{|f_{xx_v}(\lambda)|^2}{f_{xx}(\lambda)} \right], \qquad -\pi < \lambda \leq \pi$$

and is called *lagged coherence* (of lag u) between $\{X_t\}$ and $\{Y_t\}$.

(a) Show that if in (3.107) the "quadratic" term $\sum_{k=-\infty}^{\infty} b_k X_v(t-k)$ corresponding to the lag process $\{X_v(t)\}$ is missing, then

$$f_{\epsilon\epsilon}(\lambda) = f_{yy}[1 - S_1(\lambda)], \qquad -\pi < \lambda \leq \pi$$

Hence conclude that $0 \leq S_1(\lambda) \leq 1$, for all $\lambda \in (-\pi, \pi]$. Argue that when $S_1(\lambda) \equiv 1$, we have a linear system without noise, and when $S_1(\lambda) \equiv 0$, the output is all noise.

(b) With obvious notation, argue that

$$E[\xi_x(d\lambda)\overline{\xi_y(d\omega)}] = \begin{cases} f_{xy}(\lambda)d\lambda, & \text{if } \lambda = \omega \\ 0, & \text{if } \lambda \neq \omega \end{cases} \tag{3.108}$$

Hence, another proof that $S_1(\lambda)$ is a frequency domain squared correlation is implied by the Cauchy–Schwarz inequality,

$$|E[\xi_x(d\lambda)\overline{\xi_y(d\lambda)}]|^2 \leq E[|\xi_x(d\lambda)|^2]E[|\xi_y(d\lambda)|^2]$$

(c) Show that for any lag v [18],

$$0 \leq S_1(\lambda) \leq S_2(\lambda; v) \leq 1, \qquad -\pi < \lambda \leq \pi$$

(d) Show that

$$f_{\epsilon\epsilon}(\lambda) = f_{yy}[1 - S_2(\lambda; v)], \qquad -\pi < \lambda \leq \pi$$

Hence conclude that a *sensible way to estimate the unknown lag v is to maximize $S_2(\lambda; u)$ over u so as to minimize the noise variance.*

(e) Argue that when $S_1(\lambda)$ is small while $S_2(\lambda; v)$ is high for all λ, the "quadratic" term $\sum_{k=-\infty}^{\infty} b_k X_v(t-k)$ is indispensable.

25. *Spectral density estimation*. Consider the stationary $AR(2)$ process,

$$Z_t = \phi_1 Z_{t-1} + \phi_2 Z_{t-2} + \epsilon_t, \qquad t = 0, \pm 1, \pm 2, \ldots$$

where $\{\epsilon_t\}$ is white Gaussian noise. The spectral density is

$$f(\lambda) = \frac{\sigma_\epsilon^2}{2\pi} \cdot \frac{1}{|1 - \phi_1 e^{-i\lambda} - \phi_2 e^{-i2\lambda}|^2}, \qquad \lambda \in [-\pi, \pi] \tag{3.109}$$

Choose any ϕ_1, ϕ_2 such that $\phi_2 + \phi_1 < 1$, $\phi_2 - \phi_1 < 1$, $-1 < \phi_2 < 1$ (this guarantees stationarity [2, p. 58]), and simulate a time series of length $N = 1000$ from the process. The simulation can be carried out as follows. First generate IID $\epsilon_t \sim \mathcal{N}(0, \sigma_\epsilon^2)$. By fixing $Z_0 = Z_{-1} = 0$, we have

$$\begin{aligned} Z_1 &= \epsilon_1 \\ Z_2 &= \phi_1 Z_1 + \epsilon_2 \\ Z_3 &= \phi_1 Z_2 + \phi_2 Z_1 + \epsilon_2 \\ Z_4 &= \phi_1 Z_3 + \phi_2 Z_2 + \epsilon_4 \\ &\vdots \end{aligned}$$

Based on $Z_1, Z_2, \ldots, Z_{1000}$, estimate the spectral density $f(\lambda)$ by the three following procedures, and then compare the estimates relative to the true density (3.109).

(a) *A parametric approach: AR spectral estimate.* The *AR spectral estimate* is obtained by replacing ϕ_1, ϕ_2 by their estimates obtained by solving the Yule–Walker equations,

$$\begin{aligned}\hat{\rho}_1 &= \phi_1 + \phi_2\hat{\rho}_1 \\ \hat{\rho}_2 &= \phi_1\hat{\rho}_1 + \phi_2\end{aligned}$$

where $\hat{\rho}_1, \hat{\rho}_2$ are sample estimates of ρ_1, ρ_2, respectively.

(b) *A nonparametric approach: Window estimate.* Let $\bar{Z} = 1/N\sum_{t=1}^{N} Z_t$ be the sample average, and define an autocovariance estimate by

$$\hat{R}_k = \frac{1}{N}\sum_{t=1}^{N-k}(Z_t - \bar{Z})(Z_{t+k} - \bar{Z})$$

We refer to a weighting function $w(u)$ as a *lag window* if $w(u) = w(-u)$, $|w(u)| \leq 1$ for $|u| \leq 1$, and is 0 otherwise. The *window estimate* is given by

$$\hat{f}(\lambda) = \frac{1}{2\pi}\sum_{k=-M}^{M} w\left(\frac{k}{M}\right)\cos(\lambda k)\hat{R}_k \tag{3.110}$$

where $M(<< N)$ is the *truncation point* whose choice determines the variability and smoothness of the spectral estimate. There is a whole slew of lag windows to choose from. A popular choice is known as the *Blackman–Tukey* or *Tukey–Hanning window*,

$$w(u) = \begin{cases} \frac{1}{2}(1 + \cos(\pi u)), & |u| \leq 1 \\ 0, & |u| > 1 \end{cases}$$

Another popular choice is the *Parzen window*,

$$w(u) = \begin{cases} 1 - 6u^2 + 6|u|^3, & |u| \leq \frac{1}{2} \\ 2(1 - |u|)^3 & \frac{1}{2} \leq |u| \leq 1 \\ 0, & |u| > 1 \end{cases}$$

For $N = 1000$, a choice of $M = 50$ is quite reasonable.

(c) *Smoothed sample spectral density.* Define the *sample spectral density* by

$$J(\lambda) = \frac{1}{2\pi N}\left|\sum_{t=1}^{N} e^{-it\lambda}Z_t\right|^2$$

A spectral estimate is obtained by smoothing $J(\lambda)$,

$$\tilde{f}(\lambda) = \frac{1}{n}\sum_j J(\lambda_j)$$

where the $\lambda_j = 2\pi j/N$ are n values symmetric about λ, and n controls the variability and smoothness of the estimate. The estimate is sometimes called *smoothed periodogram*, and is essentially of the type (3.110).

26. *A criterion for mean square convergence* [1, p. 415]. Suppose the sequence of random variables $\{X_n\}$ satisfies as $m, n \to \infty$,

$$E[X_n X_m] \to \text{positive constant}$$

Show that $\{X_n\}$ converges in mean square to some random variable X.

27. *Stability*. A filter is said to be *stable* if a bounded input produces a bounded output. Show that a sufficient stability condition is the absolute summability of h_j.

28. *Construction of filters* [19, sec. 4.3]. Discuss the validity of the following rules of filter design.

$$\begin{aligned} \text{high-pass}(Z_t) &= Z_t - \text{low-pass}(Z_t) \\ \text{band-pass}(Z_t) &= \text{high-pass}(\text{low-pass}(Z_t)) \\ &= \text{low-pass}(\text{high-pass}(Z_t)) \end{aligned}$$

Also, a *notch* filter is obtained from the operation,

$$\mathcal{L}(Z_t) = Z_t - \text{band-pass}(Z_t)$$

29. *Inverse of a linear filter*. Let $\{X_t\}, t = 0, \pm 1, \ldots$, be weakly stationary with mean 0, and let $\mathcal{L}$ be a linear filter with absolutely summable impulse response, and transfer function $H(\lambda)$ such that $H(\lambda) \neq 0$. Show that the inverse $\mathcal{L}^{-1}$ defined by the condition $\mathcal{L}^{-1}(\mathcal{L}(X_t)) = X_t$, that is, $\mathcal{L}^{-1}\mathcal{L}$ is the identity filter, is well defined with transfer function $1/H(\lambda)$.

REFERENCES

[1] Anderson, T.W., *The Statistical Analysis of Time Series*, New York: Wiley, 1971.

[2] Box, G.E.P. and G.M. Jenkins, *Time Series Analysis, Forecasting and Control* (rev. ed.), Holden-Day: San Francisco, 1976.

[3] Brillinger, D.R., *Time Series: Data Analysis and Theory*, Holt, Rinehart & Winston: New York, 1975.

[4] Brockwell, P.J. and R.A. Davis, *Time Series: Theory and Methods* (2nd ed.), Springer-Verlag: New York, 1991.

[5] Chambers, D. and E. Slud, "Central limit theorems for nonlinear functionals of stationary Gaussian processes," *Probability Theory Rel. Fields*, Vol. 80, pp. 323–346, 1989.

[6] Doob, L.J. *Stochastic Processes*, New York: Wiley, 1953.

[7] Gaposhkin, V.F., "Criteria for the strong law of large numbers for some classes of second-order stationary processes and homogeneous random fields," *Theory Prob. Appl.*, Vol. 22, pp. 286–310, 1977.

[8] Gray, R.M. and L.D. Davisson, *Random Processes: A Mathematical Approach for Engineers*, Englewood Cliffs, N.J.: Prentice Hall, 1986.

[9] Grenander, U. and M. Rosenblatt, *Statistical Analysis of Stationary Time Series*, New York: Wiley, 1957.

[10] Hamming, R.W., *Digital Filters*, Englewood Cliffs, N.J.: Prentice Hall, 1983.

[11] He, S. and B. Kedem, "Higher order crossings of an almost periodic random sequence in noise," *IEEE Trans. Inform. Theory*, Vol. IT-35, No. 2, pp. 360–370, Mar. 1989.

[12] Heyde, C.C. and E. Seneta, "Estimation theory for growth and immigration rates in multiplicative processes," *J. Appl. Prob.*, Vol. 9, pp. 235–256, 1972.

[13] Hinich, M., "Testing for Gaussianity and linearity of a stationary time series," *J. Time Ser. Anal.*, Vol. 3, pp. 169–176, 1982.

[14] Hooghiemstra, G. and M. Keane, "Calculation of the equilibrium distribution for a solar energy storage model," *J. Appl. Prob.*, Vol. 22, pp. 852–864, 1985.

[15] Houdré, C., "On the spectral SLLN and pointwise ergodic theorem in L^α, *Ann. Prob.*, Vol. 20, pp. 1731–1753, 1992.

[16] Hurd, H.L., "Representation of strongly harmonizable periodically correlated processes and their covariances," *J. Multivar. Anal.*, Vol. 29, pp. 53–67, 1989.

[17] Kedem, B. and L.S. Chiu, "On the lognormality of rain rate," *Proc. Natl. Acad. Sci. U.S.A.*, Vol. 84, pp. 901–905, Feb. 1987.

[18] Kedem-Kimelfeld, B., "Estimating the lags of lag processes," *J. Am. Stat. Assoc.*, Vol. 70, pp. 603–605, Sept. 1975.

[19] Koopmans, L.H., *The Spectral Analysis of Time Series*, New York: Academic Press, 1974.

[20] Otnes, R.K. and L. Enochson, *Applied Time Series Analysis*, New York: Wiley, 1978.

[21] Papoulis, A., *Probability, Random Variables, and Stochastic Processes* (2nd ed.), New York: McGraw-Hill, 1984.

[22] Priestley, M.B., *Spectral Analysis and Time Series*, Vols. 1,2, London: Academic Press, 1981.

[23] Priestley, M.B., *Non-Linear and Non-Stationary Time Series Analysis*, New York: Academic Press, 1988.

[24] Rosenblatt, M., *Stationary Sequences and Related Fields*, Boston, Mass.: Birkhäuser, 1985.

[25] Schuster, A., "On the investigation of hidden periodicities with application to a supposed 26-day period of meteorological phenomena," *Terr. Magnet.*, Vol. 3, pp. 13–41, 1898.

[26] Shiryayev, A.N., *Probability*, New York: Springer-Verlag, 1984.

[27] Slutsky, E.E. "The summation of random causes as the source of cyclic processes," (Russian) *Prob. Econ. Cond.*, Vol. 3, No. 1. (English Transl.) *Econometrica*, Vol. 5, pp. 105–146, 1937.

[28] Winnicki, J., *A Unified Theory for the Branching Process with Immigration*, Doctoral Dissertation, Department of Mathematics, Univ. of Maryland, College Park, 1986.

[29] Wold, H.O.A., *Bibliography on Time Series and Stochastic Processes*, Cambridge, Mass.: M.I.T. Press, 1965.

[30] Yaglom, A.M., *Correlation Theory of Stationary and Related Random Functions,* Vols. I,II. New York: Springer-Verlag, 1987.

[31] Yule, G.U., "On a method of investigating periodicities in disturbed series with special reference to Wolfer's sunspot numbers," *Philos. Trans. R. Soc. London*, Ser. A, Vol. 226, pp. 267–298, 1927.

4

Zero-Crossings and Autocorrelation

The number of zero-crossings observed in a finitely long real-valued time series may be viewed as a measure of the oscillation exhibited by the time series. This is so because, in general, the more pronounced the oscillation, the higher is the expected number of zero-crossings, and conversely, fewer zero-crossings are expected when the time series is rather "smooth" and slowly varying. The number of zero-crossings or its expected value are obviously not the only measures of oscillation. The autocorrelation sequence, and in particular the first order autocorrelation ρ_1, can also serve as measures of oscillation. When ρ_1 is close to 1, neighboring observations are highly correlated, and in light of Theorem 2.2 and Example 2.10, we expect rather smooth realizations. Similarly, when ρ_1 is close to -1, neighboring observations are negatively correlated and a more pronounced oscillation is expected. Both being measures of oscillation, it is no wonder that ρ_1 and the expected number of zero-crossings are closely related in general, and that this relationship can be made precise under some conditions. Whatever the true relationship is, though, it must be an *inverse* relationship, reflecting the fact that slowly varying time series are associated with higher correlations, but lower expected zero-crossing rates, while the situation is reversed for markedly oscillating series. See Figure 4.1.

In this chapter we are going to investigate the relationship between ρ_1 and the expected number of zero-crossings. This relationship is of fundamental importance because it implies a formal connection between the spectrum and the number of zero-crossings. Oddly enough, we can derive explicit formulas connecting ρ_1 and the expected number of zero-crossings only in some special, but important, cases. The general case is as of yet an unsolved challenging problem.

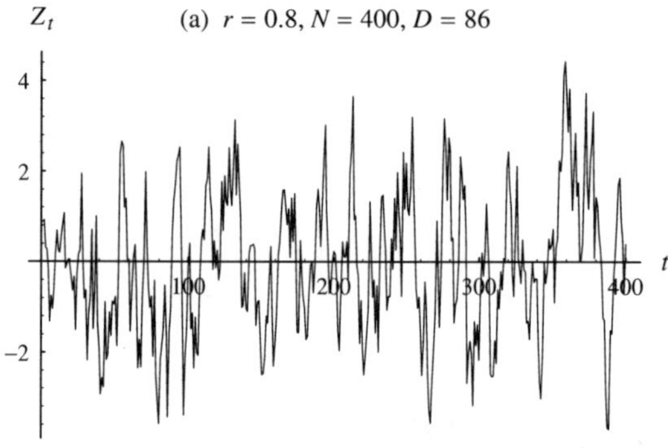

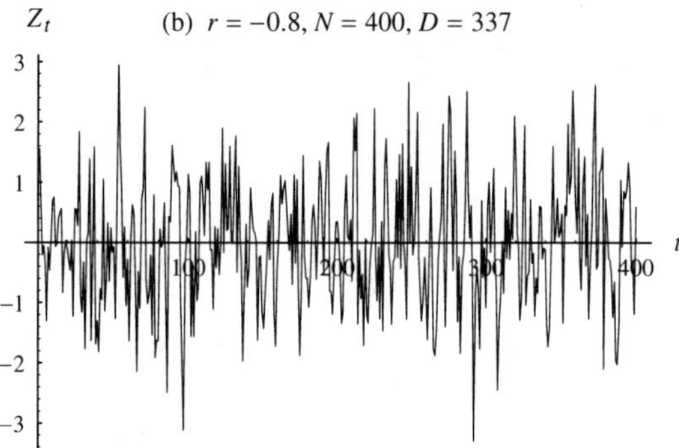

Figure 4.1: *First order autocorrelation* ρ_1 *and number of zero-crossings* D *in* $Z_t = rZ_{t-1} + \zeta_t$, $t = 1, \ldots, 400$. $\{\zeta_t\}$ *is white noise with mean* 0 *and variance* 1. (a) $\rho_1 = r = 0.8$, $D = 86$. (b) $\rho_1 = r = -0.8$, $D = 337$.

An explicit and tractable relationship between correlations and zero-crossings is useful for estimation purposes. When such a relationship exists, every parameter that can be estimated from the first order sample autocorrelation, is automatically also estimable from zero-crossings, and vice versa. However, counting zero-crossings is simpler and faster.

Remark. From now on, whenever we deal with zero-crossings and functions thereof, the underlying process and its filtered versions are assumed real-valued. By this, however, we do not rule out suitable complex analogs of zero-crossing counts that may lead to a parallel development taking after the real case.

4.1 ZERO-CROSSINGS IN DISCRETE TIME

Let $Z_1, Z_2, \ldots, Z_N$ be a zero-mean stationary real-valued time series. The zero-crossing count in discrete time is defined as the number of symbol changes in the corresponding clipped binary time series.

Define a (clipped) binary time series $X_1, X_2, \ldots, X_N$ by the nonlinear transformation,

$$X_t = \begin{cases} 1, & \text{if } Z_t \geq 0 \\ 0, & \text{if } Z_t < 0 \end{cases}$$

Clearly, $\{X_t\}$ is a stationary series because $\{Z_t\}$ is. The number of zero-crossings, denoted by D, is defined in terms of $\{X_t\}$,

$$D = \sum_{t=2}^{N} [X_t - X_{t-1}]^2 \tag{4.1}$$

If we let d_t be the indicator[1] $I_{[\cdot]}$ of the event $[X_t \neq X_{t-1}]$, then D can also be expressed as a sum of indicators,

$$D = \sum_{t=2}^{N} I_{[X_t \neq X_{t-1}]} \tag{4.2}$$

$$= \sum_{t=2}^{N} d_t \tag{4.3}$$

Evidently, D depends on the series size N, and its possible values are

$$0, 1, 2, \ldots, N-1$$

The upper bound of $N-1$ is due to the fact that the maximum possible number of symbol changes from 0 to 1 and from 1 to 0 in $X_1, \ldots, X_N$ is $N-1$. For example, to realize 11 symbol changes (the maximum possible) in Figure 4.2 we need $X = (010101010101)$ or $X = (101010101010)$, but $X = (000\cdots0)$ or $X = (111\cdots1)$ gives $D = 0$; hence, $0 \leq D \leq 11$.

The (observed) zero-crossing rate, denoted by $\hat{\gamma}$, is defined by the ratio

$$\hat{\gamma} = \frac{D}{N-1} \tag{4.4}$$

Two useful parameters associated with the binary time series $\{X_t\}$ are the probability, p, of a 1 at time t, and the conditional probability, λ_1, of a 1 at time t, given a 1 at time $t-1$,

$$p = P(X_t = 1)$$

and

$$\lambda_1 = P(X_t = 1 | X_{t-1} = 1) \tag{4.5}$$

It follows that

$$E[D] = 2p(N-1)(1-\lambda_1)$$

---

[1] $I_{[A]}$ is the indicator of the event A. It is equal to 1 if A occurs, and it is 0 otherwise.

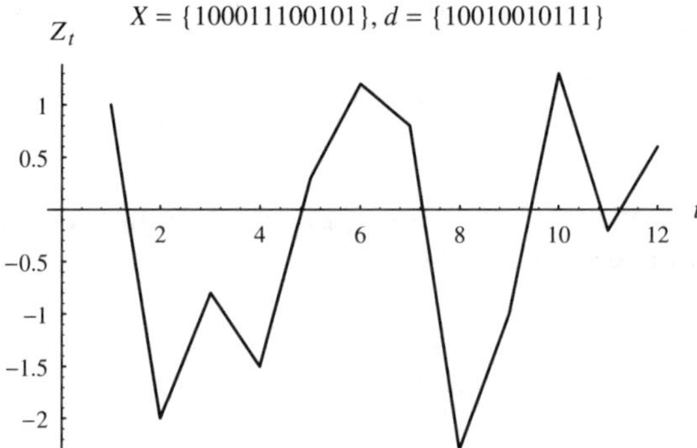

Figure 4.2: *A time series of size* 12 *with* 6 *zero-crossings* ($N = 12,\ D = 6,$ $\hat{\gamma} = 6/11$*), and the corresponding binary series* X_t *and* d_t*.*

and this depends on N as it should. However, the expected zero-crossing rate, denoted by γ,

$$\gamma = E[\hat{\gamma}] = \frac{E[D]}{N-1} = 2p(1-\lambda_1) \tag{4.6}$$

is independent of N.

The definition of D facilitates the study of the relationship problem between ρ_1 and $E[D]$, or equivalently between ρ_1 and γ.

4.2 THE COSINE FORMULA

When $\{Z_t\}$ is a stationary random sinusoid, or a stationary Gaussian process, there is a neat explicit formula connecting ρ_1 and $E[D]$,

$$\rho_1 = \cos\left(\frac{\pi E[D]}{N-1}\right) \tag{4.7}$$

The formula (4.7) is referred to as the *cosine formula*. It can be generalized to the entire class of processes whose finite dimensional distributions are elliptically symmetric. Most importantly, the cosine formula vindicates our intuition regarding the inverse relationship between the expected number of zero-crossings and the correlation between neighboring observations.

4.2.1 The Case of a Pure Sinusoid

In the case of a pure sinusoid, the relationship between ρ_1 and $E[D]$ is obtained by first observing that the frequency of the sinusoid can be estimated from zero-crossings.

Theorem 4.1. *Suppose* $\{Z_t\}, t = 0, \pm 1, \pm 2, \ldots,$ *is a stationary random sinusoid:*

$$Z_t = A\cos(\omega t) + B\sin(\omega t)$$

where A, B are uncorrelated random variables with zero means and variance σ^2 each. Let ρ_k be the autocorrelation sequence of $\{Z_t\}$. Then the cosine formula (4.7) holds.

Proof. Note that for a random stationary sinusoid

$$\rho_1 = \cos(\omega)$$

Now, as in [7, p. 29], suppose we observe in the time series C complete cycles but not $C+1$. Then ignoring end effects we have

$$\frac{N}{C+1} \leq PERIOD = \frac{2\pi}{\omega} \leq \frac{N}{C}$$

Therefore

$$\frac{2\pi C}{N} \leq \omega \leq \frac{2\pi(C+1)}{N} \tag{4.8}$$

Again, ignoring end effects, substitute

$$D = 2C$$

and (4.8) becomes

$$\frac{\pi D}{N} \leq \omega \leq \frac{\pi D}{N} + \frac{2\pi}{N} \leq \frac{\pi D}{N-1} + \frac{2\pi}{N}$$

or, by subtraction and simple manipulation,

$$\frac{-2\pi}{N} \leq \omega - \frac{\pi D}{N-1} \leq \frac{2\pi}{N}$$

Therefore with probability one, as $N \longrightarrow \infty$,

$$\frac{\pi D}{N-1} \longrightarrow \omega$$

By bounded convergence then, as $N \longrightarrow \infty$,

$$E[\frac{\pi D}{N-1}] = \frac{\pi E[D]}{N-1} \longrightarrow \omega$$

But the zero-crossing rate is independent of N, and it follows that

$$\frac{\pi E[D]}{N-1} = \omega$$

and so, we finally have

$$\rho_1 = \cos(\omega) = \cos\left(\frac{\pi E[D]}{N-1}\right)$$

□

4.2.2 The General Ellipsoidal Case

Suppose the finite dimensional distributions of $\{Z_t\}$ are governed by the ellipsoidal density

$$f(\mathbf{x}) = |\mathbf{\Sigma}|^{-1/2}\psi(\mathbf{x}'\mathbf{\Sigma}^{-1}\mathbf{x}) \tag{4.9}$$

that was discussed in Theorem 2.4. We shall say that a stochastic process is an ellipsoidal process if its finite dimensional distributions are elliptically symmetric. Clearly, Gaussian processes are ellipsoidal, and thus constitute a special but most important case. By appealing to Theorem 2.4, we have:

Theorem 4.2. *[10] Let $\{Z_t\}$, $t = 0, \pm 1, \pm 2, \ldots$, be a stationary ellipsoidal process with mean zero, variance one, and autocorrelation ρ_k. Then the cosine formula (4.7) holds.*

Proof. From Theorem 2.4, and the definition of D,

$$\begin{aligned} E[D] &= (N-1)(1 - 2P(Z_k \geq 0, Z_{k-1} \geq 0)) \\ &= (N-1)\left(\frac{1}{2} - \frac{1}{\pi}\sin^{-1}\rho_1\right) \end{aligned}$$

or

$$\begin{aligned} \cos\left(\frac{\pi E[D]}{N-1}\right) &= \cos\left(\frac{\pi}{2} - \sin^{-1}\rho_1\right) \\ &= \sin(\sin^{-1}\rho_1) \\ &= \rho_1 \end{aligned}$$

□

Corollary 4.1. *Let $\{Z_t\}$, $t = 0, \pm 1, \pm 2, \ldots$, be a a zero-mean stationary Gaussian process with autocorrelation ρ_k. Then,*

$$\rho_1 = \cos\left(\frac{\pi E[D]}{N-1}\right)$$

Thus the cosine formula (4.7) holds for stationary random sinusoids and for stationary Gaussian processes. It provides a means for a generalization to some non-Gaussian processes.

A closely related fact is an extension of the arcsine formula that gives the orthant probability for three random variables (see Example 2.15).

Theorem 4.3. *[10] Let $\{Z_t\}$, $t = 0, \pm 1, \pm 2, \ldots$, be a stationary ellipsoidal process with mean zero, variance one, and autocorrelation ρ_k. Then,*

$$P(Z_i \geq 0, Z_j \geq 0, Z_k \geq 0) = \frac{1}{8} + \frac{1}{4\pi}[\sin^{-1}\rho_{i-j} + \sin^{-1}\rho_{j-k} + \sin^{-1}\rho_{i-k}]$$

Proof. Define $Y_t = 1 - X_t, t = 0, \pm 1, \ldots,$. Then,

$$\begin{aligned} P(Z_i \geq 0, Z_j \geq 0, Z_k \geq 0) &= E[X_i X_j X_k] \\ &= E[(1-Y_i)(1-Y_j)(1-Y_k)] \\ &= 1 - \frac{3}{2} + E[Y_i Y_j] + E[Y_j Y_k] + E[Y_i Y_k] \\ &\quad - E[Y_i Y_j Y_k] \end{aligned}$$

However, by the symmetry of the ellipsoidal distribution, $(Z_{n_1}, \ldots, Z_{n_k})$ has the same distribution as $(-Z_{n_1}, \ldots, -Z_{n_k})$, and therefore,

$$2E[X_i X_j X_k] = -\tfrac{1}{2} + \tfrac{3}{4} + \tfrac{1}{2\pi}[\sin^{-1}\rho_{i-j} + \sin^{-1}\rho_{j-k} + \sin^{-1}\rho_{i-k}]$$

□

As noted in Example 2.15, the extension of Theorem 4.3 to higher order orthant probabilities, including the fourth order, is rather problematic [9], a fact that makes the evaluation of the variance of D quite cumbersome. We shall come to this point again at a later chapter.

The cosine formula elucidates the connection between the expected number of zero-crossings and the spectrum. From the definition of ρ_1 and by the Wiener–Khintchine theorem follows the relationship

$$\cos\left(\frac{\pi E[D]}{N-1}\right) = \frac{\int_{-\pi}^{\pi} \cos(\omega)\, dF(\omega)}{\int_{-\pi}^{-\pi} dF(\omega)} \tag{4.10}$$

to which we refer as the *zero-crossing spectral representation*. It shows that $E[D]$ contains spectral information, and as such plays an important role in this book. When F is continuous at the origin, an assumption we shall adopt throughout the book, symmetry implies that (4.10) can be expressed in terms of positive frequencies, as in the expression

$$\cos\left(\frac{\pi E[D]}{N-1}\right) = \frac{\int_0^{\pi} \cos(\omega)\, dF(\omega)}{\int_0^{\pi} dF(\omega)} \tag{4.11}$$

An immediate consequence of (4.11) is that the normalized expected number of zero-crossings $\pi E[D]/(N-1)$ is a *weighted average* of the spectral mass distributed over the interval $[0, \pi]$. Two extreme cases can illustrate this point vividly. First, suppose the process $\{Z_t\}$ is white noise. Then the spectrum is continuous with a flat density that is equal to some constant over the entire interval $[0, \pi]$. In this case, (4.11) gives

$$\cos\left(\frac{\pi E[D]}{N-1}\right) = 0$$

But $\cos(x)$ is strictly monotone over $[0, \pi]$ so that

$$\frac{\pi E[D]}{N-1} = \frac{\pi}{2}$$

as is well expected, since for white noise the spectral mass is evenly distributed over $[0, \pi]$. Thus, in the absence of any dominant region, $\pi E[D]/(N-1)$ falls in "the middle." On the other hand, suppose a certain frequency band becomes dominant, that is, carries more power than other bands, $\pi E[D]/(N-1)$ will tend to "land" there. In the extreme case when the process is a pure sinusoid with frequency ω_0, all the power is given to ω_0 so that,

$$\begin{aligned} F(\omega+) - F(\omega-) &> 0, \qquad \omega = \omega_0 \\ &= 0, \qquad \omega \neq \omega_0 \end{aligned}$$

and thus,

$$\cos\left(\frac{\pi E[D]}{N-1}\right) = \cos(\omega_0)$$

or, by the monotonicity of $\cos(x)$ in $[0, \pi]$,

$$\frac{\pi E[D]}{N-1} = \omega_0$$

We conclude that when a certain frequency becomes dominant, it attracts the expected normalized number of zero-crossings $\pi E[D]/(N-1)$, and the latter will land at or near that frequency. This tendency of $\pi E[D]/(N-1)$ to admit values in a dominant spectral band, when it exists, is referred to as the *dominant frequency principle*. It is a well-established empirical fact (e.g., see [12, p. 137]) which is manifested through the zero-crossing spectral representation (4.11). This principle can be recognized, time and again, as the basis for many of our theoretical results.

Example 4.1: On the Dominant Frequency Principle

The dominant frequency principle is independent of any distributional assumption (such as the Gaussian assumption), and is a general fact regarding weighted sums of sinusoids. An indication of this fact is Example 1.1. To bring this to light again, here is another indication. Consider the deterministic sum

$$Z_t = A\cos(0.8t) + B\cos(1.25t), \; t = 1, 2, \ldots, 200$$

where A, B are constants. Because Z_t is not random, $D = E[D]$. When $A = B = 1$, no frequency is dominant. We counted $D = 67$ zero-crossings, and $\pi D/199 = 1.058$, which is between 0.8 and 1.25 as expected. When $A = 0.8, B = 1$, the frequency 1.25 is dominant, and we counted $D = 79$ zero-crossings. This gives $\pi D/199 = 1.247$, which is close to the dominant frequency in agreement with the dominant frequency principle. When $A = 1.5, B = 1$, the dominancy shifts toward the lower frequency 0.8. For this case $D = 51$, and $\pi D/199 = 0.805$, close to the dominant frequency.

■

4.2.3 The Mixed Spectrum Case

By Corollary 4.1, the cosine formula (4.7) holds for any stationary zero-mean Gaussian process regardless of the spectrum type. As a consequence we can write down immediately the zero-crossing rate of a sum of random sinusoids plus any colored continuous spectrum noise.

Consider the mixed spectrum zero-mean Gaussian process,

$$Z_t = \sum_{j=1}^{p} \{A_j \cos(\omega_j t) + B_j \sin(\omega_j t)\} + \zeta_t \tag{4.12}$$

where, $t = 0, \pm 1, \pm 2, \ldots$, the A's and B's are all uncorrelated normal random variables, $E(A_j) = E(B_j) = 0$, and $\mathrm{Var}(A_j) = \mathrm{Var}(B_j) = \sigma_j^2$, and $\{\zeta_t\}$ is colored stationary Gaussian noise with mean 0 and variance σ_ζ^2, independent of the A's and B's. The noise is assumed to possess an absolutely continuous spectral distribution function $F_\zeta(\omega)$ with spectral density $f_\zeta(\omega)$, $\omega \in [-\pi, \pi]$. Without loss of generality we shall always assume that the frequencies are ordered fixed constants in $(0, \pi]$,

$$0 < \omega_1 < \omega_2 < \cdots < \omega_p \leq \pi$$

It is convenient to define a process $\{Y_t\}$ with spectral distribution function F_y by

$$Y_t = \sum_{j=1}^{p} \{A_j \cos(\omega_j t) + B_j \sin(\omega_j t)\}$$

so that

$$Z_t = Y_t + \zeta_t$$

Then from the representation of ρ_1 in (4.10), and remembering that $F(\omega) = F_y(\omega) + F_\zeta(\omega)$, we have a formula frequently used in this book (see Section 3.3),

$$\rho_1 = \cos\left(\frac{\pi E[D]}{N-1}\right) = \frac{\sum_{j=1}^{p} \sigma_j^2 \cos(\omega_j) + \int_{-\pi}^{\pi} \cos(\omega)\, dF_\zeta(\omega)}{\sum_{j=1}^{p} \sigma_j^2 + \int_{-\pi}^{\pi} dF_\zeta(\omega)} \tag{4.13}$$

and the expected zero-crossing rate is given by

$$\frac{E[D]}{N-1} = \frac{1}{\pi} \cos^{-1}\left[\frac{\sum_{j=1}^{p} \sigma_j^2 \cos(\omega_j) + \int_{-\pi}^{\pi} \cos(\omega)\, dF_\zeta(\omega)}{\sum_{j=1}^{p} \sigma_j^2 + \int_{-\pi}^{\pi} dF_\zeta(\omega)}\right] \tag{4.14}$$

From this it follows that, when, say, $\sigma_1^2 \to \infty$, and everything else stays fixed, ω_1 becomes dominant, and as expected

$$\frac{\pi E[D]}{N-1} \to \omega_1$$

When $\{\zeta_t\}$ is white Gaussian noise, (4.13) reduces to

$$\rho_1 = \cos\left(\frac{\pi E[D]}{N-1}\right) = \frac{\sum_{j=1}^{p} \sigma_j^2 \cos(\omega_j)}{\sum_{j=1}^{p} \sigma_j^2 + \sigma_\zeta^2} \tag{4.15}$$

Without any noise at all, the formula further reduces to

$$\rho_1 = \cos\left(\frac{\pi E[D]}{N-1}\right) = \frac{\sum_{j=1}^{p} \sigma_j^2 \cos(\omega_j)}{\sum_{j=1}^{p} \sigma_j^2} \tag{4.16}$$

and this implies that

$$\omega_1 \leq \frac{\pi E[D]}{N-1} \leq \omega_p$$

That is, the normalized expected zero-crossing rate falls between the lowest and highest frequencies.

Returning to the general mixed spectrum case in (4.12), suppose a time-invariant linear filter $\mathcal{L}$ with transfer function $H(\omega)$ is applied to $\{Z_t\}$. Denote by D_H the zero-crossing count in the filtered time series, $\mathcal{L}(Z)_1, \mathcal{L}(Z)_2, \ldots, \mathcal{L}(Z)_N$. It follows from (4.13) that

$$\cos\left(\frac{\pi E[D_H]}{N-1}\right) = \frac{\sum_{j=1}^{p} \sigma_j^2 |H(\omega_j)|^2 \cos(\omega_j) + \int_{-\pi}^{\pi} \cos(\omega)|H(\omega)|^2 \, dF_\zeta(\omega)}{\sum_{j=1}^{p} \sigma_j^2 |H(\omega_j)|^2 + \int_{-\pi}^{\pi} |H(\omega)|^2 \, dF_\zeta(\omega)} \tag{4.17}$$

The expected zero-crossing rate of the filtered process $\{\mathcal{L}(Z)_t\}$ is thus given by

$$\frac{E[D_H]}{N-1} = \frac{1}{\pi}\cos^{-1}\left[\frac{\sum_{j=1}^{p} \sigma_j^2 |H(\omega_j)|^2 \cos(\omega_j) + \int_{-\pi}^{\pi} \cos(\omega)|H(\omega)|^2 \, dF_\zeta(\omega)}{\sum_{j=1}^{p} \sigma_j^2 |H(\omega_j)|^2 + \int_{-\pi}^{\pi} |H(\omega)|^2 \, dF_\zeta(\omega)}\right] \tag{4.18}$$

In Section 4.4 we will show that if the filter is high-pass, then $E[D] \leq E[D_H]$. The inequality is reversed for a low-pass filter.

The expected zero-crossing rate of a sum of non-Gaussian random sinusoids is complicated in the absence of an appropriate "cosine formula." Some special cases in continuous time are investigated in [5], [25], and [29].

Sum of Two Non-Gaussian Sinusoids

A way to "transform" a sinusoidal Gaussian process is by amplitude conditioning, leaving the phase random over an interval of length 2π. We consider briefly the zero-crossing rate of an interesting special case treated in [5].

Define a continuous time process $V(t)$ by

$$V(t) = A_1 \cos(2\pi f_1 t + \phi_0) + A_2 \cos(2\pi f_2 t + \psi_0) \tag{4.19}$$

where A_1, A_2 are *fixed* positive amplitudes, ϕ_0, ψ_0 are independently and uniformly distributed in $(-\pi, \pi)$, and where the frequencies f_1, f_2, (in cycles per second) are incommensurable. As such, the process $\{V(t)\}$ may be thought of as being obtained from a Gaussian process which has the same form, but for which A_1, A_2 are independent Rayleigh random variables. It has been shown in [5] that the expected zero-crossing rate (per second) is given by

$$E[D_c] = \frac{4}{\pi} f_1 \cos^{-1}\left[\frac{f_1}{A_2}\left(\frac{A_2^2 - A_1^2}{f_1^2 - f_2^2}\right)^{1/2}\right] + \frac{4}{\pi} f_2 \sin^{-1}\left[\frac{f_2}{A_1}\left(\frac{A_2^2 - A_1^2}{f_1^2 - f_2^2}\right)^{1/2}\right] \tag{4.20}$$

where $f_1 > f_2$, and $A_1 < A_2 < A_1 f_1/f_2$. By letting A_2 vary as a Rayleigh random variable, one obtains the expected zero-crossing rate of the sum of non-Gaussian and Gaussian random sinusoids. The result, reported in [5], [25], is quite involved.

4.2.4 Explosive Oscillation

The cosine formula (4.7) paves the way for an estimation based on zero-crossing counts and rates. The following example serves as an illustration of this fact.

Let $\mathcal{B}$ be the shift operator, $\mathcal{B}Z_t = Z_{t-1}$, and consider the second order autoregressive process $\{Z_t\}$, $t = 0, \pm 1, \ldots$, defined by the second order stochastic difference equation,

$$(1 - \alpha\mathcal{B})(1 - \beta\mathcal{B})Z_t = \zeta_t \tag{4.21}$$

where $\{\zeta_t\}$, $t = 0, \pm 1, \ldots$, is Gaussian white noise with mean zero and variance σ_ζ^2. For real-valued α, β, such that $|\alpha| < 1$, $|\beta| < 1$, and for complex-valued α, β, such that $\alpha = \overline{\beta} = \rho\exp(i\theta)$, with $|\rho| < 1$, $\{Z_t\}$ is a zero-mean stationary ergodic Gaussian process. The first order autocorrelation is given by

$$\rho_1 = \frac{\alpha + \beta}{1 + \alpha\beta} \tag{4.22}$$

so that by the cosine formula (Corollary 4.1),

$$\gamma = \frac{E[D]}{N-1} = \frac{1}{\pi}\cos^{-1}\left(\frac{\alpha + \beta}{1 + \alpha\beta}\right) \tag{4.23}$$

By ergodicity, as $N \to \infty$,

$$\hat{\gamma} = \frac{D}{N-1} \to \frac{1}{\pi}\cos^{-1}\left(\frac{\alpha + \beta}{1 + \alpha\beta}\right) \tag{4.24}$$

with probability one. Therefore, by bounded convergence, also

$$\hat{\gamma} = \frac{D}{N-1} \to \frac{1}{\pi}\cos^{-1}\left(\frac{\alpha + \beta}{1 + \alpha\beta}\right) \tag{4.25}$$

in mean square. In particular, for $\alpha = \overline{\beta} = \rho\exp(i\theta)$ such that $|\rho| < 1$, and $|\rho| \to 1$, $\alpha = \overline{\beta} \to \exp(i\theta)$, (4.25) gives

$$\hat{\gamma} = \frac{D}{N-1} \to \frac{1}{\pi}\cos^{-1}(\cos(\theta)) = \frac{\theta}{\pi} \tag{4.26}$$

in mean square.

Apparently, the situation is not that simple when $\alpha = \overline{\beta} = \rho\exp(i\theta)$, and $|\rho| = 1$ exactly (case of "unit roots"). In this case, $\{Z_t\}$ is an explosive process and hence no longer stationary. The explosive nature of $\{Z_t\}$ is illustrated in Figure 4.3. However, the figure also shows the remarkable regularity of the zero-crossings by $\{Z_t\}$. This fact emboldens us to believe that (4.26) also holds in the explosive case. As it turns out this is indeed the case [11], but the proof of this fact is surprisingly technical and long, and will not be reproduced here. The interested reader is referred to [11] for the mechanics of the proof.

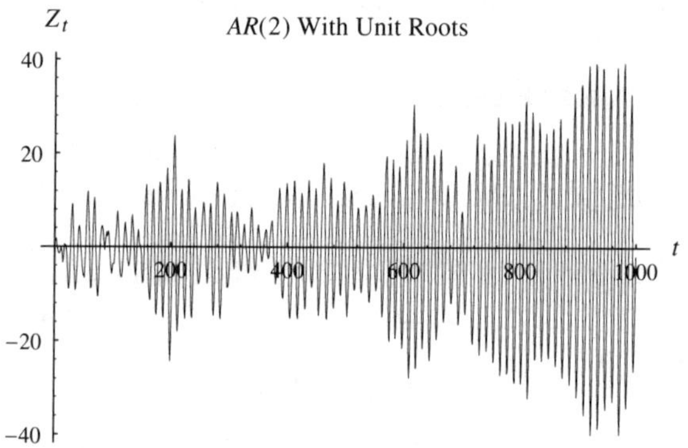

Figure 4.3: *Explosive $AR(2)$ with unit roots* $\exp(\pm i\pi/6)$.

Because of the high degree of regularity of the zero-crossings in the explosive case, the convergence in (4.26) is very fast, and is independent of the magnitude of the noise component, or even its presence altogether.

4.3 TRANSFORMED GUASSIAN PROCESSES

In this section we present a method developed in [3] for generating "cosine formulas" for some discrete time non-Gaussian processes. The underlying general idea is to transform a stationary Gaussian process by a monotone transformation that preserves the zero-crossing count (by fixing the origin), but gives a different correlation, or equivalently spectral, structure. The continuous time analogs can be obtained as limiting cases. The idea of studying level-crossings of monotone transformations of continuous time Gaussian processes occurred already in [21].

A great deal of related work on level-crossings, work that also reviews and summarizes much of the voluminous literature on this topic, can be found in [1], [6], [8], [19], [29].

4.3.1 The Discrete Time Case

Let $\varphi(x)$ be a strictly monotone (increasing or decreasing) real-valued differentiable function defined over the real line. Let $\{Z_t\}, t = 0, \pm 1, \ldots$, be a zero-mean stationary Gaussian process with unit variance and autocorrelation $\rho_z(k)$. Define a new process $\{Y_t\}$, with autocorrelation $\rho_y(k)$, by

$$Y_t = \varphi(Z_t) - \varphi(0) \tag{4.27}$$

Clearly, $\{Y_t\}$ is not necessarily Gaussian, and its mean need not be 0. The thing to observe is that with probability one *a zero-crossing occurs in* $\{Y_t\}$ *if and only if a zero-crossing occurs in* $\{Z_t\}$. That is, the zero-crossing count in $Y_1, \ldots, Y_N$ is equal to the zero-crossing count D in $Z_1, \ldots, Z_N$, with probability one. But if $\varphi(x)$ is nonlinear, then the finite dimensional distributions of $\{Y_t\}$ are different from those of $\{Z_t\}$. This implies, in particular, that the correlation structure in the two processes is different and hence we can expect different "cosine formulas."

In this connection the so-called Price's theorem (see [23, p. 339]) comes in very handy. In the context of our problem, it states that for Gaussian $\{Z_t\}$,

$$\frac{\partial E[Y_k Y_{k-1}]}{\partial \rho_z(1)} = E\left(\frac{\partial^2 Y_k Y_{k-1}}{\partial Z_k \partial Z_{k-1}}\right) \tag{4.28}$$

The method for generating the desired relationship between $\rho_y(1)$ and $E[D]$ consists of the following steps.

- Denote the zero-crossing count in $Z_1, \ldots, Z_N$ by D.
- Define
$$Y_k = \varphi(Z_k) - \varphi(0), \qquad k = 1, \ldots, N$$
The zero-crossing count in $Y_1, \ldots, Y_N$ also equals D.
- Apply Price's theorem (4.28) to obtain a functional relationship between $\rho_y(1)$ and $\rho_z(1)$ by integration.
- Substitute
$$\cos\left(\frac{\pi E[D]}{N-1}\right)$$
for $\rho_z(1)$. This gives the desired "cosine formula" that relates $\rho_y(1)$ to the expected zero-crossing count in $Y_1, \ldots, Y_N$.
- Solve for $E[D]$ in terms of $\rho_y(1)$ to obtain an expression for the zero-crossing rate of $\{Y_t\}$.

Example 4.2: A Symmetric Uniform Process

Let $\Phi(x)$ be the distribution function of the standard normal distribution, and define

$$\varphi(x) = \Phi(x)$$

Then $\varphi(0) = 1/2$ and

$$Y_t = \Phi(Z_t) - \frac{1}{2} \tag{4.29}$$

For each t, Y_t is uniformly distributed on the interval $[-1/2, 1/2]$. Write Z_1, Z_2 for Z_k, Z_{k-1}, respectively. Observe that because $\{Z_t\}$ is Gaussian, $Z_1 + Z_2$ is stochastically independent of $Z_1 - Z_2$, and that $Z_1 + Z_2$ is distributed as $N(0, 2(1 + \rho_z(1)))$ while $Z_1 - Z_2$ is distributed as $N(0, 2(1 - \rho_z(1)))$. Also, recall the elementary fact that the square of a standard normal random variable has a chi-square distribution with one degree of freedom. Then Price's theorem yields

$$\begin{aligned}\frac{\partial\rho_y(1)}{\partial\rho_z(1)} &= \frac{6}{\pi}E[\exp(-\frac{1}{2}(Z_1^2+Z_2^2))] \\ &= \frac{6}{\pi}E[\exp(-\frac{1}{4}(Z_1+Z_2)^2)]E[\exp(-\frac{1}{4}(Z_1-Z_2)^2)] \\ &= \frac{6}{\pi}M_{\chi^2}\left(\frac{-1-\rho_z(1)}{2}\right)M_{\chi^2}\left(\frac{-1+\rho_z(1)}{2}\right) \\ &= \frac{6}{\pi}\frac{1}{\sqrt{2^2-\rho_z^2(1)}} \end{aligned} \tag{4.30}$$

where $M_{\chi^2}(t) = 1/\sqrt{1-2t}$ is the moment generating function of the chi-square distribution with one degree of freedom. By integrating (4.30),

$$\rho_y(1) = \frac{6}{\pi}\sin^{-1}\frac{\rho_z(1)}{2} + C_0$$

where $C_0 = 0$ because $\rho_y(1) = 0$ if and only if $\rho_z(1) = 0$. Thus, by substituting for $\rho_z(1)$, we obtain a new "cosine formula" for a uniform process:

$$\rho_y(1) = \frac{6}{\pi}\sin^{-1}\left(\frac{1}{2}\cos\left(\frac{\pi E[D]}{N-1}\right)\right) \tag{4.31}$$

It is interesting to observe that, by series expansion of $\sin^{-1}(x)$, the new formula (4.31) is close to the cosine formula (4.7). Indeed we have

$$\rho_y(1) \approx \frac{3}{\pi}\cos\left(\frac{\pi E[D]}{N-1}\right)$$

■

Example 4.3: A cubic

Let $\varphi(x) = x^3$. Then

$$Y_t = Z_t^3 \tag{4.32}$$

Observe that for each t, Y_t has a symmetric probability density function. By Price's theorem,

$$\begin{aligned}\frac{\partial E[Y_k Y_{k-1}]}{\partial\rho_z(1)} &= 9E[Z_k^2 Z_{k-1}^2] \\ &= 9(2\rho_z^2(1)+1)\end{aligned}$$

The constant of integration is 0 and,

$$\rho_y(1) = \frac{2}{5}\rho_z^3(1) + \frac{3}{5}\rho_z(1) \tag{4.33}$$

The "cosine formula" now takes the form

$$\rho_y(1) = 0.9\cos\left(\frac{\pi E[D]}{N-1}\right) + 0.1\cos\left(\frac{3\pi E[D]}{N-1}\right) \tag{4.34}$$

■

4.3.2 Extension to Continuous Time

The preceding idea of transforming a Gaussian process monotonically proves useful also for continuous time processes. Accordingly, we derive the expected zero-crossing rate (per unit time) of a process obtained by applying a monotone transformation to a continuous time stationary Gaussian process.

Our first task is to derive the expected zero-crossing rate for a sufficiently smooth ellipsoidal process in continuous time, $\{Z(t)\}, t \in [0,1]$.

Let $\{Z(t)\}$, $-\infty < t < \infty$, be a stationary zero-mean and unit-variance ellipsoidal process, possessing a correlation function $\rho_z(t) = \rho(t)$ that is twice differentiable at 0. This implies the existence of second order spectral moments,

$$-\rho_z''(0) = \int_{-\infty}^{\infty} \lambda^2 \, dF_z(\lambda) < \infty \tag{4.35}$$

where $F_z(\lambda)$ is the spectral distribution function of $\{Z(t)\}$. Restrict attention to the time interval $[0,1]$, and assume that for $\Delta > 0$, the probability of more than a single crossing in $(t, t+\Delta]$ is negligible as $\Delta \to 0$. Define the sampled time series,

$$Z_k \equiv Z((k-1)\Delta), \qquad k = 1, 2, \ldots, N$$

such that

$$(N-1)\Delta = 1 \tag{4.36}$$

The interval $(0,1]$ is now partitioned into $N-1$ subintervals each of length Δ. This rescaling is needed for studying the expected zero-crossing rate in continuous time by letting $\Delta \to 0$ [33]. Clearly, the sampled process $\{Z_k\}$ is strictly stationary with correlation function ρ_k, say. Note that,

$$\rho_1 = \rho(\Delta) \tag{4.37}$$

As before, the corresponding binary time series is given by

$$X_k = \begin{cases} 1, & \text{if } Z_k \geq 0 \\ 0, & \text{if } Z_k < 0 \end{cases}$$

and the number of zero-crossings, now denoted more suggestively by D_N, is given in terms of X_k:

$$D_N = \sum_{t=2}^{N} [X_k - X_{k-1}]^2 \tag{4.38}$$

By the cosine formula (4.7), monotone convergence, and l'Hopital's rule then, the expected zero-crossing rate per unit time, denoted by $E[D_c]$, is given by

$$\begin{aligned} E[D_c] &= \lim_{N \to \infty} E[D_N] \\ &= \lim_{\Delta \to 0} \frac{1}{\pi\Delta} \cos^{-1}(\rho(\Delta)) \\ &= \frac{1}{\pi}\sqrt{-\rho''(0)} \end{aligned} \tag{4.39}$$

It follows that if the process is a pure sinusoid in continuous time with frequency ω, then

$$E[D_c] = \frac{1}{\pi}\sqrt{-\rho''(0)} = \frac{\omega}{\pi}$$

That is, two crossings per cycle, as is well expected.

A different method for proving (4.39) is to use the interesting fact that an ellipsoidal process is a random multiple of a Gaussian process [30],[31] (see Problem 18).

The formula (4.39) is a special case of what is known as *Rice's formula*, derived by Rice [24] under the Gaussian assumption. The more general form of Rice's formula which gives the expected level-crossing rate is given by (4.49) below; see also (4.52) in Problem 11. We find that Rice's formula (4.39) holds more generally for ellipsoidal processes, and that the discrete and continuous time formulas (4.7), and (4.39) are closely related, the latter being the limit, as $\Delta \to 0$, of the former.

Note that in deriving (4.39) the cosine formula (4.7) has been used in its inverse form

$$E[D_N] = \frac{N-1}{\pi}\cos^{-1}(\rho_1)$$

where we substituted $\rho(\Delta)$ for ρ_1 and $1/\Delta$ for $N-1$.

Now, specializing to a Gaussian $\{Z(t)\}$ with mean 0 and variance 1, let $\varphi(x)$ be as before a strictly monotone real-valued differentiable function defined over the real line, and define a new process $\{Y(t)\}$, $-\infty < t < \infty$, with autocorrelation $\rho_y(t)$, by

$$Y(t) = \varphi(Z(t)) - \varphi(0) \tag{4.40}$$

It is understood that $\rho_y(t)$ is twice differentiable at 0. For $\Delta > 0$, Price's theorem gives

$$\begin{aligned}\frac{\partial\rho_y(\Delta)}{\partial\rho_z(\Delta)} &= \frac{1}{\sigma_y^2}E[\varphi'(Z(t))\varphi'(Z(t+\Delta))]\\ &= \frac{1}{\sigma_y^2}E[\varphi'(Z(t))E(\varphi'(Z(t+\Delta))|Z(t))]\end{aligned}$$

Observe that as $\Delta \to 0$,

$$E[\varphi'(Z(t+\Delta))|Z(t)] \to \varphi'(Z(t))$$

Therefore, by iterated expectation, as $\Delta \to 0$,

$$\frac{d\rho_y}{d\rho_z} \to \frac{E[\varphi'(Z(t))]^2}{\sigma_y^2} \tag{4.41}$$

Now, by the chain rule,

$$\frac{d^2\rho_y}{d\Delta^2} = \frac{d^2\rho_y}{d\rho_z^2}\left(\frac{d\rho_z}{d\Delta}\right)^2 + \frac{d\rho_y}{d\rho_z}\frac{d^2\rho_z}{d\Delta^2}$$

and by (4.41), as $\Delta \to 0$, we obtain an equation that relates the second order spectral moments,

$$\rho_y''(0) = \frac{E[\varphi'(Z(t))]^2}{\sigma_y^2}\rho_z''(0) \tag{4.42}$$

From this and Rice's formula (4.39), we obtain the expected zero-crossing rate per unit time for $Y(t)$ in (4.40),

$$E[D_c] = \frac{1}{\pi}\sqrt{\frac{\text{Var}[\varphi(Z(t))]}{E[\varphi'(Z(t))]^2}}\sqrt{-\rho_y''(0)} \tag{4.43}$$

In particular, Rice's formula (4.39) is obtained from (4.43) as a special case by setting $\varphi(x) = x$. The modified Rice's formula (4.43) was introduced in [3].

It is interesting to note that because $E[D_c]$ is the same for both $\{Z(t)\}$ and $\{Y(t)\}$, the generalized formula (4.43) gives the change in the second order spectral moment (4.35) effected in going from $\{Z(t)\}$ to $\{Y(t)\}$. To clarify this point we appeal to Chernoff's inequality, which states that when $\varphi(x)$ is sufficiently smooth (see Problem 10 in Chapter 2),

$$\text{Var}[\varphi(Z(t))] \leq E[\varphi'(Z(t))]^2$$

Thus, for a sufficiently smooth $\varphi(x)$, Chernoff's inequality implies an increase in the second spectral moment: $\sqrt{-\rho_z''(0)} \leq \sqrt{-\rho_y''(0)}$. This, however, does not come as a surprise since in general memoryless nonlinear transformations tend to whiten the output in the sense that its spectrum occupies a greater bandwidth than the input spectrum [32]. Hence, the last remark together with the generalized Rice's formula (4.43) provide an intuitive verification of Chernoff's inequality.

The increase in the second spectral moment can be quite appreciable as can be seen from the next example.

Example 4.4: A Cubic Revisited

Let $\varphi(x) = x^3$. Then

$$Y(t) = Z^3(t) \tag{4.44}$$

and (4.43) gives

$$E[D_c] = \frac{1}{\pi}\sqrt{\frac{5}{9}}\sqrt{-\rho_y''(0)} \tag{4.45}$$

One can check that the same formula can be obtained via equation (4.33) replacing the 1s by Δ's and letting $\Delta \to 0$, as is done in [3].

Thus, going from $\{Z(t)\}$ to $\{Y(t)\}$ we have an 80% increase in the second spectral moment. Evidently, any attempt to replace (4.45) by Rice's formula (4.39) may result in gross error.

■

4.3.3 Level-Crossing Rate

Rice's formula for zero-crossings (4.39) admits a straightforward generalization in terms of level-crossings of monotone transformations of Gaussian processes [21].

To see this, let $\{Z(t)\}$, $-\infty < t < \infty$, $E[Z(t)] = 0$, $\text{Var}[Z(t)] = 1$, be a stationary Gaussian process with autocorrelation $\rho_z(t)$, assumed twice differentiable at $t = 0$. Consider the memoryless transformation

$$Y(t) = \varphi(Z(t))$$

where $\varphi(x)$ is strictly monotone increasing and continuously differentiable. Unlike (4.40), here we do not subtract $\varphi(0)$. Fix u, and denote the expected number of level-crossings per unit time (expected level-crossing rate) of $Y(t)$ corresponding to level u by $E[D_y(u)]$. When the mean square derivative $Y'(t)$ is well defined, then (see Problem 11)

$$E[D_y(u)] = \int_{-\infty}^{\infty} |v| g(u, v)\, dv \tag{4.46}$$

where $g(u, v)$ is the joint density of $(Y(t), Y'(t))$ for each fix t. However, for any fixed level u,

$$P\left(Y(t) \le u, \frac{Y(t+h) - Y(t)}{h} \le v\right) =$$

$$P[Z(t) \le \varphi^{-1}(u), Z(t+h) \le \varphi^{-1}(hv + \varphi(Z(t)))] =$$

$$\int_{-\infty}^{\varphi^{-1}(u)} \int_{-\infty}^{\varphi^{-1}(hv+\varphi(x))} \frac{\exp\left\{[-1/2(1-\rho_z^2(h))][x^2 - 2\rho_z(h)xy + y^2]\right\}\, dy dx}{2\pi\sqrt{1-\rho_z^2(h)}} =$$

$$\int_{-\infty}^{\varphi^{-1}(u)} \frac{1}{\sqrt{2\pi}} e^{-x^2/2} \Phi\left[\frac{\varphi^{-1}(hv + \varphi(x)) - \rho_z(h)x}{\sqrt{1-\rho_z^2(h)}}\right] dx$$

where $\Phi(x)$ is the distribution function of the standard normal distribution, and where $(y - \rho_z(h)x)^2 + x^2(1 - \rho_z^2(h))$ was substituted for $x^2 - 2\rho_z(h)xy + y^2$. Partial differentiation with respect to u, v, and then taking the limit as $h \to 0$ together with l'Hopital's rule (twice) give the joint density of $(Y(t), Y'(t))$,

$$g(u, v) = \frac{\exp\left\{-\frac{1}{2}\left[(\varphi^{-1}(u))^2 + (v^2/(-\rho_z''(0))[\varphi'(\varphi^{-1}(u))]^2)\right]\right\}}{2\pi\sqrt{-\rho_z''(0)}[\varphi'(\varphi^{-1}(u))]^2} \tag{4.47}$$

$u > \varphi(-\infty)$, $-\infty < v < \infty$. Substitution of (4.47) in (4.46) gives [21]

$$E[D_y(u)] = \frac{1}{\pi}\sqrt{-\rho_z''(0)} \exp\left\{-\frac{1}{2}[\varphi^{-1}(u)]^2\right\}, \qquad u > \varphi(-\infty) \tag{4.48}$$

The particular case $\varphi(x) = x$ yields Rice's original result for the expected level-crossing rate per unit time of a zero-mean, unit variance, stationary Gaussian process $\{Z(t)\}$ in continuous time,

$$E[D_z(u)] = \frac{1}{\pi}\sqrt{-\rho_z''(0)}\exp\left\{-\frac{1}{2}u^2\right\}, \qquad u > -\infty \tag{4.49}$$

Another particular case of interest is that of the lognormal process $\varphi(Z(t)) = \exp\{Z(t)\}$. Then [22],

$$E[D_y(u)] = \frac{1}{\pi}\sqrt{-\rho_z''(0)}\exp\left\{-\frac{1}{2}\log^2 u\right\}, \qquad u > 0 \tag{4.50}$$

Recall that in (4.43) the rate was given in terms of the autocorrelation of the transformed process $\varphi(Z(t)) - \varphi(0)$, while in (4.48) the rate is given in terms of the autocorrelation of the original process $Z(t)$. Also, in the previous section use was made of Price's formula, a fact that resulted in some saving in computation, while here straightforward integration was used.

From (4.47) we obtain the curious fact that, for each fixed t, the conditional distribution of $Y'(t)$ given that $Y(t) = u$ is normal with mean 0 and variance $(-\rho_z''(0))[\varphi'(\varphi^{-1}(u))]^2$. However, this is quite expected if one writes $Y'(t) = \varphi'(\varphi^{-1}(u))Z'(t)$, noting that $Z'(t)$ is normal with mean 0 and variance $-\rho_z''(0)$.

There have been more attempts to study the expected level-crossing rate of other non-Gaussian processes. Without getting into details, we mention that in recent years there has been a growing interest in the class of so-called stationary, harmonizable, symmetric, α-stable stochastic processes whose marginal distributions have fatter and longer tails than that of the Gaussian distribution. In its standard form, the characteristic function of the marginal distribution is given by $\exp(-|t|^\alpha)$, $0 < \alpha \leq 2$ (for $\alpha = 2$ the process becomes Gaussian). Let $\{Z(t)\}$ be such a process corresponding to an $\alpha \in (0, 2)$. Then [2]

$$\lim_{u\to\infty} u^\alpha E[D_z(u)] = C$$

where C is a constant which depends on α. Examples indicate that the approximation

$$E[D_z(u)] \sim Cu^{-\alpha}$$

holds even for relatively low values of the level u [2].

4.4 EFFECT OF MONOTONE GAIN

Consider a linear filter with gain $|H(\omega)|^2$. Intuitively, when the gain is monotone increasing, the filter becomes a high-pass filter and we expect to see an increase in the number of zero-crossings. Conversely, fewer zero-crossings are expected when the gain is monotone decreasing, because then the filter acts as a low-pass filter. Surprisingly enough, a direct proof of this common sense is far from being trivial. In order to vindicate this intuition, we appeal to the autocorrelation and the cosine formula. As it turns out, the notion of stochastic ordering introduced in Chapter 2 comes into full play in describing the effect of high-pass and low-pass filtering on the autocorrelation.

The autocorrelation of a sinusoid is not affected by a filter. However, as we shall see, the converse holds as well when the gain is strictly monotone. In other words, if the autocorrelation is not affected by a filter with a strictly increasing or decreasing gain, the process must be a sinusoid.

The following theorem describes the effect of a linear filter with a monotone gain on the first-order autocorrelation ρ_1. In the proof, $f(\omega)$ denotes the spectral density function of $\{Z_t\}$. For convenience, we assume that $f(\omega)$ is a generalized density so that it may contain δ-functions corresponding to the points of jump in the spectral distribution function $F(\omega)$.

Theorem 4.4. *[17] Let $|H(\omega)|^2$ be the squared gain of a linear filter acting on a real-valued zero-mean stationary process $\{Z_t\}, t = 0, \pm 1, \pm 2, \ldots$, and let $\rho_1(H)$ be the first order autocorrelation in the filtered process. Then we have:*

(a) *If $|H(\omega)|^2$ is monotone increasing in $[0, \pi]$, then*

$$\rho_1 \geq \rho_1(H)$$

When the gain is monotone decreasing the inequality is reversed.

(b) *Assume that $|H(\omega)|^2$ is strictly monotone. Then*

$$\rho_1 = \rho_1(H)$$

if and only if $\{Z_t\}$ is a pure sinusoid with frequency $\cos^{-1}(\rho_1)$, with probability one.

Proof. The proof of part (a) makes use of properties of stochastic ordering. Accordingly, we define for $\lambda \in [0, \pi]$ the *probability* distribution functions,

$$\tilde{F}(\lambda) \equiv \frac{\int_0^\lambda f(\omega)\, d\omega}{\int_0^\pi f(\omega)\, d\omega}$$

and

$$\tilde{F}_H(\lambda) \equiv \frac{\int_0^\lambda |H(\omega)|^2 f(\omega)\, d\omega}{\int_0^\pi |H(\omega)|^2 f(\omega)\, d\omega}$$

Suppose $|H(\omega)|^2$ is monotone increasing. Then, using $\int_0^\pi = \int_0^\lambda + \int_\lambda^\pi$,

$$\tilde{F}(\lambda) - \tilde{F}_H(\lambda)$$

$$= \frac{\int_0^\pi |H(\omega)|^2 f(\omega)\, d\omega \int_0^\lambda f(\omega)\, d\omega - \int_0^\pi f(\omega)\, d\omega \int_0^\lambda |H(\omega)|^2 f(\omega)\, d\omega}{\int_0^\pi f(\omega)\, d\omega \int_0^\pi |H(\omega)|^2 f(\omega)\, d\omega}$$

$$= \frac{\int_\lambda^\pi |H(\omega)|^2 f(\omega)\, d\omega \int_0^\lambda f(\omega)\, d\omega - \int_\lambda^\pi f(\omega)\, d\omega \int_0^\lambda |H(\omega)|^2 f(\omega)\, d\omega}{\int_0^\pi f(\omega)\, d\omega \int_0^\pi |H(\omega)|^2 f(\omega)\, d\omega}$$

$$\geq \frac{|H(\lambda)|^2 \int_\lambda^\pi f(\omega)\, d\omega \int_0^\lambda f(\omega)\, d\omega - |H(\lambda)|^2 \int_\lambda^\pi f(\omega)\, d\omega \int_0^\lambda f(\omega)\, d\omega}{\int_0^\pi f(\omega)\, d\omega \int_0^\pi |H(\omega)|^2 f(\omega)\, d\omega}$$

$$= 0$$

and therefore for all $\lambda \in [0, \pi]$

$$\tilde{F}(\lambda) \geq \tilde{F}_H(\lambda)$$

Thus, if X, Y are two random variables such that

$$X \sim \tilde{F}_H, \;\; Y \sim \tilde{F}$$

then X is stochastically larger than Y, and this means, by Theorem 2.1, that

$$E[g(X)] \geq E[g(Y)]$$

for every nonnegative increasing function g defined on $[0, \pi]$. In particular, for the nonnegative increasing function

$$g(\omega) = 1 - \cos(\omega), \qquad 0 \leq \omega \leq \pi$$

we obtain,

$$1 - \int_0^\pi \cos(\omega)\, d\tilde{F}_H(\omega) \geq 1 - \int_0^\pi \cos(\omega)\, d\tilde{F}(\omega)$$

But this means that $\rho_1(H) \leq \rho_1$.

In the same way we can also prove the reversed inequality when the gain is monotone decreasing.

To prove (b), observe that when $\{Z_t\}$ is a sinusoid with frequency ω_0 then

$$\rho_1 = \rho_1(H) = \cos(\omega_0)$$

Conversely, suppose $\rho_1 = \rho_1(H)$, but that $\{Z_t\}$ is not a sinusoid. Without loss of generality assume that $|H(\omega)|^2$ is strictly monotone increasing in $[0, \pi]$. Then, because of the assumption that the process is not a pure sinusoid, the numerator I, say, of the difference $\rho_1 - \rho_1(H)$ satisfies

$$I = \int\int_{0 \leq \lambda < \omega \leq \pi} [|H(\omega)|^2 - |H(\lambda)|^2][\cos(\lambda) - \cos(\omega)]\, dF(\lambda)\, dF(\omega) > 0 \quad (4.51)$$

However, this contradicts the fact that $\rho_1 - \rho_1(H) = 0$.

□

The formula (4.51) is an integration trick; it can be easily verified in the purely discrete spectrum case. Similar results have been proved in [26] and [27] under various conditions on the correlation function.

In the Gaussian case, due to the cosine formula, which is defined by a strictly decreasing transformation, the implication concerning the expected number of zero-crossings is now clear: for zero-crossings the inequality in part (a) of the theorem is reversed, while equality still persists in part (b).

Corollary 4.2. *Suppose that the process $\{Z_t\}$ is Gaussian, and let $\gamma(H)$ be the zero-crossing rate in the filtered process.*

(a) *If $|H(\omega)|$ is monotone increasing in $[0, \pi]$, then*

$$\gamma \leq \gamma(H)$$

The inequality is reversed if $|H(\omega)|$ is monotone decreasing.

(b) *Assume $|H(\omega)|$ is strictly monotone. Then*

$$\gamma = \gamma(H)$$

if and only if $\{Z_t\}$ is a pure sinusoid with probability one. The frequency of the sinusoid is given by $\pi\gamma$.

An important special case is the difference operator. Because its gain is strictly increasing, we expect an increase in the number of zero-crossings observed in a differenced process. However, if the expected number of zero-crossings is not altered by differencing, the process must be a pure sinusoid [16], [26], [27]. In other words, discounting an end effect, *a zero-mean stationary Gaussian process in discrete time is, with probability one, a pure sinusoid with random amplitude and phase if and only if the expected number of zero-crossings is equal to the expected number of peaks and troughs in a time series from the process.*

4.5 PROBLEMS AND COMPLEMENTS

1. Suppose $\{Z_t\}$, $t = 0, \pm 1, \ldots,$ is a zero-mean stationary process. Define $\{Y_t\}$ by the summation $Y_t = Z_t + Z_{t+1}$. Determine the gain of this filter and show that $\rho_1 \leq \rho(H)$. What is the relationship between γ and $\gamma(H)$?
2. Consider the $AR(1)$ filter

$$\mathcal{L}_\alpha \equiv 1 + \alpha\mathcal{B} + \alpha^2\mathcal{B}^2 + \cdots + , \qquad \alpha \in (-1, 1)$$

where $\mathcal{B}$ is the backward shift. Show that the transfer function of this filter is given by

$$H(\omega; \alpha) = \frac{1}{1 - \alpha \exp(-i\omega)}$$

Show that the corresponding gain is monotone.

3. *Random curve crossings [15].* Suppose $\{Z_t\}$, $t = 0, \pm 1, \ldots,$ is a zero-mean stationary bounded process: $|Z_t| \leq A$, A a positive constant. Let $\{U_t\}$, $t = 0, \pm 1, \ldots,$ be a process of independent random variables such that for each t, U_t is uniformly

distributed in $(-A, A)$. Assume that the two processes $\{Z_t\}$ and $\{U_t\}$ are independent. Let $Y_t = 1$ if $Z_t \geq U_t$, and $Y_t = 0$ otherwise (see Figure 4.4.) Show that

$$E[Z_t Z_{t-1}] = A^2 \left(1 - \frac{2E[C]}{N-1}\right)$$

where C is the number of symbol changes in $Y_1, \ldots, Y_N$. Conclude that the number of symbol changes in $Y_1, \ldots, Y_N$ leads to an unbiased estimate of $E[Z_t Z_{t-1}]$.

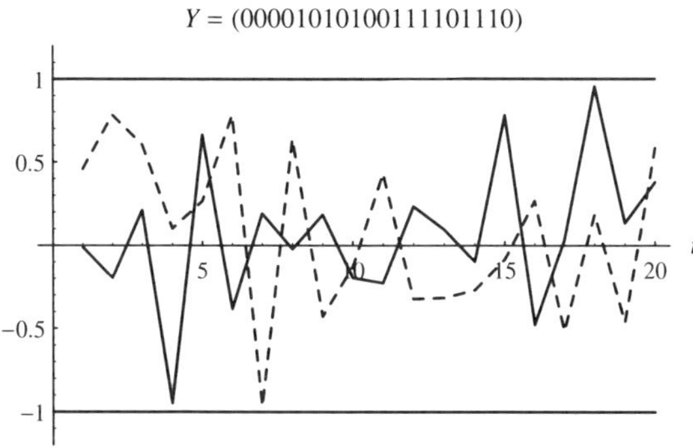

Figure 4.4: *Random curve crossings: mutual crossings of a time series Z_t with $\rho_1 = 0$ (solid line) and a random curve U_t (dashed line). $A = 1, N = 20, C = 10$, and $A^2\,(1 - (2C/N - 1)) = -0.0526$ estimates ρ_1.*

4. *Multiple level clipping [20]*. Let $\{Z_t\}, t = 0, \pm 1, \ldots$, be a stationary Gaussian process. For each fixed t assume that $Z_t \sim N(0,1)$. Define a clipped process $\{Y_t\}$, $t = 0, \pm 1, \ldots$, by the nonlinear transformation,

$$Y_t = \begin{cases} 1, & \text{if } Z_t > u \\ 0, & \text{if } -v \leq Z_t \leq u \\ -1, & \text{if } Z_t < -v \end{cases}$$

where u, v are arbitrary positive constants. Show that ρ_1 can be obtained from $E[Y_t Z_{t+1}]$. That is, show that

$$E[Y_t Z_{t+1}] = \frac{\rho_1}{\sqrt{2\pi}} (\exp \frac{-u^2}{2} + \exp \frac{-v^2}{2})$$

Hence, conclude that $E[Y_t Z_{t+1}]$ can be obtained from $E[D]$, u and v. (Hint: Use iterated expectation, $E[Y_t Z_{t+1}] = E[E[Y_t Z_{t+1} | Y_t]]$.)

5. *Zero-crossing rate of a uniform process in continuous time [3]*. Consider the continuous time version $Y(t)$ of the uniform process discussed in Example 4.2. Use formula (4.43) to show that the expected zero-crossing rate per unit time of $Y(t)$ is given by

$$E[D_c] = \frac{1}{\sqrt{2\pi\sqrt{3}}} \sqrt{-\rho_y''(0)}$$

6. *Zero-crossing rate of a lognormal process in continuous time [3].* Let $Z(t)$ be a zero-mean stationary Gaussian process in continuous time. Define

$$Y(t) = \exp(Z(t)) - 1$$

Use formula (4.43) to show that the expected zero-crossing rate per unit time of $Y(t)$ is given by

$$E[D_c] = \frac{1}{\pi}\sqrt{\frac{e-1}{e}}\sqrt{-\rho_y''(0)}$$

7. *Zero-crossing rate of an odd power of a Gaussian process [3].* Let $Z(t)$ be a zero-mean, unit-variance, stationary Gaussian process in continuous time. For an odd positive integer n, define

$$Y(t) = Z^n(t)$$

Use formula (4.43) to show that the expected zero-crossing rate per unit time of $Y(t)$ is given by

$$E[D_c] = \frac{1}{\pi}\sqrt{\frac{2n-1}{n^2}}\sqrt{-\rho_y''(0)}$$

Conclude that the second spectral moment of $Y(t)$ increases with n. This is in agreement with the fact that memoryless transformations of Gaussian processes tend to invoke a greater spectral bandwidth for the output [32].

8. *A formal Stieltjes–Sheppard-type orthant probability [10].* Suppose $\{Z_t\}$, $t = 0, \pm 1, \ldots$, is a zero-mean stationary process such that $P(Z_t \geq 0) = 1/2$. Define,

$$r_1 \equiv \cos\left(\frac{\pi E[D]}{N-1}\right)$$

Show that,

$$P(Z_t \geq 0, Z_{t-1} \geq 0) = \frac{1}{4} + \frac{1}{2\pi}\sin^{-1}(r_1)$$

Argue that in the Gaussian case, r_1 can be replaced by ρ_1.

9. *Symbol changes in a Markov chain [14].* Recall the definition of λ_1 in (4.5). Let $\{X_t\}$, $t = 1, \ldots, N$ be a binary 0–1 stationary Markov chain, and denote by D the number of symbol changes from 0 to 1 and from 1 to 0 observed in $X_1, \ldots, X_N$. If $E[X_t] = 1/2$, and d is the value of D, show that

$$P(X_1 = x_1, \ldots, X_N = x_N) = \frac{1}{2}\lambda_1^{(N-1)-d}(1-\lambda_1)^d$$

and that this implies that D has the binomial distribution $b(N-1, 1-\lambda_1)$.

10. *D and X_t are uncorrelated* [18]. Suppose $\{Z_t\}$, $t = 0, \pm 1, \ldots$, is a zero-mean stationary Gaussian process, and recall the definition of $\{X_t\}$ and D. Show that $\text{Cov}(D, X_t) = 0$ uniformly in t, and that this implies $\text{Cov}(D, \sum_{t=1}^{N} X_t) = 0$.

11. *Expected level-crossing rate [24].* (a) Suppose $\Delta > 0$. Argue that for a sufficiently smooth stationary process $X(t)$ in continuous time such that $P(X(t) = u) = 0$, the expected number of level-crossings per unit time of level u is given by the limit

$$\lim_{\Delta \to 0} \frac{P(\text{A level-crossing in}(t, t+\Delta])}{\Delta}$$

(b) Conclude that the expected level-crossing rate per unit time corresponding to level u is the limit

$$\lim_{\Delta\to 0} \frac{\int_0^\infty \int_{u-\Delta\dot{x}}^{u} p(x,\dot{x})\, dx\, d\dot{x}}{\Delta} + \frac{\int_{-\infty}^0 \int_u^{u-\Delta\dot{x}} p(x,\dot{x})\, dx\, d\dot{x}}{\Delta} =$$

$$\int_0^\infty \dot{x}p(u,\dot{x})\, d\dot{x} - \int_{-\infty}^0 \dot{x}p(u,\dot{x})\, d\dot{x} = \int_{-\infty}^\infty |\dot{x}|p(u,\dot{x})\, d\dot{x}$$

where $p(x,\dot{x})$ is the joint probability density of $(X(t), X'(t))$.

(c) Use the previous expression to show that the expected number of crossings of level u in $(0,T]$ by a zero-mean stationary Gaussian process with autocovariance $\psi(t)$ is given by

$$\frac{T}{\pi}\sqrt{-\frac{\psi''(0)}{\psi(0)}} \exp\left\{\frac{-u^2}{2\psi(0)}\right\} \tag{4.52}$$

Argue that this expectation is infinite when the autocorrelation is of the form $\rho(t) = \exp(-|t|)$. (Hint: In the Gaussian case $X(t) \sim \mathcal{N}(0,\psi(0))$ and $X'(t) \sim \mathcal{N}(0,-\psi''(0))$ are independent.)

(d) Conclude from (4.52) that in the zero-mean stationary Gaussian case as in (c), the expected number of local maxima per unit time is given by

$$\frac{1}{2\pi}\sqrt{-\frac{\psi^{(4)}(0)}{\psi''(0)}}$$

provided the fourth order derivative $\psi^{(4)}(0)$ is finite.

12. *Mixed spectrum [13],[24].* Use the spectral moment form of Rice's formula

$$E[D_c] = \frac{1}{\pi}\sqrt{-\frac{\psi''(0)}{\psi(0)}} = \frac{1}{\pi}\left[\frac{\int_{-\infty}^\infty \omega^2 dF(\omega)}{\int_{-\infty}^\infty dF(\omega)}\right]^{1/2}$$

to derive the continuous time analog of (4.18).

13. *Expected number of upcrossings.* For any continuous time process $X(t)$ with a well-defined mean square derivative $X'(t)$, let $N_u(0,T]$ denote the number of upcrossings of level u by $X(t)$ in $(0,T]$. Show that

$$E\{N_u(0,T]\} = \int_0^T \int_0^\infty vp(u,v,t)\, dv\, dt$$

where $p(u,v,t)$ is the joint probability density of $(X(t), X'(t))$.

14. *Upcrossings of a random sinusoid [19].* Let A, B be independent standard normal random variables. Define a zero-mean stationary Gaussian process in continuous time by

$$Z(t) = A\cos(\omega t) + B\sin(\omega t), \qquad -\infty < t < \infty$$

where $\omega > 0$.

(a) Show that

$$E\{N_u(0,T]\} = \frac{\omega T}{2\pi} \exp\left(-\frac{u^2}{2}\right)$$

(b) Let M^* be the maximum of the process in $[0,T]$ for $0 < T < \pi/\omega$. For $u > 0$ show that

$$P(M^* \leq u) = \Phi(u) - E\{N_u(0,T]\}$$

(Hint: $P(M^* > u) = P(Z(0) > u) + P(Z(0) \leq u, N_u(0,T] \geq 1)$ and consider the probability of the event $\{Z(0) > u, N_u(0,T] \geq 1\}$ for $0 < T < \pi/\omega$.)

15. *Expected length of excursions above a fixed level [21].* Define the monotone increasing transformation $Y(t) = \varphi(X(t))$, where $X(t)$ is a stationary zero-mean unit-variance Gaussian process in continuous time with autocorrelation $\rho_x(t)$. Let $N_u(0,T]$ be as above, and define

$$Z_u(0,T] = \int_0^T I_{[Y(t)\geq u]}\, dt$$

where $I_{[\mathcal{A}]}$ is the indicator of the event $\mathcal{A}$. (a) Show that the mean length of an excursion above level u by $Y(t)$ in $(0,T]$ is given by the ratio

$$\frac{E\{Z_u(0,T]\}}{E\{N_u(0,T]\}} = \frac{2\pi[1-\Phi(\varphi^{-1}(u))]}{\sqrt{-\rho_x''(0)}\exp\left\{-\frac{1}{2}[\varphi^{-1}(u)]^2\right\}}$$

(b) Show that when $\varphi^{-1}(u)$ is sufficiently large,

$$\frac{E\{Z_u(0,T]\}}{E\{N_u(0,T]\}} \approx \sqrt{\frac{2\pi}{-\rho_x''(0)}} \cdot \frac{1}{\varphi^{-1}(u)}$$

(Hint: Use the standard normal tail estimate $1 - \Phi(x) \sim \phi(x)/x$.)

16. *Upcrossings of lognormal process [22].* Let $\{Z(t)\}$, $-\infty < t < \infty$, be a zero-mean unit-variance stationary Gaussian process, and let $Y(t) = \exp\{Z(t)\}$. Use the level-crossings formulas (4.49), (4.50) to show that when $u > 0.56$ the expected number of *upcrossings* per unit time of level u by $Y(t)$ exceeds that of $Z(t)$.

17. *Sampling rate and zero-crossings [28].* Let $\{X(t)\}$, $-\infty < t < \infty$, be a zero-mean stationary Gaussian process with autocovariance $\psi(t)$ and spectral density

$$f(\omega) = \begin{cases} \dfrac{2}{\pi}, & \dfrac{-\pi}{2} < \omega < \dfrac{\pi}{2} \\ 0, & \text{otherwise} \end{cases}$$

Define

$$X_\Delta(n) = X(n\Delta), \qquad n = 0, \pm 1, \pm 2, \ldots$$

A zero-crossing by $X(t)$ occurs in $(n\Delta, (n+1)\Delta]$ iff

$$I_{[X_\Delta(n)X_\Delta(n+1)<0]} = 1$$

where $I_{[\mathcal{A}]}$ is the indicator of the event $\mathcal{A}$. Fix T, and let $\Delta > 0$ be such that $T = k\Delta$. Assume the probability of more than one zero-crossing in $(n\Delta, (n+1)\Delta]$ is negligible. Clearly, a loss in zero-crossings occurs in going from $X(t)$ to $X_\Delta(n)$.

(a) Find the expected number of zero-crossings by $X_\Delta(n)$ in $(0, T]$.
(b) Find the corresponding number for $X(t)$ in $(0, T]$.
(c) Show that the ratio of your answer in (a) to that in (b) is

$$g\left(\frac{1}{\Delta}\right) = \frac{\sqrt{12}}{\Delta}\left\{\frac{1}{2} - \frac{1}{\pi}\sin^{-1}\left(\frac{2}{\pi\Delta}\sin\left(\frac{\pi\Delta}{2}\right)\right)\right\}$$

(d) Plot $g(1/\Delta)$ as a function of $1/\Delta$.
(e) Conclude that already by sampling at twice the Nyquist rate, that is, $\Delta = 1$ in the present case, we can expect to recover more than [illegible]% of the zero-crossings.
(f) Repeat steps (a)–(e) with the spectral density (because of symmetry it suffices to define it over the positive support)

$$f(\omega) = \begin{cases} \dfrac{2\alpha}{\pi}, & 0 \le \omega < \dfrac{\pi}{2} \\ \dfrac{2(1-\alpha)}{\pi}, & \dfrac{\pi}{2} \le \omega \le \pi \end{cases}$$

$0 \le \alpha \le 1$. Conclude that for the specific process at hand, from a zero-crossing point of view, sampling at twice the Nyquist rate is a sensible rule.

18. *Zero-crossing rate of an ellipsoidal process [31].* Define a strictly stationary zero-mean ellipsoidal process in continuous time $Z(t)$ by the product $Z(t) = AY(t)$, where A is a random variable uniformly distributed in the interval $(0, 1)$, independently of a zero-mean stationary Gaussian process $Y(t)$. Use this representation to show that the expected number of zero-crossings per unit time for $Z(t)$ is given by Rice's formula

$$\frac{1}{\pi}\left[-\frac{R''(0)}{R(0)}\right]^{1/2}$$

where $R(t)$ is the autocovariance of $Z(t)$. (Hint: The autocorrelation of $Z(t)$ is equal to that of $Y(t)$.)

19. *Highest frequency in a random trigonometric polynomial [4].* Consider the random trigonometric polynomial of degree n in $\cos(t)$,

$$Z_n(t) = \sum_{k=1}^{n} A_k \cos(kt), \qquad t \in (0, 2\pi]$$

where the A_k are independent $\mathcal{N}(0, 1)$ random variables. Show that

$$\frac{\pi \times \{\text{maximum number of real zeros in } (0, 2\pi]\}}{2\pi}$$

cannot exceed the highest frequency n. (Hint: $Z_n(t)$ can be expressed in terms of a polynomial (what degree ?) in $x = e^{it}$.)

REFERENCES

[1] Abrahams, J., "A survey of recent progress on level-crossing problems for random processes," in *Communications and Networks, A Survey of Recent Advances*, I.F. Blake and H.V. Poor, Eds., pp. 6–25, New York: Springer-Verlag, 1986.

[2] Adler, R.J., G. Samorodnitsky, and T. Gadrich. "The expected number of level crossings for stationary, harmonisable, symmetric, stable processes," *Ann. Appl. Probab.*, Vol. 3, No. 2, 1993.

[3] Barnett, J. and B. Kedem, "Zero-crossing rates of functions of Gaussian processes," *IEEE Trans. Inform. Theory*, Vol. IT-37, pp. 1188–1194, No. 4, July 1991.

[4] Bharucha-Reid, A.T. and M. Sambandham, *Random Polynomials*, New York: Academic Press, 1986.

[5] Blachman, N.M., "Zero-crossing rate for the sum of two sinusoids or a signal plus noise," *IEEE Trans. Inform. Theory*, Vol. IT-21, No. 6, pp. 671–675, Nov. 1975.

[6] Blake, I.F. and W.C. Lindsey, "Level crossing problems for random processes", *IEEE Trans. Inform. Theory*, Vol. IT-19, No. 3, pp. 295–315, May 1973.

[7] Bloomfield, P., *Fourier Analysis of Time Series: An Introduction*, New York: Wiley, 1976.

[8] Cramér, H. and M.R. Leadbetter, *Stationary and Related Stochastic Processes*, New York: Wiley, 1967.

[9] David, F.N., "A note on the evaluation of the multivariate normal integral," *Biometrika*, Vol. 40, pp. 458–459, 1953.

[10] He, S. and B. Kedem, "On the Stieltjes-Sheppard orthant probability formula," Tech. Rep. TR-89-69, Department of Mathematics, Univ. of Maryland, College Park, 1989.

[11] He, S. and B. Kedem, "The zero-crossing rate of autoregressive processes and its link to unit roots," *J. Time Ser. Anal.*, Vol. 11, pp. 201–213, 1990.

[12] Hess, W., *Pitch Determination of Speech Signals*, Berlin: Springer-Verlag, 1983.

[13] Ito, M.R. and R.W. Donaldson, "Zero-crossing measurements for analysis and recognition of speech sounds," *IEEE Trans. Audio Electroacoust.*, Vol. AU-19, No. 3, pp. 235–242, Sept. 1971.

[14] Kedem, B., *Binary Time Series*, New York: Dekker, 1980.

[15] Kedem, B., "Some graphical considerations in time series analysis," *IEEE Trans. Pattern Anal. Machine Intell.*, Vol. PAMI-4, No. 5, pp. 493–499, Sept. 1982.

[16] Kedem, B., "On the sinusoidal limit of stationary time series," *Ann. Stat.*, Vol. 12, No. 2, pp. 665–674, 1984.

[17] Kedem, B. and T. Li, "Monotone gain, first-order autocorrelation, and expected zero-crossing rate," *Ann. Stat.*, Vol. 19, No. 3, pp. 1672–1676, 1991.

[18] Kedem, B. and G. Reed, "On the variance of higher order crossings with special reference to a fast white noise test," *Biometrika*, Vol. 73, No. 1, pp. 143–149, 1986.

[19] Leadbetter, M.R, Lindgren, G., and H. Rootzén, *Extremes and Related Properties of Random Sequences and Processes*, New York: Springer-Verlag, 1983.

[20] Okamoto, M.B. and K. Iwase, "A statistical index representing a gross feature of several climatic time series by hard limiting," *J. Meteoron. Soc. Japan*, Vol. 60, No. 2, pp. 726–727, Apr. 1982.

[21] Orsingher, E., "Level crossings of transformations of stationary Gaussian processes," *Metron*, Vol. 37, pp. 81–100, 1979.

[22] Orsingher, E., "Upcrossings and conditional behavior of the lognormal process," *Boll. Un. Mat. Ital.*, Vol. 18-B, pp. 1017–1034, 1981.

[23] Papoulis, A., *Probability, Random Variables, and Stochastic Processes* (2nd ed.), New York, McGraw-Hill, 1984.

[24] Rice, S.O., "Mathematical analysis of random noise," *Bell Syst. Tech. J.*, Vol. 23, pp. 282–332, 1944, and Vol. 24, pp. 46–156, 1945, reprinted in *Selected Papers on Noise and Stochastic Processes*," N. Wax, Ed., New York: Dover, 1954.

[25] Rice, S.O., "Statistical properties of a sine wave plus random noise," *Bell Syst. Tech. J.*, Vol. 27, pp. 109–157, Jan. 1948.

[26] Romanovsky, V.I., "Sur la loi sinusoidale limite," *Rend. Circ. Mat. Palermo*, Vol. 56, pp. 82–111, 1932.

[27] Slutsky, E.E., "The summation of random causes as the source of cyclic processes," *Prob. Econ. Cond.*, Vol. 3, No. 1, 1927 (in Russian). (English translation in *Econometrica*, Vol. 5, pp. 105–146, Apr. 1937.)

[28] Tick, L.J and P. Shaman, "Sampling rates and appearance of stationary Gaussian processes," *Technometrics*, Vol. 8, pp. 91–106, Feb. 1966.

[29] Tikhonov, V.N., "Characteristics of overshoots in random processes (review)," *Radio Eng. Electron. Phys.*, Vol. 9, pp. 295–320, Mar. 1964.

[30] Wise, G.L., "A comment on *Zero-crossing rates of functions of Gaussian processes*," *IEEE Trans. Inform. Theory*, Vol. IT-38, No. 1, p. 213, Jan. 1992.

[31] Wise, G.L. and N.C. Gallagher, "On spherically invariant random processes," *IEEE Trans. Inform. Theory*, Vol. IT-24, No. 1, pp. 118–120, Jan. 1978.

[32] Wise, G.L., A.P. Traganitis, and J.B. Thomas, "The effect of a memoryless nonlinearity on the spectrum of a random process," *IEEE Trans. Inform. Theory*, Vol. IT-23, No. 1, pp. 84–89, Jan. 1977.

[33] Ylvisaker, D.N., "The expected number of zeros of a stationary Gaussian process," *Ann. Stat.*, Vol. 36, No. 3, pp. 1043–1046, 1965.

5

Higher Order Crossings and Correlations

In the previous chapter, the number of zero-crossings is suggested as a measure of oscillation displayed by a time series. Pursuing this idea a bit further, we observe that the number of zero-crossings in a time series can be used as a measure of oscillation (or its change) brought about by the application of a linear filter. By extension, we can apply to a time series a family of filters, obtain the corresponding family of zero-crossing counts, and thus provide a summary of the oscillation "history" observed in the time series and in its filtered versions. The resulting family of zero-crossing counts is referred to as *higher order crossings* or simply as HOC. Thus, HOC are zero-crossing counts observed in a time series and in its filtered versions. HOC from stationary time series are related to the autocorrelation $\rho_k,\ k = 1, 2, \dots$, and hence, also to the spectrum. As we shall see, in the Gaussian case, there are (expected) HOC families and also HOC sequences that determine the spectrum up to a constant.

In this and subsequent chapters, we shall be concerned with developing a certain systematic approach to the analysis of stationary time series based on HOC by exploring the fruitful connection between time-invariant linear filtering and zero-crossing counts. The idea that zeros of filtered signals are of relevance in applications is a natural signal processing idea. For example, in the continuous time case, in order to determine the expected number of extrema per unit time, one simply finds the expected zero-crossing rate in the derivative of a random signal [13]. Upcrossings of differenced time series are discussed in [12], while in [2, ch. 9], a filter is applied prior to some counting procedures. There are other suitable references in the engineering and scientific literature. A more systematic study of HOC has been initiated in [10] and [11], using repeated differencing, and later extended in a sequence of papers, some of which were reviewed in [7].

Given the proximity between the zero-crossing count and the first order autocorrelation noted in the previous chapter, and in line with the above definition of higher order crossings, we refer to a sequence or family of first order autocorrelations obtained by applying to a time series a family of filters as *higher order correlations* or simply HOC again. The precise type of HOC under consideration will be clear from the context. In the Gaussian case, these two types of HOC are essentially equivalent due to the cosine formula. The qualification is needed because the cosine formula only relates theoretical rather than observed quantities.

The HOC idea can be reincarnated in many different ways. A useful example is discussed in Section 7.4.

5.1 PARAMETRIC FAMILIES OF ZERO-CROSSING COUNTS

The mathematical formulation of the idea behind HOC can be outlined in four steps as follows.

1. We start with a zero-mean real-valued process in discrete time

$$\{Z_t\}, \qquad t = 0, \pm 1, \pm 2, \ldots$$

2. Let

$$\{\mathcal{L}_\theta(\cdot),\ \theta \in \Theta\}$$

be a parametric family of filters indexed by θ. The parameter space Θ may be an interval or a countable set in one or more dimensions. That is, θ may be a vector.

3. Fix $\theta \in \Theta$ and let D_θ be the number of zero-crossings in discrete time observed in the time series

$$\mathcal{L}_\theta(Z)_1, \mathcal{L}_\theta(Z)_2, \ldots, \mathcal{L}_\theta(Z)_N$$

of length N from the filtered process

$$\{\mathcal{L}_\theta(Z)_t\}, \qquad t = 0, \pm 1, \pm 2, \ldots$$

4. The corresponding HOC family is given by

$$\{D_\theta, \theta \in \Theta\}$$

As in the previous chapter, D_θ is defined from the clipped process,

$$X_t(\theta) = \begin{cases} 1, & \text{if } \mathcal{L}_\theta(Z)_t \geq 0 \\ 0, & \text{if } \mathcal{L}_\theta(Z)_t < 0 \end{cases}$$

$t = 0, \pm 1, \pm 2, \ldots$, by counting symbol changes in $X_1(\theta), \ldots, X_N(\theta)$:

$$D_\theta = \sum_{t=2}^{N} [X_t(\theta) - X_{t-1}(\theta)]^2$$

When, for some fixed θ_0, $\mathcal{L}_{\theta_0}(\cdot)$ corresponds to the identity filter, then D_{θ_0} is the number of zero-crossings in the original unfiltered series $Z_1, \ldots, Z_N$, that we sometimes denote simply by D. The expected higher order zero-crossing rate is given by

$$\gamma(\theta) \equiv \frac{E[D_\theta]}{N-1}$$

and the corresponding observed rate is defined, as before, by the ratio,

$$\hat{\gamma}(\theta) \equiv \frac{D_\theta}{N-1}$$

Let $H(\omega; \theta)$ be the transfer function associated with $\mathcal{L}_\theta(\cdot)$, and assume the original process $\{Z_t\}$ is a zero-mean stationary Gaussian process. Then the zero-crossing spectral representation, discussed in the previous chapter, has the form

$$\cos[\pi\gamma(\theta)] = \cos\left(\frac{\pi E[D_\theta]}{N-1}\right) = \frac{\int_{-\pi}^{\pi} \cos(\omega)|H(\omega;\theta)|^2 \, dF(\omega)}{\int_{-\pi}^{\pi} |H(\omega;\theta)|^2 \, dF(\omega)} \tag{5.1}$$

where $F(\omega)$ is the spectral distribution function of the original process $\{Z_t\}$. Symmetry implies the equivalent, and at times more convenient form,[1]

$$\cos[\pi\gamma(\theta)] = \cos\left(\frac{\pi E[D_\theta]}{N-1}\right) = \frac{\int_{0}^{\pi} \cos(\omega)|H(\omega;\theta)|^2 \, dF(\omega)}{\int_{0}^{\pi} |H(\omega;\theta)|^2 \, dF(\omega)} \tag{5.2}$$

Thus, by changing the parameter θ we can effect a desired change in the expected zero-crossing count by modifying the spectrum of the process judiciously through linear filtering. This observation is crucial. As it turns out, the representation (5.1) and its equivalent form (5.2) enable us to understand the effect of filtering on zero-crossings through the spectrum even in the general non-Gaussian case. Now, because the expected (normalized) zero-crossing rate $\pi\gamma(\theta)$ takes values in $[0, \pi]$, it is easy to relate it directly by (5.2) to specific frequency bands and discrete frequencies in the spectral support. Thus when θ corresponds to a high-pass (low-pass) filter, $\pi\gamma(\theta)$ moves to the right (left) so that by varying θ intelligently we can force or "push" $\pi\gamma(\theta)$ to land on any desired frequency. This mechanism allows the construction of HOC sequences useful in spectrum analysis.

Note that because θ is allowed to be a vector, the spectral representations (5.1) and (5.2) also cover the case of sequential filtering via the product of the

---

[1]We always assume the absence of the dc component.

transfer functions. In other words, if $\mathcal{L}_{\theta_j}(\cdot)$ is a linear filter with transfer function $H(\omega;\theta_j)$, and if we let $\theta = (\theta_1, \theta_2, \ldots, \theta_m)$, then the zero-crossing spectral representation corresponding to the sequential filter

$$\mathcal{L}_\theta = \mathcal{L}_{\theta_m} \cdots \mathcal{L}_{\theta_2}\mathcal{L}_{\theta_1}$$

can be obtained with

$$H(\omega;\theta) = H(\omega;\theta_m)\cdots H(\omega;\theta_2)H(\omega;\theta_1)$$

Similar remarks apply to other types of combinations of linear filters.

It is a curious fact that although $\{E[D_\theta],\ \theta \in \Theta\}$ is a time domain sequence, its normalization $\{\pi E[D_\theta]/(N-1),\ \theta \in \Theta\}$ may be viewed as a sequence in the frequency domain because it admits values in the spectral support. In this sense, HOC constitute a domain that intersects with both the time and frequency domains. In general, $\{E[D_\theta], \theta \in \Theta\}$ may be interpreted as yet another type of "spectrum" that relates directly to oscillation irrespective of stationarity. In this respect, $2(N-1)/E[D_\theta]$, where θ corresponds to the identity operation, may be viewed as the basic "period" of the process.

5.1.1 The Ideal Bandpass Case

For a typical application of the zero-crossing spectral representation (5.1), consider the HOC from a family of ideal band-pass filters operating on Gaussian white noise in discrete time. Suppose in the interval $[0,\pi]$ the band-pass extends from the frequency α to the frequency β radians per unit time. That is, the spectral density of the filtered white noise is constant over the interval $(-\beta,-\alpha) \cup (\alpha,\beta) \subset (-\pi,\pi]$, and is equal to 0 otherwise in the interval $(-\pi,\pi]$. By denoting the corresponding HOC by $D_{(\alpha,\beta)}$, we obtain from (5.1),

$$\frac{\pi E[D_{(\alpha,\beta)}]}{N-1} = \cos^{-1}\left[\frac{\sin(\beta)-\sin(\alpha)}{\beta-\alpha}\right] \tag{5.3}$$

so that as $\beta \to \alpha$,

$$\frac{\pi E[D_{(\alpha,\beta)}]}{N-1} \to \frac{\pi E[D_{(\alpha,\alpha)}]}{N-1} = \cos^{-1}(\cos(\alpha)) = \alpha$$

as is well expected from the dominant frequency principle. When $\alpha = 0$, (5.3) gives

$$\frac{\pi E[D_{(0,\beta)}]}{N-1} = \cos^{-1}\left[\frac{\sin(\beta)}{\beta}\right]$$

The continuous time analog of (5.3) was derived by Rice [13]. Accordingly, from Rice's formula, the expected number of zero-crossings per unit time is

$$2\left[\frac{1}{3}\frac{f_\beta^3 - f_\alpha^3}{f_\beta - f_\alpha}\right]^{1/2} \tag{5.4}$$

where now the passband extends from f_α to f_β cycles per unit time.

Really, (5.3) and (5.4) are not that different. To see this, use a second order Taylor series approximation in (5.3) to obtain

$$\frac{E[D_{(\alpha,\beta)}]}{N-1} \approx \frac{1}{\pi}\left[\frac{1}{3}\frac{\beta^3-\alpha^3}{\beta-\alpha}\right]^{1/2}$$

To switch to cycles per unit time (ranging from 0 to 0.5) we write, $\alpha = 2\pi f_\alpha$, and $\beta = 2\pi f_\beta$, so that,

$$\frac{E[D_{(\alpha,\beta)}]}{N-1} \approx 2\left[\frac{1}{3}\frac{f_\beta^3-f_\alpha^3}{f_\beta-f_\alpha}\right]^{1/2}$$

For $f_\alpha = 0$ we obtain Rice's useful result that

$$\frac{E[D_{(0,\beta)}]}{N-1} \approx \frac{2}{\sqrt{3}}f_\beta \doteq 1.155 f_\beta$$

the approximation being better for small $f_\beta \in [0, 0.5]$. In terms of radians per unit time, the result is

$$\frac{E[D_{(0,\beta)}]}{N-1} \approx \frac{1}{\pi\sqrt{3}}\beta \doteq 0.1838\beta$$

The corresponding approximation for the normalized zero-crossing rate is given by

$$\frac{\pi E[D_{(0,\beta)}]}{N-1} \approx \frac{1}{\sqrt{3}}\beta \doteq 0.5774\beta$$

which is more than half the highest frequency in the spectral support.

5.2 HIGHER ORDER CORRELATIONS

The higher order correlation corresponding to $\mathcal{L}_\theta(\cdot)$ is denoted by $\rho_1(\theta)$. It is *always* given by

$$\rho_1(\theta) = \frac{\int_{-\pi}^{\pi} \cos(\omega)|H(\omega;\theta)|^2\, dF(\omega)}{\int_{-\pi}^{\pi} |H(\omega;\theta)|^2\, dF(\omega)} \tag{5.5}$$

In the Gaussian case we also have,

$$\rho_1(\theta) = \cos[\pi\gamma(\theta)] = \cos\left(\frac{\pi E[D_\theta]}{N-1}\right) \tag{5.6}$$

When $\{Z_t(\theta)\}$ is complex, the notion of HOC from zero-crossings loses its ordinary meaning, but $\rho_1(\theta)$ can still be defined as the real part of the complex correlation,

$$\rho_1(\theta) = \frac{\Re\{E[Z_t(\theta)\overline{Z_{t+1}(\theta)}]\}}{E|Z_t(\theta)|^2}$$

As a rule, we shall always assume that the original process $\{Z_t\}$ is real. However, the filtered process $\{Z_t(\theta)\}$ may be complex. In this chapter only real filters are considered, so that $\{Z_t(\theta)\}$ is real for all θ.

Although in general there exists a close association between $\rho_1(\theta)$ and $E[D_\theta]$ (we saw examples of this in Chapter 4), there are several advantages to each over the other. For example, zero-crossings have a certain advantage in the presence of outliers. This is so because, in the presence of extreme values, the sample autocorrelation may resemble the autocorrelation of white noise even though the process itself is far from being white noise. This can easily be verified when a time series contains a single large outlier. In this case the sample autocorrelation tends to zero (see Problem 1). The zero-crossing count from the unfiltered original data, however, is left intact. Another example where zero-crossings have an advantage is when the process is strictly stationary but has no moments. In this case, $\rho_1(\theta)$ cannot be defined in the usual sense. The expected zero-crossing count, however, is well defined because D_θ possesses all moments, and thus provides a firsthand feel of the oscillation regardless of the existence of moments. An advantage of $\rho_1(\theta)$ is that it can be related to the spectrum, when it is well defined, without regard to whether the process is Gaussian or not. For $E[D_\theta]$ to do the same, a "cosine formula" is needed.

A few words about our terminology. "HOC from zero-crossings" were defined in terms of the *observed* $\{D_\theta, \theta \in \Theta\}$. The family of *expected* HOC is $\{E[D_\theta], \theta \in \Theta\}$. In contrast, "HOC from correlation" are defined in terms of the "*expected*" $\{\rho_1(\theta), \theta \in \Theta\}$, while the corresponding family of *observed* HOC is given by $\{\hat{\rho}_1(\theta), \theta \in \Theta\}$, where $\hat{\rho}_1(\theta)$ is any sample estimator of $\rho_1(\theta)$. It is convenient to refer to all these quantities, rather loosely, as HOC. In general, no confusion is likely to arise from this practice. However, when necessary, the adjectives "expected" and "observed" will be added, and the HOC under consideration (from zero-crossings or correlation) will be identified. In short, for $\theta \in \Theta$, we refer to either of $D_\theta, E[D_\theta], \rho_1(\theta), \hat{\rho}_1(\theta)$, as HOC.

We shall now obtain a general useful representation for $\rho_1(\theta)$ in terms of the autocorrelation ρ_k [8].

Let $\{Z_t\}, t = 0, \pm 1, \pm 2, \ldots$, be a zero-mean stationary process with autocovariance R_k, autocorrelation ρ_k, and generalized spectral density $f(\omega)$ that may contain δ-functions corresponding to the points of jump in the spectral distribution function. Let $\mathcal{L}_\theta$ be a parametric family of linear time-invariant filters with impulse response $\{h_n(\theta)\}_{n=-\infty}^{\infty}$ and transfer function $H(\omega;\theta)$, where

$$H(\omega;\theta) \equiv \mathcal{FT}(h_n(\theta)) = \sum_n \exp(-in\omega) h_n(\theta) \tag{5.7}$$

Wherever applicable, we always make the regularity assumptions that

$$\int_{-\pi}^{\pi} |H(\omega;\theta)|^2 f(\omega)\, d\omega < \infty$$

We define the filtered process by the convolution,

$$Z_t(\theta) \equiv \mathcal{L}_\theta(Z)_t = \sum_n h_n(\theta) Z_{t-n} \tag{5.8}$$

In other words

$$Z_t(\theta) \equiv h_t(\theta) \otimes Z_t \tag{5.9}$$

where $\otimes$ denotes convolution. Denote the autocovariance and autocorrelation of $\{Z_t(\theta)\}$ by $R_k(\theta)$, and $\rho_k(\theta)$, respectively. The next result shows how to compute the autocorrelation of $\{Z_t(\theta)\}$ from ρ_k.

Lemma 5.1. *Let $\mathcal{J}_\theta$ be a time invariant linear filter, indexed by θ, whose impulse response $\{g_n(\theta)\}_{n=-\infty}^{\infty}$ is given by the convolution*

$$g_n(\theta) = h_n(\theta) \otimes h_{-n}(\theta)$$

Then, for $n = 0, \pm 1, \pm 2, \ldots,$

$$\rho_n(\theta) = \frac{\mathcal{J}_\theta(\rho)_n}{\mathcal{J}_\theta(\rho)_0} \tag{5.10}$$

and

$$\mathcal{J}_\theta(\rho)_n = g_n(\theta) + \sum_{t=1}^{\infty} (g_{n+t}(\theta) + g_{n-t}(\theta))\rho_t \tag{5.11}$$

Proof. The Fourier transform ($\mathcal{FT}$) of a convolution gives

$$\begin{aligned} \mathcal{FT}(\mathcal{J}_\theta(R)_n) = \mathcal{FT}(g_n(\theta) \otimes R_n) &= \mathcal{FT}(h_n(\theta) \otimes h_{-n}(\theta) \otimes R_n) \\ &= H(\omega;\theta)\overline{H(\omega;\theta)} 2\pi f(\omega) \end{aligned}$$

By taking the inverse Fourier transform we obtain

$$\mathcal{J}_\theta(R)_n = \int_{-\pi}^{\pi} \exp(in\omega) |H(\omega;\theta)|^2 f(\omega)\, d\omega = R_n(\theta)$$

Therefore

$$\rho_n(\theta) = \frac{R_n(\theta)}{R_0(\theta)} = \frac{\mathcal{J}_\theta(R)_n}{\mathcal{J}_\theta(R)_0} = \frac{R_0 \mathcal{J}_\theta(\rho)_n}{R_0 \mathcal{J}_\theta(\rho)_0}$$

□

This fact will be used later in connection with the "*α-filter.*"

5.3 PROPERTIES OF HOC FROM DIFFERENCES

Let ∇ be the difference operator defined by

$$\nabla Z_t \equiv Z_t - Z_{t-1}$$

and define

$$\mathcal{L}_\theta \equiv \nabla^{\theta-1}, \qquad \theta \in \{1, 2, 3, \ldots\}$$

with $\mathcal{L}_1 \equiv \nabla^0$ being the identity filter. The corresponding HOC

$$D_1, D_2, D_3, \ldots$$

are called the *simple* HOC. They are always evaluated from differenced records of the same length N. Thus, D_1 is the number of zero-crossings in $Z_1, \ldots, Z_N$, D_2 is the number of zero-crossings in $\nabla Z_1, \ldots, \nabla Z_N$, D_3 is the number of zero-crossings in $\nabla^2 Z_1, \ldots, \nabla^2 Z_N$, and so on. The transfer function of $\mathcal{L}_\theta$ is given by

$$H(\omega; \theta) = (1 - \exp(-i\omega))^{\theta-1}$$

and the squared gain is

$$|H(\omega; \theta)|^2 = (2\sin(\omega/2))^{2(\theta-1)}$$

Substituting this in (5.1) gives the spectral representation for simple HOC in the Gaussian case,

$$\rho_1(\theta) = \cos\left(\frac{\pi E[D_\theta]}{N-1}\right) = \frac{\int_{-\pi}^{\pi} \cos(\omega)(\sin(\omega/2))^{2(\theta-1)}\, dF(\omega)}{\int_{-\pi}^{\pi} (\sin(\omega/2))^{2(\theta-1)}\, dF(\omega)} \tag{5.12}$$

for $\theta = 1, 2, 3, \ldots$.

Unlike the process $\{Z_t\}$ which is defined over all the integers, in practice we only have finite time series and lose an observation with each difference. To avoid end effects we must index the data by moving to the right. Thus, if it is desired to evaluate k higher order crossings, the index $t = 1$ is given to the kth or a later observation. The following example clarifies this point.

Example 5.1: Evaluating Simple D_1, D_2, D_3

Suppose the given record is

1 6 1 7 8 9 2 3 0 7

Subtracting the sample average, 4.4, we obtain the centered series,

−3.4 1.6 − 3.4 2.6 3.6 4.6 − 2.4 − 1.4 − 4.4 2.6

To evaluate D_1, D_2, D_3, we let $Z_1 = -3.4$, $Z_2 = 2.6$, reserving the first −3.4 and 1.6 for Z_{-1} and Z_0, respectively. We can now evaluate $Z_t, \nabla Z_t, \nabla^2 Z_t$, $t = 1, 2, \ldots, 8$ as

Z	:	−3.4	2.6	3.6	4.6	−2.4	−1.4	−4.4	2.6
∇Z	:	−5.0	6.0	1.0	1.0	−7.0	1.0	−3.0	7.0
$\nabla^2 Z$	:	−10.0	11.0	−5.0	0.0	−8.0	8.0	−4.0	10.0

The corresponding binary time series are

$$\begin{array}{lcllllllll} X(1) & : & 0 & 1 & 1 & 1 & 0 & 0 & 0 & 1 \\ X(2) & : & 0 & 1 & 1 & 1 & 0 & 1 & 0 & 1 \\ X(3) & : & 0 & 1 & 0 & 1 & 0 & 1 & 0 & 1 \end{array}$$

Then $D_1 = 3, D_2 = 5, D_3 = 7$, and are all evaluated from records of length 8. Note that by our definition, a shift from a negative value to 0 is counted as a crossing.

When facing shrinking records resulting from other types of filtering, similar remarks apply as well.

■

From the definition of $\{X_t(j)\}, j = 1, 2, \ldots$, it follows that, if $X_{t-1}(j) \neq X_t(j)$, then $X_t(j+1) = X_t(j)$. This is shown clearly in the previous example. Thus, every symbol change in $\{X_t(j)\}$ leads to at least one symbol change in $\{X_t(j+1)\}$. This means that we should have $D_{j+1} \geq D_j$. This is indeed what happens in practice for moderate and large record lengths N, provided j is not too large relative to N. However, in finite data records we may have an end effect at the beginning of a time series, and this must be compensated for. What we really have is the sure inequality

$$D_{j+1} \geq D_j - 1 \tag{5.13}$$

Taking expectations and dividing by $N - 1$ throughout in (5.13), we obtain

$$\frac{E[D_{j+1}]}{N-1} \geq \frac{E[D_j]}{N-1} - \frac{1}{N-1} \tag{5.14}$$

Upon letting $N \to \infty$, and recalling the fact that $\gamma(j) = E[D_j]/(N-1)$ is independent of N under (strict) stationarity, we have the expected inequality,

$$\gamma(j+1) = \frac{E[D_{j+1}]}{N-1} \geq \frac{E[D_j]}{N-1} = \gamma(j) \tag{5.15}$$

or

$$E[D_{j+1}] \geq E[D_j] \tag{5.16}$$

Because a difference is a high-pass filter with a strictly monotone gain, the inequality (5.16) is a verification of part (a) of Corollary 4.2, but without recourse to the Gaussian assumption. By part (b) of that corollary, if in addition the process is Gaussian, equality in (5.16) for $j = 1$ holds if and only if the process is a pure sinusoid with probability one. Also, recall that the maximum number of symbol changes cannot exceed $N - 1$. We have thus established the following monotonicity property of the simple HOC [5].

Theorem 5.1. *Suppose $\{Z_t\}$, $t = 0, \pm 1, \ldots$, is a zero-mean stationary process. Then, regardless of spectrum type,*

(a)

$$0 \leq E[D_1] \leq E[D_2] \leq \cdots \leq N - 1$$

(b) *If in addition the process is Gaussian, then*

$$\frac{E[D_1]}{N-1} = \frac{E[D_2]}{N-1}$$

if and only if the process is a pure sinusoid with probability one. The frequency of the sinusoid is given by $\pi E[D_1]/(N-1)$.

Example 5.2: Monotonicity of Simple HOC

Consider a zero-mean stationary Gaussian process defined by a first order autoregressive process

$$Z_t = \phi Z_{t-1} + \zeta_t \tag{5.17}$$

where $\{\zeta_t\}$ is Gaussian white noise. Two time series of length $N = 200$ from the process, one corresponding to $\phi = 0.5$ and the other to $\phi = -0.5$, were generated by a computer simulation and are shown in Figure 5.1, together with the observed HOC. It is seen that even for these relatively short series, the first few D_j tend to increase. Since $\rho_1 = \phi$, the series with $\phi = -0.5$ is more oscillatory and we would expect a higher D_1, as is indeed the case.

■

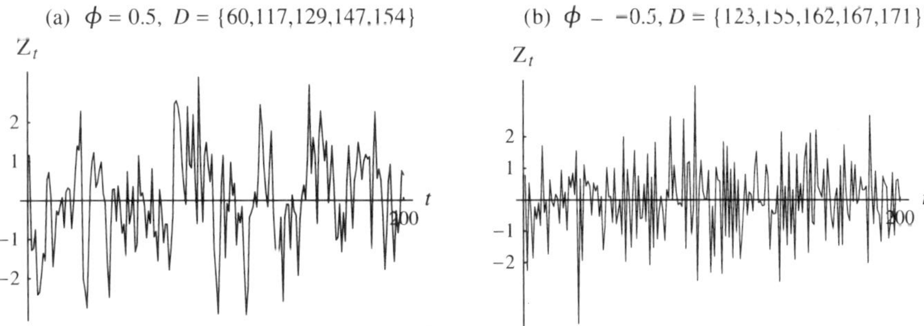

Figure 5.1: *Simple HOC from a first order autoregressive process with parameter* (*a*) $\phi = 0.5$, (*b*) $\phi = -0.5$. $N = 200$.

Thus, the sequence of expected zero-crossing rates $\{E[D_j]/(N-1)\}$, $j = 1, 2, 3, \ldots$, is a monotone increasing sequence, and is bounded by 1. From a theorem in calculus we know therefore that the sequence must converge. But converge to what? To answer this question we appeal to the dominant frequency principle. Suppose the highest positive frequency in the spectral support is ω^*. That is, no power is given to frequency bands that lie entirely above ω^*. Now, as mentioned before, the difference operator ∇ is a high-pass filter and as such it amplifies high frequencies. But a second difference ∇^2 amplifies high frequencies even more, and ∇^3 even more so, and so on. It follows by induction that if we apply the filter ∇^j to $\{Z_t\}$, and let $j \to \infty$, the power is pushed all the way toward ω^*. As a result ω^* becomes dominant, and in accordance with the dominant frequency principle, $\{D_j/(N-1)\}$, and hence also $\{E[D_j]/(N-1)\}$, should converge to ω^* in some sense, as $j \to \infty$. In the Gaussian case there is an easy proof of this fact.

Theorem 5.2. *Suppose* $\{Z_t\}$, $t = 0, \pm 1, \ldots$, *is a zero-mean stationary Gaussian process, and let* ω^* *be the highest positive frequency in the spectral support. Then, regardless of spectrum type, the sequence of simple expected normalized HOC* $\{\pi E[D_j]/(N-1)\}$ *converges to the highest frequency,*

$$\frac{\pi E[D_j]}{N-1} \to \omega^* \tag{5.18}$$

as $j \to \infty$.

Proof. Note that if we define $\nu_j(\cdot)$, $j = 0, 1, 2, \ldots$, by

$$\nu_j(d\omega) = \frac{\sin^{2j}(\omega/2)\, dF(\omega)}{\int_{-\pi}^{\pi} \sin^{2j}(\lambda/2)\, dF(\lambda)} \tag{5.19}$$

where F is the spectral distribution function of $\{Z_t\}$,then (5.12) can be written more compactly as

$$\cos\left(\frac{\pi E[D_{j+1}]}{N-1}\right) = \int_{-\pi}^{\pi} \cos(\omega)\nu_j\ (d\omega)$$

One can verify that $\nu_j(\Lambda)$ is a probability defined over subsets Λ in $[0, \pi]$. In the appendix to this chapter we show that,

$$\nu_j \Rightarrow \frac{1}{2}\delta_{-\omega^*} + \frac{1}{2}\delta_{\omega^*},\ j \to \infty \tag{5.20}$$

where δ_u is the unit point mass at u. That is, $\delta_u(\Lambda) = 1$ if $u \in \Lambda$, and is 0 otherwise. This implies, as $j \to \infty$,

$$\cos\left(\frac{\pi E[D_{j+1}]}{N-1}\right) \to \frac{1}{2}(\cos(-\omega^*) + \cos(\omega^*)) = \cos(\omega^*)$$

Now use the fact that $\cos(x)$, $x \in [0, \pi]$, is monotone.

□

In Chapter 6 we will show that (5.18) also holds in some sense for the *observed* sequence $\{\pi D_j/(N-1)\}$ (that is without "E"). It is simpler to show this for the important special case $\omega^* = \pi$ which we consider next.

Suppose the highest frequency in the spectral support is π, a case often met in practice due to the frequent presence of wide-band ambient noise. Then, as we keep on differencing, more and more spectral weight is given to π and the oscillation in the repeatedly differenced series tends to maximum possible oscillation. There are at least two ways to describe this rather intuitive fact. The first is to consider the limit of $\rho_1(j+1)$, and the second is to consider the limit of the process $\{X_t(j)\}$, as j increases. Accordingly, from the definition (5.19) of ν_j, we can rewrite (5.5) as

$$\rho_1(j+1) = \int_{-\pi}^{\pi} \cos(\omega)\nu_j\ (d\omega)$$

for $j = 0, 1, 2, \ldots$. But since ν_j converges as in (5.20) with π replacing ω^*, we have for $j = 0, 1, 2, \ldots,$

$$\rho_1(j+1) \to \cos(\pi) = -1 \tag{5.21}$$

See Problem 16 for a verification of (5.21) in the special case of white noise. Now, (5.21) implies that if Z_t is positive, then Z_{t-1} tends to be negative and vice versa. It follows that the binary process $\{X_t(j)\}$ converges, as $j \to \infty$, to a *degenerate* binary process in which a 1 is followed by a 0 and vice versa. So, realizations in $\{X_t(j)\}$ for large j tend to have the form

$$\cdots 0101010101010101 \cdots$$

or

$$\cdots 1010101010101010 \cdots$$

each with equal probability. This degenerate state is a mode that signifys extreme oscillation. It is independent of the original process $\{Z_t\}$ as long as π is in the spectrum. Clearly, this is quite expected in light of Theorem 5.2, because when $\omega^* = \pi$ the theorem tells us that, as $j \to \infty$,

$$\frac{E[D_j]}{N-1} \to 1 \tag{5.22}$$

and this can happen only in extreme oscillation. In the appendix to this chapter it is shown that the convergence in (5.22) also holds with probability one for the random sequence $D_j/(N-1)$ and not just for the corresponding sequence of expected values, provided *both* N and j increase indefinitely. We summarize this discussion in the next theorem, called the *higher order crossings theorem* or HOCT for short [11].

Theorem 5.3. *Let* $\{Z_t\}$, $t = 0, \pm 1, \ldots,$ *be a zero-mean stationary process, and assume that* π *is included in the spectral support. Then,*

(a)

$$\{X_t(j)\} \Rightarrow \begin{cases} \cdots 01010101 \cdots, & \textit{with probability 1/2} \\ \cdots 10101010 \cdots, & \textit{with probability 1/2} \end{cases}$$

as $j \to \infty$.

(b) $\lim_{j\to\infty} \lim_{N\to\infty}(D_j/N-1) = 1$, *with probability 1.*

The higher order crossings theorem has a graphical aspect to it in the sense that D_1, D_2, D_3 are features that can be discerned with a casual eye examination of the data. This is so since D_1 is the number of zero-crossings, and except for an end effect, D_2 is the number of peaks and troughs, while D_3 is the number of inflection points in the graph of the original time series. The intuitive interpretation of D_k for $k \geq 4$ is more difficult. We shall have many opportunities to see that only the first few D_k are useful for discrimination purposes precisely because the discrimination power in the simple HOC diminishes rather fast as k increases. Thus, the theorem supports the common sense idea that easily discernible visual features are more potent for discrimination and classification purposes than features associated with D_k, for large k, that we cannot "see."

Example 5.3: Demonstration of HOCT

The theorem is demonstrated using two time series of length $N = 1000$ from the first order autoregressive process (5.17) with $\phi = 0.5$ and $\phi = 0.8$. The results are displayed in Figure 5.2. The rightmost column gives the D_j, $j = 1, \ldots, 16$, but the binary arrays should be thought of as sections from the infinite binary arrays obtained from the binary processes $\{X_t(j)\}, j = 1, 2, 3, \ldots$. We can see that the initial rate of convergence toward the degenerate state $\cdots 01010101 \cdots$ is rather fast, but that the rate slows down to a great extent already for moderately large j.

■

j	$\{X\}$	D	$\{X\}$	D
01	100100000111111	207	011100001111111	331
02	001100110111111	513	011000111010010	567
03	001100110101101	659	010001111010010	690
04	011001100101101	715	010001101010110	738
05	011001001101001	745	010101001010110	774
06	010011001101001	773	010101011010100	798
07	010011011101011	807	010101011010100	810
08	010110011001011	823	010101010010101	829
09	010110010001010	829	010101010010101	837
10	010100110101010	849	010101010110101	848
11	010100110101010	855	010101010110101	854
12	010101110101010	865	010101010100101	858
13	010101000101010	875	010101010100101	858
14	010101010101010	883	010101010101101	862
15	010101010101010	885	010101010101101	868
16	010101010101010	893	010101010101001	870
	$\phi = 0.8$		$\phi = 0.5$	

Figure 5.2: *Demonstration of the higher order crossings theorem using a first order autoregressive process with parameter* $\phi = 0.8, 0.5$. $N = 1000$.

Example 5.4: The Rate of Increase of Simple HOC

It is interesting to compare the rate of increase displayed by the simple HOC from several processes. Figure 5.3 gives the graphs of D_j versus j, for $j = 1, 2, \ldots, 10$, and $N =$

1000, generated again from the first order autoregressive process (5.17). The parameter ϕ varies from 0.75 to -0.75. From the figure we can see that the initial rate of increase in the simple HOC differs from process to process. However, as j continues to increase, the rate of increase tapers off considerably and the D_j themselves become quite similar. This demonstrates the fact that discrimination based on D_j for *large* j is in general a fruitless effort, since different processes yield almost the same D_j. Normally, only the very first few D_j are useful in discrimination between processes.

■

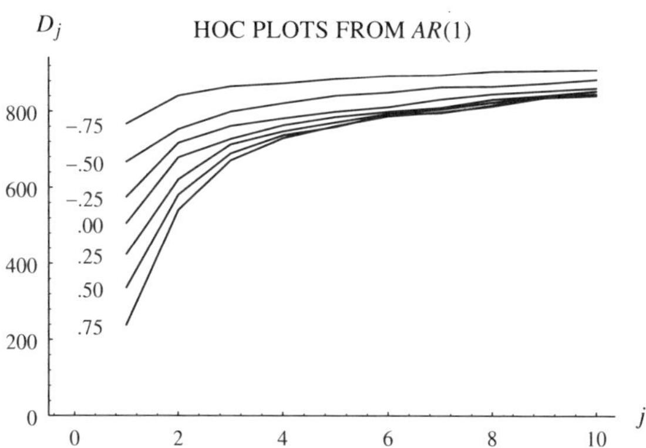

Figure 5.3: *Plots of* D_j *versus* $j = 1, 2, \ldots, 10$, *from a first order autoregressive process with parameter* ϕ *ranging from* -0.75 *to* 0.75. $N = 1000$.

Example 5.5: HOC from Speech Data

Figure 5.4 shows the graphs of twenty simple HOC obtained from the utterances of the English words "zero," "one," and "two." The original time series were sampled at a rate of 8000 samples per second, and we used $N = 2000$ samples in obtaining $D_1, D_2, \ldots, D_{20}$. Again, the D_j are monotone as expected, and exhibit a fast initial growth. The plots suggest that successful discrimination can be achieved from fewer than twenty D_j. In many cases in practice, six or even as few as three D_j suffice for effective discrimination purposes. It is useful to note that the zero-crossing count D_1 is almost the same for the three cases. What distinguishes between the different utterances are not zero-crossings but *higher* order crossings.

■

Our last result in this section says that the sequences $\{E[D_j]\}$, and $\{\rho_j\}$, $j = 1, 2, 3, \ldots$, are equivalent under the Gaussian assumption. That is, each sequence can be obtained from the other. To see this, we first observe that the repeatedly differenced process $\{\nabla^k Z_t\}$ admits the representation

$$\nabla^k Z_t = \sum_{j=0}^{k} \binom{k}{j} (-1)^j Z_{t-j} \tag{5.23}$$

for $k = 0, 1, 2, \ldots$. Recall that R_k denotes the autocovariance at lag k, and

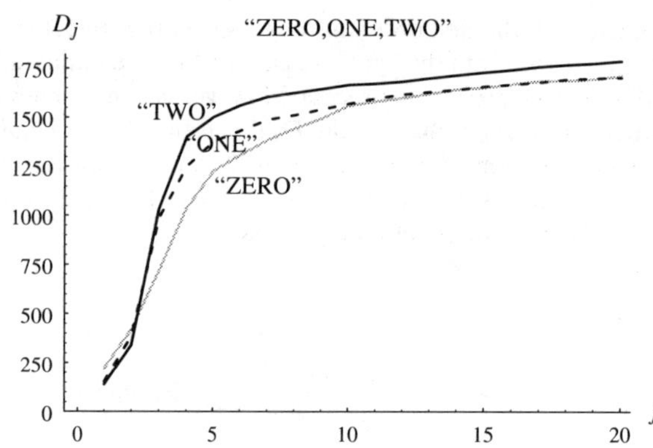

Figure 5.4: *Plots of D_j versus $j = 1, 2, \ldots, 20$, from the utterances of "zero," "one" (dashed), and "two" (dark). $N = 2000$.*

assume the mean of the process is 0. Then

$$\begin{aligned} \mathrm{Var}[\nabla^k Z_t] = & \\ & R_0\left[\binom{k}{0}^2+\binom{k}{1}^2+\cdots+\binom{k}{k}^2\right] \\ -R_1&\left[\binom{k}{1}\binom{k}{0}+\binom{k}{2}\binom{k}{1}+\cdots+\binom{k}{k}\binom{k}{k-1}\right] \\ -R_{-1}&\left[\binom{k}{0}\binom{k}{1}+\binom{k}{1}\binom{k}{2}+\cdots+\binom{k}{k-1}\binom{k}{k}\right] \\ & +\cdots+(-1)^k R_k+(-1)^k R_{-k} \end{aligned} \tag{5.24}$$

and

$$\begin{aligned} E[\nabla^k Z_t\, \nabla^k Z_{t-1}] = & \\ -R_0&\left[\binom{k}{0}\binom{k}{1}+\binom{k}{1}\binom{k}{2}+\cdots+\binom{k}{k-1}\binom{k}{k}\right] \\ +R_1&\left[\binom{k}{0}\binom{k}{0}+\binom{k}{1}\binom{k}{1}+\cdots+\binom{k}{k}\binom{k}{k}\right] \\ +R_{-1}&\left[\binom{k}{0}\binom{k}{2}+\binom{k}{1}\binom{k}{3}+\cdots+\binom{k}{k-2}\binom{k}{k}\right] \\ & +\cdots+(-1)^k R_{k+1} \end{aligned} \tag{5.25}$$

For positive integers a, b, n we have the combinatorial identities

$$\binom{a}{0}\binom{b}{n}+\binom{a}{1}\binom{b}{n-1}+\cdots+\binom{a}{n}\binom{b}{0} = \binom{a+b}{n} \tag{5.26}$$

and

$$\binom{b}{a} = \binom{b}{b-a} \tag{5.27}$$

By forming the ratio of $E[\nabla^k Z_t \nabla^k Z_{t-1}]$ to $\text{Var}[\nabla^k Z_t]$, dividing the resulting numerator and denominator by $R_0 = \text{Var}[Z_t]$, and by invoking the combinatorial identities (5.26) and (5.27), we finally obtain the *long formula*:

$$\rho_1(k+1) = \cos\left(\frac{\pi E[D_{k+1}]}{N-1}\right) = \tag{5.28}$$

$$\frac{-\binom{2k}{k-1} + \rho_1\left[\binom{2k}{k} + \binom{2k}{k-2}\right] - \rho_2\left[\binom{2k}{k-1} + \binom{2k}{k-3}\right] + \cdots + (-1)^k \rho_{k+1}}{\binom{2k}{k} - 2\rho_1\binom{2k}{k-1} + 2\rho_2\binom{2k}{k-2} - \cdots + (-1)^k 2\rho_k}$$

Note that the long formula can also be derived from Lemma 5.1 (see Problem 4) and that it is not that long after all, because it can be written compactly as

$$\rho_1(k+1) = \cos\left(\frac{\pi E[D_{k+1}]}{N-1}\right) = \frac{\nabla^{2k}\rho_{k-1}}{\nabla^{2k}\rho_k} \tag{5.29}$$

where now ∇ operates on ρ_k. The long formula can be viewed as a recursion for obtaining the autocorrelations $\rho_1, \rho_2, \ldots$, from $E[D_1], E[D_2], \ldots$. The converse is obviously true as well. For example, from $E[D_1], E[D_2]$ we can determine ρ_1 using the cosine formula, and

$$\rho_2 = -1 + 2\cos\left(\frac{\pi E[D_1]}{N-1}\right) - 2\cos\left(\frac{\pi E[D_2]}{N-1}\right)\left[1 - \cos\left(\frac{\pi E[D_1]}{N-1}\right)\right] \tag{5.30}$$

Conversely, from ρ_1, ρ_2 we obtain $E[D_1]$ from the inverse form of the cosine formula, and the inverse form of

$$\cos\left(\frac{\pi E[D_2]}{N-1}\right) = \frac{-1 + 2\rho_1 - \rho_2}{2(1-\rho_1)} \tag{5.31}$$

In general, the vectors $(\rho_1, \rho_2, \ldots, \rho_k)$ and $(E[D_1], E[D_2], \ldots, E[D_k])$ determine each other. Thus, the two sequences $\{\rho_k\}$ and $\{E[D_k]\}$, $k = 1, 2, 3, \ldots$, are equivalent. However, whereas ρ_k is not necessarily monotone, the expected simple HOC sequence $\{E[D_k]\}$ is always monotone increasing, a fact that makes it a "more standardized" tool in comparing two or more time series.

Now, recall that $\rho_1, \rho_2, \ldots$ are the Fourier coefficients of the normalized (total power is 1) spectral distribution function, and as such determine it uniquely. We have therefore established the following important fact.

Theorem 5.4. *For a zero-mean stationary Gaussian process, the sequence of expected simple HOC $\{E[D_k]\}$ determines the spectrum up to a constant.*

The preceding discussion can be summarized by the suggestive symbolism,

$$\{E[D_k]\} \Leftrightarrow \{\rho_k\} \Leftrightarrow \overline{F(\omega)} \tag{5.32}$$

where $\overline{F(\omega)} = F(\omega)/\text{Var}[Z_t]$ is the normalized spectrum. The equivalence relation (5.32) is another indication that zero-crossings of filtered processes contain useful spectral information, and it may be viewed as a ramification of the Wiener–Khintchine relationship. The simple HOC are by no means the only HOC that provide useful spectral information.

Two-Dimensional HOCT

Suppose[2] we have a two-dimensional process where each two-dimensional index (t_1, t_2) is associated with a single random value Z_{t_1,t_2}. Such a process is called a *random field*, but it is simpler to think of this as a two-dimensional "picture" or image. What happens if we keep differencing (horizontally and vertically) and clipping indefinitely? The answer is given in the next paragraph.

The higher order crossings theorem 5.3 can be extended to two dimensions as follows. Let Z_{t_1,t_2}, $t_1, t_2 = 0, \pm 1, \pm 2, \ldots$, be a two-dimensional zero-mean stationary random field, with a two-dimensional spectral density $f(\lambda_1, \lambda_2)$ supported over the rectangle $-\pi < \lambda_1, \lambda_2 \leq \pi$. Define,

$$(1 - \mathcal{B}_{t_1})Z_{t_1,t_2} = Z_{t_1-1,t_2}$$

$$(1 - \mathcal{B}_{t_2})Z_{t_1,t_2} = Z_{t_1,t_2-1}$$

Also, define the clipping operator $\mathcal{U}$ by,

$$\mathcal{U}Y_t = \begin{cases} 1, & \text{if } Y_t \geq 0 \\ 0, & \text{if } Y_t < 0 \end{cases}$$

The two-dimensional HOCT states that the two-dimensional binary process $\mathcal{U}(1-\mathcal{B}_{t_1})^{k-1}(1-\mathcal{B}_{t_2})^{k-1}Z_{t_1,t_2}$, $k = 1, 2, 3, \ldots$, converges, as $k \to \infty$, to a degenerate infinite binary "checkerboard":

$$\mathbf{a} = \begin{array}{cccccccccccc} \cdot & \cdot & \cdot & \cdot & \cdot & \cdot & \cdot & \cdot & \cdot & \cdot & \cdot & \cdot \\ \cdot & \cdot & \cdot & \cdot & \cdot & \cdot & \cdot & \cdot & \cdot & \cdot & \cdot & \cdot \\ \cdot & \cdot & \cdot & \cdot & \cdot & \cdot & \cdot & \cdot & \cdot & \cdot & \cdot & \cdot \\ \cdot & \cdot & \cdot & 0 & 1 & 0 & 1 & 0 & 1 & \cdot & \cdot & \cdot \\ \cdot & \cdot & \cdot & 1 & 0 & 1 & 0 & 1 & 0 & \cdot & \cdot & \cdot \\ \cdot & \cdot & \cdot & 0 & 1 & 0 & 1 & 0 & 1 & \cdot & \cdot & \cdot \\ \cdot & \cdot & \cdot & 1 & 0 & 1 & 0 & 1 & 0 & \cdot & \cdot & \cdot \\ \cdot & \cdot & \cdot & 0 & 1 & 0 & 1 & 0 & 1 & \cdot & \cdot & \cdot \\ \cdot & \cdot & \cdot & 1 & 0 & 1 & 0 & 1 & 0 & \cdot & \cdot & \cdot \\ \cdot & \cdot & \cdot & 0 & 1 & 0 & 1 & 0 & 1 & \cdot & \cdot & \cdot \\ \cdot & \cdot & \cdot & \cdot & \cdot & \cdot & \cdot & \cdot & \cdot & \cdot & \cdot & \cdot \\ \cdot & \cdot & \cdot & \cdot & \cdot & \cdot & \cdot & \cdot & \cdot & \cdot & \cdot & \cdot \\ \cdot & \cdot & \cdot & \cdot & \cdot & \cdot & \cdot & \cdot & \cdot & \cdot & \cdot & \cdot \end{array}$$

[2]This subsection may be omitted upon first reading.

Let $\mathbf{a}'$ be the array $\mathbf{a}$ shifted to the right once. The complete statement is that as $k \to \infty, \mathcal{U}(1-\mathcal{B}_{t_1})^{k-1}(1-\mathcal{B}_{t_2})^{k-1}Z_{t_1,t_2}$ converges (weakly) to $\mathbf{a}$, or to $\mathbf{a}'$, each with probability $1/2$. The proof is entirely analogous to the one-dimensional case [4].

As in the one-dimensional case, the number of horizontal and vertical symbol changes (i.e., the two-dimensional analog of simple HOC) tends to increase with k, and in fact the initial rate of convergence is relatively fast. For example, consider the stationary wave model

$$Z_{t_1,t_2} = 0.4[Z_{t_1-1,t_2} + Z_{t_1,t_2-1}] + \zeta_{t_1,t_2}$$

where ζ_{t_1,t_2} is white Gaussian noise with mean 0 and variance 1. A simulation gives the number of horizontal and vertical symbol changes (from 0 to 1 and vice versa) in a 15×15 subarray from $\mathcal{U}(1-\mathcal{B}_{t_1})^{k-1}(1-\mathcal{B}_{t_2})^{k-1}Z_{t_1,t_2}$, for $k = 1, 2, \ldots, 8$, as $111, 229, 282, 305, 315, 326, 333, 340$, respectively. Clearly, the maximum number of symbol changes is $2(15 \times 14) = 420$. Thus, more than 80% of the maximum number of possible symbol changes was obtained with $k = 8$. This is quite typical.

5.4 HOC FROM REPEATED SUMMATION

The counterparts of the simple HOC are higher order crossings obtained by the repeated application of the summation operator $1 + \mathcal{B}$,

$$(1+\mathcal{B})Z_t = Z_t + Z_{t-1}$$

The HOC sequence is generated from the family of filters,

$$\mathcal{L}_\theta = (1+\mathcal{B})^{\theta-1}, \qquad \theta \in \{1, 2, 3, \ldots\}$$

with $\mathcal{L}_1 = (1+\mathcal{B})^0$ being the identity filter. It is convenient to denote this HOC sequence by $\{{}_\theta D\}$. By this notation ${}_1D = D_1$, however, whereas the simple HOC D_j tend to increase, the ${}_jD$ tend to decrease. Thus, the counterpart of Theorem 5.1 is the following fact (see Problem 10).

Theorem 5.5. *Suppose* $\{Z_t\}$, $t = 0, \pm 1, \ldots$, *is a zero-mean stationary process. Then, regardless of spectrum type,*

(a)

$$N - 1 \geq E[{}_1D] \geq E[{}_2D] \geq \cdots \geq 0$$

(b) *If in addition the process is Gaussian, then*

$$\frac{E[{}_1D]}{N-1} = \frac{E[{}_2D]}{N-1}$$

if and only if the process is a pure sinusoid with probability one. The frequency of the sinusoid is given by $\pi E[{}_1D]/(N-1)$.

The proof follows the same pattern as that of Theorem 5.1. Similarly, the analog of Theorem 5.2 is:

Theorem 5.6. *Suppose* $\{Z_t\}$, $t = 0, \pm 1, \ldots$, *is a zero-mean stationary Gaussian process, and let* ${}_*\omega$ *be the lowest positive frequency in the spectral support. Then, regardless of spectrum type, the sequence of expected normalized HOC* $\{\pi E[{}_jD]/(N-1)\}$ *converges to the lowest frequency,*

$$\frac{\pi E[{}_jD]}{N-1} \rightarrow {}_*\omega \tag{5.33}$$

as $j \rightarrow \infty$.

We leave the proof of this fact as an exercise (Problem 11). Evidently, (5.33) is another manifestation of the dominant frequency principle which provides a way to estimate the lowest frequency in the spectral support. We can also obtain a "long formula" as in (5.28) by replacing in (5.28) all the minus signs by plus signs, which establishes the equivalence of these HOC and the normalized spectrum under the Gaussian assumption.

Example 5.6: Crossings from Repeated Summation

Consider the first order autoregressive process (5.17). With $N = 1000$, and $\phi = 0.8$, the first ten ${}_jD$ from a computer simulation are:

$$212, 154, 126, 118, 108, 92, 88, 84, 80, 78$$

With $\phi = -0.8$, the first ten ${}_jD$ are:

$$805, 447, 325, 260, 210, 194, 178, 161, 157, 153$$

Thus the ${}_jD$ decrease as is well expected.

■

5.4.1 Discrete Spectrum Analysis by HOC

As an application of Theorems 5.2 and 5.6, we consider the problem of locating the frequencies in a purely discrete spectrum. This is a typical application of our HOC methods, and we shall discuss it again in greater detail at a later chapter. For the time being we consider the stationary process that admits the representation,

$$Z_t = \sum_{j=1}^{p} \{A_j \cos(\omega_j t) + B_j \sin(\omega_j t)\} \tag{5.34}$$

where, $t = 0, \pm 1, \pm 2, \ldots$, the A's and B's are all uncorrelated, $E(A_j) = E(B_j) = 0$, and $\mathrm{Var}(A_j) = \mathrm{Var}(B_j) = \sigma_j^2$. This is an ideal situation of a pure signal where no noise is involved. Without loss of generality, we assume that the frequencies are ordered fixed constants,

$$0 < \omega_1 < \omega_2 < \cdots < \omega_p < \pi$$

The problem is to determine all the frequencies $\omega_1, \omega_2, \ldots, \omega_p$ from expected HOC sequences, without prior knowledge of the number of frequencies p. Note that here the lowest positive frequency is $_*\omega = \omega_1$, while the highest is $\omega^* = \omega_p$, the next highest frequency is ω_{p-1}, and so on.

The solution runs as follows. First we invoke Theorem 5.6 and obtain the lowest positive frequency ω_1 from the convergence of

$$\frac{\pi E[{}_1D]}{N-1}, \frac{\pi E[{}_2D]}{N-1}, \frac{\pi E[{}_3D]}{N-1} \cdots$$

Then, by invoking Theorem 5.2, we obtain the highest frequency ω_p from the convergence of

$$\frac{\pi E[D_1]}{N-1}, \frac{\pi E[D_2]}{N-1}, \frac{\pi E[D_3]}{N-1} \cdots$$

and filter it out by an ideal low-pass filter. Then, ω_{p-1} is the highest frequency in the filtered process. Theorem 5.2 is invoked again, and we obtain ω_{p-1}. The procedure is repeated until the "next highest frequency" coincides with ω_1, that has already been determined earlier, and we stop. As a byproduct we also obtain p, by counting the number of determined frequencies. We summarize this procedure in the form of an algorithm.

Algorithm 5.1
Determination of $\omega_1, \cdots, \omega_p, \& p$

1. Obtain the lowest positive frequency $_*\omega \equiv \omega_1$ from repeated summation:
$$\frac{\pi E[{}_jD]}{N-1} \to \omega_1, \qquad j \to \infty$$
Do not filter ω_1 out.
2. Obtain the highest frequency $\omega^* \equiv \omega_p$ from repeated differencing:
$$\frac{\pi E[D_j]}{N-1} \to \omega_p, \qquad j \to \infty$$
Filter ω_p out by an ideal low-pass filter with cutoff frequency ω_p-.
3. ω_{p-1} is now the highest frequency. Repeat step 2 to determine ω_{p-1}, and filter it out with an ideal low-pass filter with cutoff frequency $\omega_{p-1}-$.
4. Repeat the procedure to obtain $\omega_{p-2}, \omega_{p-3}, \ldots$, until the next highest frequency is equal to ω_1, the lowest frequency that we already know, and stop.

As an example of this procedure, we consider the model (5.34) with the five sinusoids with amplitudes and frequencies given in Table 5.1.

We approximate an ideal low-pass filter by the sine-Butterworth filter with transfer function whose squared gain is given by

TABLE 5.1 Amplitudes and Frequencies of Five Sinusoids

j	ω_j	A_j	B_j
1	0.75	1.034	1.220
2	1.25	−0.250	1.090
3	1.60	−1.462	2.266
4	2.00	0.817	−0.501
5	2.30	0.873	−1.660

$$|H(\omega)|^2 = \left(\frac{1}{1+(\sin(\omega/2)/\sin(\omega_b/2))^{20}}\right)^{10}$$

where $\omega \in [0,\pi]$. Here $\omega_b \in [0,\pi]$ is the cutoff frequency. With $N = 450$, and with *observed* rather than expected HOC, we first determine the lowest frequency as 0.749. The other frequencies are determined in succession as 2.295, 2.001, 1.602, 1.245, and 0.756, which is fairly close to 0.749, and the search is stopped. In all cases, 16 iterations are used. That is, each frequency was obtained from a (normalized) HOC sequence of length 16. The results are given in terms of (normalized) HOC plots in Figure 5.5. It should be noted that these HOC were obtained by repeated differencing of, first, the original data, and then of the data filtered by the sine-Butterworth filter.

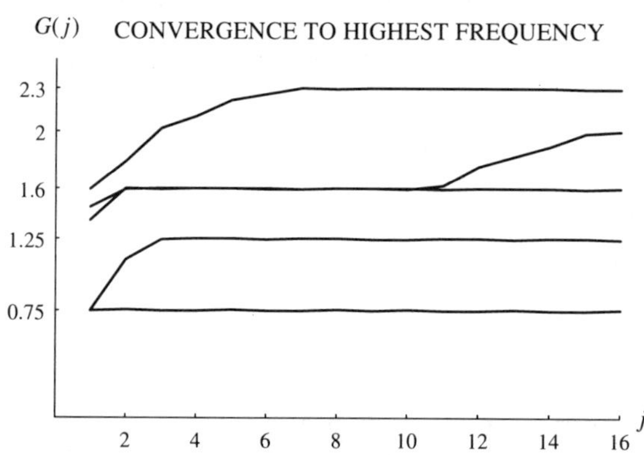

Figure 5.5: *Convergence of* $G(j) = \pi D_j/(N-1)$ *to the highest frequency before and after low-pass filtering.* $N = 450$.

5.5 THE SLUTSKY FILTER

There is no reason why we cannot combine sequential differencing and summation together to produce more versatile HOC features. Consider the family of filters

$$\mathcal{L}_{m,n} = (1-\mathcal{B})^{m-1}(1+\mathcal{B})^{n-1} \tag{5.35}$$

$m, n \in \{1, 2, 3, \ldots\}$, characterized by the squared gain

$$|H(\omega; m, n)|^2 = 2^{m+n}(1 - \cos(\omega))^m(1 + \cos(\omega))^n$$

Note that $\theta = (m, n)$ is now a vector. It is convenient to denote the corresponding HOC sequence by $\{{}_nD_m\}$ where m, n refer to the number of repeated differencing and repeated summation, respectively. When m, n tend to infinity such that $m/n = c$, where c is a constant, the frequency

$$\lambda_c = \cos^{-1}\left(\frac{1-c}{1+c}\right) \tag{5.36}$$

becomes dominant because $|H(\omega; m, n)|^2$ is unimodal with peak at λ_c (see Problem 12). It follows from the dominant frequency principle that

$$\frac{\pi E[{}_nD_m]}{N-1} \to \lambda_c \tag{5.37}$$

as $m, n \to \infty$, such that $m/n = c$.

5.6 HOC FROM THE *AR*(1) FILTER

The HOC family obtained from the family of filters

$$\mathcal{L}_\alpha \equiv 1 + \alpha\mathcal{B} + \alpha^2\mathcal{B}^2 + \cdots, \qquad \alpha \in (-1, 1) \tag{5.38}$$

plays an important role in both theory and applications, and will be considered here in detail. We refer to the filter (5.38) colloquially as the *α-filter*, and to the corresponding family of higher order crossings $\{D_\alpha\}$, $\alpha \in (-1, 1)$, as HOC from the *α-filter*. Exponential smoothing is another term used for (5.38). See Example 3.9 and the discussion about recursive filters preceding it. Observe that

$$\mathcal{L}_\alpha(Z)_t = Z_t + \alpha Z_{t-1} + \alpha^2 Z_{t-2} + \cdots \tag{5.39}$$

We can rewrite (5.39) in its shorter version by defining

$$Z_t(\alpha) = \mathcal{L}_\alpha(Z)_t, \qquad \alpha \in (-1, 1) \tag{5.40}$$

Then (5.39) becomes

$$Z_t(\alpha) = \alpha Z_{t-1}(\alpha) + Z_t, \qquad \alpha \in (-1, 1) \tag{5.41}$$

When $\{Z_t\}$ is a zero-mean stationary Gaussian process then, as we shall see, $\gamma(\alpha)$ is monotone decreasing, and we obtain the spectral representation

$$\cos[\pi\gamma(\alpha)] = \frac{\int_{-\pi}^{\pi} \cos(\omega)|H(\omega; \alpha)|^2\, dF(\omega)}{\int_{-\pi}^{\pi} |H(\omega; \alpha)|^2\, dF(\omega)} \tag{5.42}$$

where

$$|H(\omega;\alpha)|^2 = \frac{1}{1-2\alpha\cos(\omega)+\alpha^2}, \qquad \alpha \in (-1,1), \quad \omega \in [0,\pi] \quad (5.43)$$

Similarly to the expected HOC from differences, we will show that $\gamma(\alpha)$ is equivalent to the normalized spectrum.

The impulse response of the α-filter is readily seen to be

$$h_n(\alpha) = \begin{cases} \alpha^n, & n = 0, 1, 2, \ldots \\ 0, & \text{otherwise} \end{cases}$$

Define,

$$g_n(\alpha) = h_n(\alpha) \otimes h_{-n}(\alpha) = \sum_{m=-\infty}^{\infty} h_m(\alpha) h_{m-n}(\alpha)$$

Then it is not difficult to see that,

$$g_n(\alpha) = \frac{\alpha^{|n|}}{1-\alpha^2}, \qquad n = 0, \pm 1, \pm 2, \ldots$$

Now let $\mathcal{J}_\alpha(\cdot)$ be the linear filter with this impulse response. By operating on ρ_k, we obtain after some algebra,

$$\mathcal{J}_\alpha(\rho)_n = \frac{\alpha^{|n|}}{1-\alpha^2} + \frac{1}{1-\alpha^2}\sum_{t=1}^{\infty}[\alpha^{|n+t|} + \alpha^{|n-t|}]\rho_t$$

Let $\phi_1(\alpha)$ be the *correlation generating function* of $\{Z_t\}$,

$$\phi_1(\alpha) \equiv \sum_{n=0}^{\infty} \rho_n \alpha^n, \qquad |\alpha| < 1$$

It follows that

$$\mathcal{J}_\alpha(\rho)_0 = \frac{1}{1-\alpha^2}[2\phi_1(\alpha) - 1]$$

and

$$\mathcal{J}_\alpha(\rho)_1 = \frac{\alpha\phi_1(\alpha) + \alpha^{-1}\phi_1(\alpha) - \alpha^{-1}}{1-\alpha^2}$$

and therefore by Lemma 5.1

$$\rho_1(\alpha) = \frac{(1+\alpha^2)\phi_1(\alpha) - 1}{\alpha[2\phi_1(\alpha) - 1]}, \qquad |\alpha| < 1 \qquad (5.44)$$

Solving for $\phi_1(\alpha)$ in terms of $\rho_1(\alpha)$ we finally have [8]:

Theorem 5.7. *Suppose the α-filter*

$$\mathcal{L}_\alpha \equiv 1 + \alpha\mathcal{B} + \alpha^2\mathcal{B}^2 + \cdots, \qquad \alpha \in (-1,1)$$

operates on a zero-mean stationary process $\{Z_t\}$. Then the correlation generating function $\phi_1(\alpha)$ of $\{Z_t\}$ is obtained from $\rho_1(\alpha)$ by the equation

$$\phi_1(\alpha) = \frac{1 - \alpha\rho_1(\alpha)}{1 - 2\alpha\rho_1(\alpha) + \alpha^2}, \qquad |\alpha| < 1 \tag{5.45}$$

If in addition $\{Z_t\}$ is Gaussian, then for $|\alpha| < 1$,

$$\begin{aligned}\phi_1(\alpha) &= \frac{1 - \alpha\cos(\pi\gamma(\alpha))}{1 - 2\alpha\cos(\pi\gamma(\alpha)) + \alpha^2} \\ &= [1 - \alpha\cos(\pi\gamma(\alpha))]|H(\pi\gamma(\alpha);\alpha)|^2\end{aligned} \tag{5.46}$$

It follows from (5.46) that in the Gaussian case knowledge of the normalized expected zero-crossing rate $\gamma(\alpha)$ is equivalent to knowing the normalized spectral density

$$\tilde{f}(\omega) = \frac{f(\omega)}{\int_{-\pi}^{\pi} f(\lambda)\, d\lambda}$$

Corollary 5.1. *Let $\{Z_t\}$, $t = 0, \pm 1, \pm 2, \ldots,$ be a zero-mean stationary Gaussian process. For $|\alpha| < 1$ and $\omega \in [0, \pi]$, the following equivalent relations hold:*

$$\gamma(\alpha) \Leftrightarrow \phi_1(\alpha) \Leftrightarrow \rho_1(\alpha) \Leftrightarrow \tilde{f}(\omega)$$

Example 5.7: HOC from the α-Filter

(a) Gaussian White Noise

For $|\alpha| < 1$,

$$\rho_1(\alpha) = \alpha \tag{5.47}$$

$$\gamma(\alpha) = \frac{1}{\pi}\cos^{-1}(\alpha) \tag{5.48}$$

and

$$\phi_1(\alpha) \equiv 1 \tag{5.49}$$

Given any process, we find it very helpful in applications to compare its zero-crossing rate $\gamma(\alpha)$ to that of white noise given in (5.48). See Figures 5.6 and 5.7.

(b) First Order Autoregressive Process

Consider a stationary Gaussian $AR(1)$ process with parameter a_1,

$$Z_t = a_1 Z_{t-1} + \epsilon_t, \qquad t = 0, \pm 1, \ldots$$

where $|a_1| < 1$, and $\{\epsilon_t\}$ is Gaussian white noise. For $|\alpha| < 1$,

$$\rho_1(\alpha) = \frac{\alpha + a_1}{1 + a_1\alpha} \tag{5.50}$$

$$\gamma(\alpha) = \frac{1}{\pi}\cos^{-1}\left(\frac{\alpha + a_1}{1 + a_1\alpha}\right) \tag{5.51}$$

and

$$\phi_1(\alpha) = \frac{1}{1 - a_1\alpha} \tag{5.52}$$

Figure 5.6 gives the graphs of $\gamma(\alpha)$ for $a_1 = -0.6, \ldots, 0, \ldots, 0.6$. The case $a_1 = 0$ (middle curve) corresponds to white noise and is given for reference. The curves above (below) the middle one correspond to negative (positive) a_1.

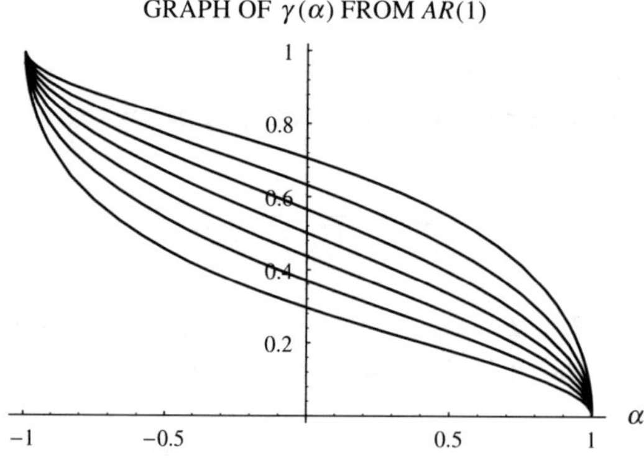

Figure 5.6: *Graph of* $\gamma(\alpha)$ *from* $AR(1)$ *with parameter* $a_1 = -0.6, -0.4, -0.2, 0, 0.2, 0.4, 0.6$. *The upper curve corresponds to* $a_1 = -0.6$, *the one below it corresponds to* $a_1 = -0.4$, *etc.*

(c) Sum of Sinusoids Plus White Noise

Consider the mixed spectrum process

$$Z_t = \sum_{j=1}^{p}(A_j \cos(\omega_j t) + B_j \sin(\omega_j t)) + \epsilon_t$$

where the A's and B's are all independent, $A_j, B_j \sim N(0, \sigma_j^2)$, and $\{\epsilon_t\}$ is white Gaussian noise with mean 0 and variance σ_ϵ^2, independent of the A's and B's. For $|\alpha| < 1$,

$$\rho_1(\alpha) = \frac{\sum_{j=1}^{p}[\sigma_j^2 \cos(\omega_j)/(1 - 2\alpha\cos(\omega_j) + \alpha^2)] + \alpha\sigma_\epsilon^2/(1-\alpha^2)}{\sum_{j=1}^{p}[\sigma_j^2/(1 - 2\alpha\cos(\omega_j) + \alpha^2)] + \sigma_\epsilon^2/(1-\alpha^2)} \tag{5.53}$$

$$\gamma(\alpha) = \frac{1}{\pi}\cos^{-1}(\rho_1(\alpha)) \tag{5.54}$$

and

$$\phi_1(\alpha) = \frac{\sum_{j=1}^{p}[\sigma_j^2(1 - \alpha\cos(\omega_j))/(1 - 2\alpha\cos(\omega_j) + \alpha^2)] + \sigma_\epsilon^2}{\sum_{j=1}^{p}\sigma_j^2 + \sigma_\epsilon^2} \tag{5.55}$$

Figure 5.7 gives the graphs of $\gamma(\alpha)$ in the case of $p = 1$ for several values of the frequency ω_1 when the signal-to-noise ratio is 20 dB (very little noise). The figure also gives the zero-crossing rate of white noise for comparison. Near the origin the graph of $\gamma(\alpha)$ is close to a horizontal line, as is expected in light of our next result.

■

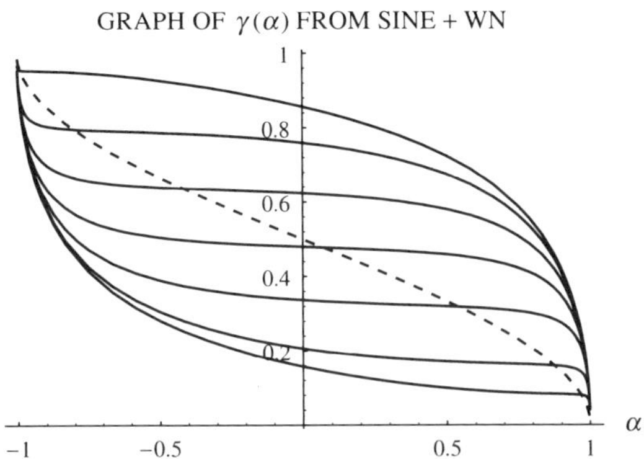

Figure 5.7: *Plots of $\gamma(\alpha)$ from a single sinusoid plus white noise; SNR=20 dB. The plots correspond to frequency ω_1 = 0.25, 0.5, 1, 1.5, 2, 2.5, 3. The intersecting dashed curve is the zero-crossing rate of white noise.*

As a function of $\alpha \in (-1, 1)$, $\gamma(\alpha)$ is *always* monotone decreasing, a fact that makes it useful in comparing different processes. When the process is a pure sinusoid, $\gamma(\alpha)$ is a constant. In the Gaussian case the converse is also true. An indication of the validity of these assertions was already given in the preceding three special cases. Before providing a formal proof we need to prove a lemma.

Lemma 5.2. *Consider the family of filters $\{\mathcal{L}_\alpha(\cdot),\ \alpha \in A\}$, and assume that for each $\alpha \in A$, $\mathcal{L}_\alpha(\cdot)$ has a well-defined inverse $\mathcal{L}^-{}_\alpha(\cdot)$. Suppose that for $\alpha \leq \beta,\ \alpha, \beta \in A$,*

$$G(\omega; \alpha, \beta) = \frac{|H(\omega; \beta)|^2}{|H(\omega; \alpha)|^2}$$

is monotone decreasing in $\omega \in [0, \pi]$. Then

$$\rho_1(\alpha) \leq \rho_1(\beta)$$

If in addition $\{Z_t\}$ is Gaussian, then also

$$E[D_\alpha] \geq E[D_\beta]$$

The inequalities are reversed when $G(\omega; \alpha, \beta)$ is monotone increasing in $\omega \in [0, \pi]$.

Proof. Define $Z_t(\alpha) \equiv \mathcal{L}_\alpha(Z)_t$. Operate on $\{Z_t(\alpha)\}$ with $\mathcal{L}_\beta \mathcal{L}^-{}_\alpha(\cdot)$. The squared gain of this sequential filter is $G(\omega; \alpha, \beta)$, which is assumed to be monotone decreasing. Therefore by Theorem 4.4,

$$\rho_1(\alpha) \leq \rho_1(G)$$

where $\rho_1(G)$ is the first order correlation in the filtered process

$$\mathcal{L}_\beta \mathcal{L}^-{}_\alpha \mathcal{L}_\alpha(Z)_t = \mathcal{L}_\beta(Z)_t = Z_t(\beta)$$

Therefore $\rho_1(G) = \rho_1(\beta)$ and

$$\rho_1(\alpha) \leq \rho_1(G) = \rho_1(\beta)$$

In the Gaussian case this entails,

$$\rho_1(\alpha) = \cos\left(\frac{\pi E[D_\alpha]}{N-1}\right) \leq \rho_1(\beta) = \cos\left(\frac{\pi E[D_\beta]}{N-1}\right)$$

and since $\cos(x)$ is monotone decreasing in $[0, \pi]$ we obtain

$$\frac{\pi E[D_\alpha]}{N-1} \geq \frac{\pi E[D_\beta]}{N-1}$$

The converse can be proved similarly.

□

We now have.

Theorem 5.8. *Let $\{Z_t\}$, $t = 0, \pm 1, \pm 2, \ldots$, be a zero mean stationary Gaussian process, and consider the HOC from the α-filter. Then*

(a) *$\gamma(\alpha)$ is monotone decreasing in $\alpha \in (-1, 1)$.*

(b) *Let $-1 < \alpha < \beta < 1$. Then, $\gamma(\alpha) = \gamma(\beta)$ if and only if $\{Z_t\}$ is a pure sinusoid with probability one.*

Proof. Apply Lemma 5.2 to a ratio in terms of the squared gain of the α-filter,

$$|H(\omega; \alpha)|^2 = \frac{1}{1 - 2\alpha\cos(\omega) + \alpha^2}$$

If we can show that

$$G(\omega; \alpha, \beta) = \frac{|H(\omega; \beta)|^2}{|H(\omega; \alpha)|^2} = \frac{1 - 2\alpha\cos(\omega) + \alpha^2}{1 - 2\beta\cos(\omega) + \beta^2}$$

is *strictly* monotone decreasing, then by Lemma 5.2 we are done. So, for $0 \leq \omega_1 < \omega_2 \leq \pi$ the numerator of the difference

$$G(\omega_2; \alpha, \beta) - G(\omega_1; \alpha, \beta)$$

is equal to

$$2(\cos(\omega_1) - \cos(\omega_2))(1 - \alpha\beta)(\alpha - \beta) < 0$$

Therefore $G(\omega; \alpha, \beta)$ is strictly decreasing. Also, the α-filter has a well-defined inverse, say $\mathcal{L}^-$ (what is it?). Therefore (a) is proved by Lemma 5.2. To prove (b) note that since our filter $\mathcal{L}_\alpha$ has a well-defined inverse that we denote by $\mathcal{L}^-$,

$\{Z_t(\beta)\}$ is obtained from $\{Z_t(\alpha)\}$ by the well-defined linear filter $\mathcal{L}_\beta\mathcal{L}^-{}_\alpha(\cdot)$ and this, we just saw, has a strictly decreasing gain that, by assumption, leaves the first order autocorrelation intact. Therefore, by Corollary 4.2, $\{Z_t(\alpha)\}$ is a pure sinusoid with probability one. By inverse filtering, this implies that $\{Z_t\}$ is also a pure sinusoid with probability one. Conversely, if the process is a pure sinusoid, we certainly have the equality $\gamma(\alpha) = \gamma(\beta)$.

□

5.7 APPENDIX TO CHAPTER 5

Convergence of ν_j in Theorem 5.2

For

$$\nu_j(d\omega) = \frac{\sin^{2j}(\omega/2)\, dF(\omega)}{\int_{-\pi}^{\pi} \sin^{2j}(\lambda/2)\, dF(\lambda)}$$

we want to show that, as $j \to \infty$,

$$\nu_j \Rightarrow \frac{1}{2}\delta_{-\omega^*} + \frac{1}{2}\delta_{\omega^*}$$

where ω^* is the highest frequency in the spectral support, and δ_u is the unit point mass at u (giving probability one to sets that contain u). So [11], *define*

$$|H(\omega)|^2 = \sin^2(\omega/2)$$

Then $|H(\omega)|^2$ is strictly increasing in $[0, \pi]$. For any $\epsilon > 0$ we have,

$$\begin{aligned}
\nu_j[0, \omega^* - \epsilon] &= \frac{\int_0^{\omega^*-\epsilon} |H(\omega)|^{2j}\, dF(\omega)}{\int_{-\omega^*}^{\omega^*} |H(\omega)|^{2j}\, dF(\omega)} \\
&\leq \frac{\int_0^{\omega^*-\epsilon} |H(\omega)|^{2j}\, dF(\omega)}{\int_{\omega^*-\epsilon/2}^{\omega^*} |H(\omega)|^{2j}\, dF(\omega)} \\
&\leq \left[\frac{|H(\omega^* - \epsilon)|^2}{|H(\omega^* - \epsilon/2)|^2}\right]^j \frac{\int_0^{\omega^*-\epsilon} dF(\omega)}{\int_{\omega^*-\epsilon/2}^{\omega^*} dF(\omega)} \\
&\to 0, \qquad j \to \infty
\end{aligned}$$

by the monotonicity of $|H(\omega)|^2$. Therefore by symmetry, as $j \to \infty$,

$$\nu_j[-\omega^* + \epsilon, \omega^* - \epsilon] \to 0$$

But ν_j is a probability such that for all j, $\nu_j[-\omega^*, \omega^*] = 1$. Therefore, as $j \to \infty$,

$$\nu_j[-\omega^*, -\omega^* + \epsilon) \cup (\omega^* - \epsilon, \omega^*] \to 1$$

By the symmetry of ν_j, it converges to a symmetric probability measure supported only at $\pm\omega^*$.

Proof of (b) in Theorem 5.3 We give the proof under the Gaussian assumption. The general case is essentially the same [11]. The underlying idea of the proof is to let N go to infinity first, and only then also let j go to infinity.

First recall the inequality of simple HOC,

$$D_j - 1 \leq D_{j+1}$$

and that D_j is a *sum* of indicators

$$D_j = d_2(j) + d_3(j) + \cdots + d_N(j)$$

where $d_t(j) = [X_t(j) - X_{t-1}(j)]^2$. Because $\{Z_t\}$ is (strictly) stationary, so is the indicator process $\{d_t(j)\}$ associated with the $(j-1)$th difference. Thus, by strict stationarity, the limit

$$\overline{D_j} \equiv \lim_{N\to\infty} \frac{D_j}{N-1}$$

exists with probability one, and is monotone in j. Therefore

$$\lim_{j\to\infty} \overline{D_j} \equiv \overline{D} \leq 1$$

exists with probability one. Also, by dominated convergence

$$E[\overline{D_j}] = \lim_{N\to\infty} \frac{E[D_j]}{N-1} = \frac{E[D_j]}{N-1}$$

because the expected zero-crossing rate is independent of N. However, because $\omega^* = \pi$, we know from Theorem 5.2 that, as $j \to \infty$,

$$E[\overline{D_j}] = \frac{E[D_j]}{N-1} \to \frac{\omega^*}{\pi} = 1$$

Collecting all these statements, and by invoking monotone convergence, we have

$$E[\overline{D}] = \lim_{j\to\infty} E[\overline{D_j}] = \lim_{j\to\infty} \lim_{N\to\infty} \frac{E[D_j]}{N-1} = \lim_{j\to\infty} \frac{E[D_j]}{N-1} = 1$$

However, since $\overline{D} \leq 1$ with probability one, the only way for its expectation to be equal to 1 is that $\overline{D}$ itself is equal to one with probability one.

□

5.8 PROBLEMS AND COMPLEMENTS

1. Consider a zero-mean time series $Z_1, Z_2, \ldots, Z_N$, such that $Z_1 > 0$, and let

$$\hat{\rho}_1 = \frac{\sum_{t=2}^{N} Z_t Z_{t-1}}{\sum_{t=2}^{N} Z_t^2}$$

be the sample estimate of ρ_1. Show that $\hat{\rho}_1 \to 0$ as $Z_1 \to \infty$, but that the zero-crossing count is left unchanged.

2. Consider the first order autoregressive process $Z_t = \phi Z_{t-1} + \epsilon_t$, $t = 0, \pm 1, \ldots$, where $\{\epsilon_t\}$ is Gaussian white noise and $|\phi| < 1$. Let $\{D_j\}$ be the sequence of simple HOC.

(a) Use the long formula (5.28) and the fact that $\rho_k = \phi^k$ to show that

$$\phi = 1 + 2\cos\left(\frac{\pi E[D_2]}{N-1}\right)$$

(b) Discuss the consistency of the estimator

$$\tilde{\phi} = 1 + 2\cos\left(\frac{\pi D_2}{N-1}\right)$$

(c) Run a computer simulation to compare the performance of $\tilde{\phi}$ with

$$\hat{\phi} = \cos\left(\frac{\pi D_1}{N-1}\right)$$

and ([3, ch. 5])

$$\breve{\phi} = \cos\left(\frac{2\pi(\text{number of } 1 - \text{runs})}{N-1}\right)$$

and the least squares estimator

$$\bar{\phi} = \frac{\sum_{t=2}^{N} Z_t Z_{t-1}}{\sum_{t=2}^{N} Z_t^2}$$

Note that $1 - runs$ refer to the runs of consecutive 1s separated by 0s in the clipped binary series $X_1, X_2, \ldots, X_N$.

(d) From part (b), conclude that ϕ can be estimated from the number of peaks only [14].

(e) Discuss the estimation of ϕ from D_3 by appealing to the long formula (5.28). That is, discuss the estimation problem from the number of "inflection" points in $Z_1, Z_2, \ldots, Z_N$.

(f) Can ϕ be estimated from D_j for any j?

3. Let $\{Z_t\}$, $t = 0, \pm 1, \pm 2, \ldots$, be a zero-mean stationary Gaussian process with autocorrelation ρ_k, and let $X_t = 1$ when $Z_t \geq 0$; otherwise $X_t = 0$. Consider the binary time series $X_1, X_2, \ldots, X_N$, and define the "HOC" sequence $\{``D_k"\}$ by,

$$``D_k" = \sum_{t=k+1}^{N} [X_t - X_{t-k}]^2$$

Recall Corollary 2.1, and show that

$$\rho_k = \cos\left(\frac{\pi E[``D_k"]}{N-k}\right)$$

4. Suppose $\{Z_t\}, t = 0, \pm 1, \pm 2, \ldots$, is a zero-mean stationary Gaussian process. Verify the long formula by applying Lemma 5.1. (Hint: Define for $j = 0, 1, \ldots, \theta$,

$$h_j(\theta) = (-1)^j \binom{\theta}{j}$$

Otherwise let $h_j(\theta) = 0$. Then define

$$g_n(\theta) = h_n(\theta) \otimes h_{-n}(\theta) = \sum_j h_j(\theta) h_{j-n}(\theta)$$

to obtain the required filter. Operate with the filter on the autocorrelation ρ_k.)

5. Show that the inverse of the α-filter is given by $1 - \alpha\mathcal{B}$.

6. Suppose $\{Z_t\}, t = 0, \pm 1, \ldots$, is a zero-mean stationary Gaussian process. Assume that π is in the spectral support. It must be then the highest frequency. Prove that the simple HOC satisfy the strict inequality $E[D_m] < E[D_{m+1}]$. (Hint: Suppose equality holds for some m, will the degenerate case addressed by the HOCT ever be achieved? Appeal for an answer to part (b) of Theorem 5.1.)

7. Under the Gaussian assumption, obtain the spectral representation of the HOC from repeated summation. Argue that the representation can be used to obtain the "long formula" corresponding to repeated summation.

8. Let

$$Z_t = \cos(0.8t) - 1.3\cos(1.25t) + \sin(2t), \qquad t = 1, 2, \ldots, 1050$$

   (a) Evaluate $D_1, D_2, \ldots, D_{30}$, from the section $Z_{851}, Z_{852}, \ldots, Z_{1050}$, of length $N = 200$. Show that $\pi D_{30}/199 \approx 2$.

   (b) Evaluate ${}_1D, {}_2D, \ldots, {}_{30}D$, from the section $Z_{851}, Z_{852}, \ldots, Z_{1050}$, of length $N = 200$. Show that $\pi {}_{30}D/199 \approx 0.8$.

9. Consider the sequence of processes $\{X_t(j)\}$, obtained by clipping

$$\{(1 + \mathcal{B})^{j-1} Z_t\}$$

$j = 1, 2, \ldots$. Under what conditions does $\{X_t(j)\}$ approach, as j increases, the degenerate states $\cdots 00000000 \cdots$ or $\cdots 11111111 \cdots$, each with probability $1/2$?

10. Prove Theorem 5.5.

11. Prove Theorem 5.6.

12. Under the Gaussian assumption, show that

$$\frac{\pi E[{}_nD_m]}{N-1} \to \lambda_c$$

as $m, n \to \infty$, such that $m/n = c$. (Hint [5]: Note that $|H|^2$ is symmetric, unimodal at λ_c, and

$$\frac{|H(\lambda_c - \epsilon; m, n)|^2}{|H(\lambda_c - \epsilon/2; m, n)|^2} \to 0, \qquad n \to \infty$$

Then for $\Lambda \subset [0, \pi]$, define,

$$\nu_n(\Lambda) = \frac{\int_\Lambda |H(\omega; m, n)|^2 \, dF(\omega)}{\int_{-\pi}^{\pi} |H(\lambda; m, n)|^2 \, dF(\lambda)}$$

and show that as $n \to \infty$

$$\nu_n \Rightarrow \frac{1}{2}\delta_{-\lambda_c} + \frac{1}{2}\delta_{\lambda_c}$$

Compare with a similar result in the appendix to this chapter.)

13. Verify part (b) of Theorem 4.4 for the α-filter by using the correlation generating function $\phi_1(\alpha)$ (5.45).

(Hint [9]: Differentiate $\phi_1(\alpha)$ using first its definition and then (5.45). Let $\alpha \to 0^+$, and equate the resulting expressions. Argue that the process must satisfy the stochastic difference equation $Z_t - 2\rho_1 Z_{t-1} + Z_{t-2} = 0$.)

14. *Verification of SLT [6].* Verify part (b) of Theorem 5.1 by considering the stationary zero-mean Gaussian $AR(2)$ process

$$Z_t = a_1 Z_{t-1} + a_2 Z_{t-2} + \epsilon_t, \qquad t = 0, \pm 1, \ldots$$

where ϵ_t is white Gaussian noise with mean 0 and variance σ_ϵ^2. Denote the autocorrelation by ρ_k. Show that as $0 \leq E[D_2] - E[D_1] \to 0$,

$$a_1 \to 2\rho_1, \qquad a_2 \to -1, \qquad \sigma_\epsilon^2 \to 0$$

in which case $Z_t - 2\rho_1 Z_{t-1} + Z_{t-2} = 0$. (Hint: Express a_1, a_2 in terms of ρ_1, ρ_2, and use the "long formula" (5.28).)

15. Suppose $\{Z_t\}, t = 0, \pm 1, \pm 2, \ldots$, is a zero-mean stationary Gaussian process, and consider the simple HOC from differences D_1, D_2, D_3. What can be said about $\{Z_t\}$

(a) when $E[D_1] = E[D_3]$?
(b) when $E[D_2] = E[D_3]$?

(Hint: What happens when $E[D_1] = E[D_2]$?)

16. *Autocorrelation of differenced white noise [1, p. 63].* Let $\{u_t\}$ be white noise with variance σ_u^2, and consider the polynomials of degree p, q in $\mathcal{B}$,

$$\mathcal{P}(\mathcal{B}) = a_0 + a_1\mathcal{B} + \cdots + a_p\mathcal{B}^p$$

$$\mathcal{Q}(\mathcal{B}) = b_0 + b_1\mathcal{B} + \cdots + b_q\mathcal{B}^q$$

(a) Prove that $E[\mathcal{P}(\mathcal{B})u_t\mathcal{Q}(\mathcal{B})u_t]$ is the coefficient of x^q in $\sigma_u^2\mathcal{P}(x)x^q\mathcal{Q}(x^{-1})$.
(b) Use this to show that

$$\text{Corr}(\nabla^k u_t, \nabla^k u_{t-j}) = \frac{\binom{2k}{k-j}}{\binom{2k}{k}}(-1)^j \to (-1)^j$$

as $k \to \infty$.

17. *Verification of Theorem 5.4.* Consider the zero-mean Gaussian $AR(2)$ process

$$Z_t = a_1 Z_{t-1} + a_2 Z_{t-2} + \epsilon_t, \qquad t = 0, \pm 1, \ldots$$

where ϵ_t is white Gaussian noise with mean 0 and variance 1. Express the spectral density of Z_t,

$$f(\omega) = \frac{1}{2\pi} \cdot \frac{1}{|1 - a_1 \exp(-i\omega) - a_2 \exp(-i2\omega)|^2}$$

in terms of $E[D_1], E[D_2]$.

REFERENCES

[1] Anderson, T.W., *The Statistical Analysis of Time Series*, New York: Wiley, 1971.

[2] Durst, F., A. Melling, and J.H. Whitelaw, *Principles and Practice of Laser-Dopler Anemometry,* London: Academic Press, 1976.

[3] Kedem, B., *Binary Time Series*, New York: Dekker, 1980.

[4] Kedem, B., "A two-dimensional higher order crossings theorem," *IEEE Trans. Inform. Theory*, Vol. IT-29, No. 1, pp. 159–161, Jan. 1983.

[5] Kedem, B., "On the sinusoidal limit of stationary time series," *Ann. Stat.*, Vol. 12, No. 2, pp. 665–674, 1984.

[6] Kedem, B., "A stochastic characterization of the sine function," *Am. Math. Mon.*, Vol. 93, No. 6, pp. 430–440, June–July 1986.

[7] Kedem, B., "Spectral analysis and discrimination by zero-crossings," *Proc. IEEE*, Vol. 74, No. 11, pp. 1477–1493, Nov. 1986.

[8] Kedem, B. and T. Li. "Higher order crossings from a parametric family of linear filters," Tech. Rep. TR-89-47, Department of Mathematics, Univ. of Maryland, College Park, 1989.

[9] Kedem, B. and T. Li, "Monotone gain, first-order autocorrelation, and expected zero-crossing rate," *Ann. Stat.*, Vol. 19, No. 3, pp. 1672–1676, 1991.

[10] Kedem, B. and E. Slud, "On goodness of fit of time series models: An application of higher order crossings," *Biometrika*, Vol. 68, No. 2, pp. 551–556, Aug. 1981.

[11] Kedem, B. and E. Slud, "Time series discrimination by higher order crossings," *Ann. Stat.*, Vol. 10, No. 3, pp. 786–794, Sept. 1982.

[12] Lomnicki, Z.A. and S.K. Zaremba, "Some applications of zero-one processes," *J. R. Stat. Soc. B*, Vol. 17, No. 2, pp. 243–255, 1955.

[13] Rice, S.O., "Mathematical analysis of random noise," *Bell Syst. Tech. J.*, Vol. 23, pp. 282–332, 1944, and Vol. 24, pp. 46–156, 1945, reprinted in *Selected Papers on Noise and Stochastic Processes*, N. Wax, Ed., New York: Dover, 1954.

[14] Stigler, S.M., "Estimating serial correlation by visual inspection of diagnostic plots," *Am. Stat.*, Vol. 40, No. 2, pp. 111–116, May 1986.

6

Statistical Properties of HOC

Assume that $Z_1, Z_2, \ldots, Z_N$ is a stationary Gaussian time series with zero mean and autocorrelation $\rho_j = E[Z_t Z_{t-j}]/E[Z_t^2]$, and let X_t be the corresponding clipped binary time series, $X_t = I_{[Z_t \geq 0]}$. Then also $d_t \equiv (X_t - X_{t-1})^2$ is binary: $d_t = 1$ if a zero-crossing in discrete time occurs between $t-1$ and t, and $d_t = 0$ otherwise. As in (4.3), the number of zero-crossings D in $Z_1, Z_2, \ldots, Z_N$, is therefore the sum of indicators

$$D = d_2 + d_3 + \cdots + d_N$$

It is convenient to study the statistical properties of D from those of the binary process $\{d_t\}$. For example, from the cosine formula

$$E[d_t] = \frac{1}{\pi} \cos^{-1} \rho_1$$

and hence

$$E[D] = \frac{(N-1)\cos^{-1} \rho_1}{\pi}$$

We shall follow the same rule as before, and suppress N in the notation for the number of zero-crossings D.

In this chapter we embark upon the study of the sampling properties of HOC. Some of our results are quite technical in nature, catering to mixed spectra and their peculiarities, but unfortunately this cannot be avoided. To ease the presentation we state the main facts in the form of theorems. The theorems cover cases that are likely to arise in practice, and are thus of prime interest.

6.1 ASYMPTOTIC NORMALITY

The exact distribution of the number of zero-crossings D is quite intractable, unless one assumes special structures for X_t. The less ambitious, but more practical, asymptotic distribution is obtainable under certain conditions on the autocorrelation. We present two such results.

Consider the moment function

$$\begin{aligned}\kappa_x(i,j,k) &\equiv E[X_t - 1/2][X_{t+i} - 1/2][X_{t+j} - 1/2][X_{t+k} - 1/2] \\ &\quad -\{R_x(i)R_x(j-k) + R_x(j)R_x(i-k) + R_x(k)R_x(i-j)\}\end{aligned}$$

This is the fourth order cumulant function of $\{X_t\}$ discussed in Chapter 3 (see also [1, p. 444]). When ρ_k is absolutely summable, so is $\kappa_x(1,-j,1-j)$ as a function of j ([6, p. 40]). The following theorem is from [6, pp. 66–68].

Theorem 6.1. *Let* $\{Z_t\}$, $t = 0, \pm1, \pm2, \ldots$, *be a zero-mean stationary Gaussian process with autocorrelation* ρ_j. *If* $\sum_{j=-\infty}^{\infty} |\rho_j| < \infty$, *then*

$$\sum_{j=-\infty}^{\infty} |\kappa_x(1,-j,1-j)| < \infty$$

and

$$\frac{D - E[D]}{\sqrt{N}} \xrightarrow{\mathcal{L}} \mathcal{N}(0, \sigma_1^2), \qquad N \to \infty$$

where

$$\sigma_1^2 = \frac{1}{\pi^2} \sum_{j=-\infty}^{\infty} \{(\sin^{-1}\rho_j)^2 + \sin^{-1}\rho_{j-1}\sin^{-1}\rho_{j+1} + 4\pi^2\kappa_x(1,-j,1-j)\}$$

The proof of Theorem 6.1 is sketched in the appendix to this chapter. The same result holds for any D_θ provided the autocorrelation of the filtered data is absolutely summable. The difficulty with Theorem 6.1 is that in general we cannot get hold of the term due to the fourth order cumulant function. There are, however, special cases where this term has a closed form. For a careful discussion of this see [1, ch. 8].

The next result gives a useful lower bound for σ_1^2, under a somewhat weaker assumption than absolute summability of the autocorrelation.

Theorem 6.2. [22] *Let* $\{Z_t\}$, $t = 0, \pm1, \pm2, \ldots$, *be a zero-mean stationary Gaussian process with autocorrelation* ρ_j, $\mathrm{Var}[Z_0] = 1$, *and square integrable spectral density* f. *Then*

$$\frac{D - E[D]}{\sqrt{N}} \xrightarrow{\mathcal{L}} \mathcal{N}(0, \sigma_1^2), \qquad N \to \infty$$

where σ_1^2 satisfies

$$\sigma_1^2 \geq \frac{4}{\pi(1-\rho_1^2)} \int_{-\pi}^{\pi} |\rho_1 - \cos(\omega)|^2 f^2(\omega)\, d\omega > 0$$

Although Theorem 6.2 supplies an elegant lower bound for the asymptotic variance of D, in practice we need to know f in order to obtain this approximation. An expression for σ_1^2 as an infinite sum is given in [4].

Continuous Time Central Limit Theorem

We mention for completeness the continuous time version of Theorem 6.2 as given in [22].

Assume that $\{Z(t)\}$, $-\infty < t < \infty$, is a zero-mean stationary Gaussian process with $\mathrm{Var}[Z(t)] = 1$, and correlation function $\rho(t)$. According to [22], if $D(T)$ denotes the number of zero-crossings in the time interval $[0, T]$, and $\rho(t), \rho''(t)$ are square integrable on the real line, then

$$\frac{1}{\sqrt{T}} \left(D(T) - T\frac{\sqrt{-\rho''(0)}}{\pi} \right) \xrightarrow{\mathcal{L}} \mathcal{N}(0, \sigma_2^2), \qquad T \to \infty$$

where

$$\sigma_2^2 \geq \frac{-\rho''(0)}{\pi} \int_{-\infty}^{\infty} \left(1 + \frac{\omega^2}{\rho''(0)}\right)^2 f^2(\omega)\, d\omega > 0$$

Similar results proved under various conditions can be found in [2], [3], [19]. In particular, it is shown in [3] under some regularity conditions, including the square integrability of $\rho(t)$ and $\rho''(t)$, that the asymptotic variance σ_2^2 is given by

$$\sigma_2^2 = \frac{1}{\pi}\sqrt{-\rho''(0)} + \frac{2}{\pi} \int_0^{\infty} \left\{ \frac{E[|Z'(0)Z'(t)| \,\big|\, Z(0) = Z(t) = 0]}{2\sqrt{1-\rho^2(t)}} + \frac{\rho''(0)}{\pi} \right\} dt$$

6.2 VARIANCE APPROXIMATION

From Theorems 6.1 and 6.2, and from the continuous time counterparts, we can see that the asymptotic variance of the number of zero-crossings is rather problematic. In this section we present ways for getting some quick practical approximations to the variance of D in discrete time.

6.2.1 A Markov Approximation

Denote the one and two step conditional probabilities of $\{X_t\}$ by

$$\lambda_j \equiv P(X_t = 1 | X_{t-j} = 1), \qquad j = 1, 2$$

Then $E[d_t] = 1 - \lambda_1$, and $\text{Var}[d_t] = \lambda_1(1 - \lambda_1)$. It follows that

$$\begin{aligned}\text{Var}[D] &= \sum_{t=2}^{N} \text{Var}[d_t] + \sum_{s \neq t} \text{Cov}[d_s, d_t] \\ &= (N-1)\lambda_1(1-\lambda_1) + 2\sum_{s<t} \text{Cov}[d_s, d_t] \qquad (6.1)\end{aligned}$$

The source of difficulty in obtaining a simple expression for $\text{Var}[D]$ is that in general, no simple expression for $\text{Cov}[d_s, d_t]$ in terms of first and second order quantities exists. This is contrary to the fact that $\text{Var}[d_t] = \text{Cov}[d_t, d_t]$ does have such an expression in terms of λ_1. Fortunately, there is a useful approximation to the whole sum $2\sum_{s<t} \text{Cov}[d_s, d_t]$, and hence also to $\text{Var}[D]$, under a certain assumption.

Experience shows that in many cases the dependence in $\{d_t\}$ is rather weak as expressed by very low correlations unless the original series $\{Z_t\}$ displays very large absolute correlations. Thus, as a first order approximation, we shall treat $\{d_t\}$ as a first order two state (0 or 1) stationary Markov chain. Recall that a process $\{Y_t\}$, $t = 0, \pm 1, \pm 2, \ldots$, with a finite or countably infinite number of possible states (possible values) is a Markov chain if

$$P(Y_{t_n} = y_n | Y_{t_{n-1}} = y_{n-1}, \ldots, Y_{t_1} = y_1) = P(Y_{t_n} = y_n | Y_{t_{n-1}} = y_{n-1})$$

where $t_1 < \cdots < t_{n-1} < t_n$. This immediately leads to a computable expression for $\text{Var}[D]$, which seems to hold surprisingly well for continuous spectrum processes [9].

Under the Markov assumption, the two parameters needed to specify the probabilistic behavior of $\{d_t\}$ are

$$p = P(d_t = 1) = 1 - \lambda_1$$

and

$$\nu_1 = P(d_t = 1 | d_{t-1} = 1) = \frac{1 - 2\lambda_1 + \lambda_2}{2(1 - \lambda_1)}$$

By using standard arguments from the theory of Markov chains (see [5, ch. 2], [13]) one obtains,

$$\text{Var}[D] \approx (N-1)pq + \frac{2pq(\nu_1 - p)}{1 - \nu_1}[(N-1) - V] \qquad (6.2)$$

where $q = 1 - p$, and

$$V = \frac{q[1 - ((\nu_1 - p)/q)^{N-1}]}{1 - \nu_1}$$

In most cases V is negligible compared with N and can be safely omitted, a fact leading to the useful approximation,

$$
\begin{aligned}
\text{Var}[D] &\approx (N-1)pq + (N-1)\frac{2pq(\nu_1 - p)}{1-\nu_1} \\
&= (N-1)pq\left[\frac{1-2p+\nu_1}{1-\nu_1}\right]
\end{aligned} \tag{6.3}
$$

Moreover, since $(N-1)\lambda_1(1-\lambda_1) = (N-1)pq$, we can see from (6.1) and (6.3) that

$$
\frac{2pq(\nu_1 - p)}{1-\nu_1}(N-1) \tag{6.4}
$$

provides the desired approximation for $2\sum_{s<t}\text{Cov}[d_s, d_t]$.

This idea can be extended to HOC sequences $\{D_\theta\}$ obtained from series of size N. For example, consider the simple HOC from repeated differences. Then $X_t^{(k)} = I_{[\nabla^{k-1}Z_t \geq 0]}$, and we let $d_t^{(k)} = (X_t^{(k)} - X_{t-1}^{(k)})^2$,

$$
\lambda_j^{(k)} \equiv P(X_t^{(k)} = 1 | X_{t-j}^{(k)} = 1), \qquad j, k = 1, 2, 3, \ldots
$$

and

$$
\nu_1^{(k)} = P(d_t^{(k)} = 1 | d_{t-1}^{(k)} = 1) = \frac{1 - 2\lambda_1^{(k)} + \lambda_2^{(k)}}{2(1-\lambda_1^{(k)})}
$$

The extension of (6.3) is given by

$$
\text{Var}[D_k] \approx (N-1)p^{(k)}q^{(k)}\left[\frac{1-2p^{(k)}+\nu_1^{(k)}}{1-\nu_1^{(k)}}\right] \tag{6.5}
$$

where $p^{(k)} \equiv P(d_t^{(k)} = 1) = 1 - \lambda_1^{(k)}$.

The transition probabilities $\lambda_1^{(k)}, \lambda_2^{(k)}$ are related to the autocorrelation by the cosine formula. From the definition

$$
\rho_j^{(k)} \equiv \text{Corr}(\nabla^k Z_t, \nabla^k Z_{t-j}), \qquad j, k = 0, 1, 2, \ldots
$$

and the cosine formula, we obtain

$$
\lambda_j^{(k)} \equiv P(X_t^{(k)} = 1 | X_{t-j}^{(k)} = 1) = \frac{1}{2} + \frac{1}{\pi}\sin^{-1}\rho_j^{(k-1)} \tag{6.6}
$$

This permits shifting from ρ's to λ's and vice versa whenever necessary. Observe that $\rho_j = \rho_j^{(0)}$.

Given the autocorrelation sequence $\rho_1, \rho_2, \ldots, \rho_K$, the approximation for $\text{Var}[D_k]$, $k = 1, 2, \ldots, K-1$, given in (6.5) can be computed by the following algorithm (see Problem 5 in Chapter 8).

Algorithm 6.1

1. Computation of $\rho_1^{(k)}$,$\rho_2^{(k)}$, $k = 0, 1, \ldots, K-1$:
   $\rho_1^{(k)} = \rho_1(k+1)$ is obtained from the long formula (5.27), and

$$\rho_2^{(k)} = -1 + 2\rho_1^{(k)} - 2\rho_1^{(k+1)}(1 - \rho_1^{(k)})$$

2. Computation of $\lambda_1^{(k)}, \lambda_2^{(k)}, k = 1, 2, \ldots, K$:

$$\lambda_1^{(k)} = \frac{1}{2} + \frac{1}{\pi}\sin^{-1}\rho_1^{(k-1)}$$
$$\lambda_2^{(k)} = \frac{1}{2} + \frac{1}{\pi}\sin^{-1}\rho_2^{(k-1)}$$

3. Computation of $p^{(k)}, \nu_1^{(k)}, k = 1, 2, \ldots, K$:

$$p^{(k)} = 1 - \lambda_1^{(k)}, \qquad \nu_1^{(k)} = \frac{1 - 2\lambda_1^{(k)} + \lambda_2^{(k)}}{2(1 - \lambda_1^{(k)})}$$

4. Substitute $p^{(k)}$, $q^{(k)} = 1 - p^{(k)}$, $\nu_1^{(k)}$ in (6.5) to obtain the Markov approximation for $\mathrm{Var}[D_k]$, $k = 1, 2, \ldots, K - 1$.

Table 6.1 illustrates the quality of approximation obtained by the use of (6.5). The table compares $\sqrt{\mathrm{Var}[D_k]}$, $k = 1, \ldots, 6$, $N = 1000$, as obtained from (6.5) (actually, the figures in the table were computed by including also the V term as in (6.2), replacing p, q, ν_1 by $p^{(k)}, q^{(k)}, \nu_1^{(k)}$, but this can be ignored as the V values were mostly around 0.9) with a sample estimate $\sqrt{\hat{\mathrm{V}}\mathrm{ar}[D_k]}$, $k = 1, \ldots, 6$, obtained from 100 independent realizations each of length $N = 1000$ generated by certain Gaussian autoregressive moving average (*ARMA*) models. As a check on the simulation, the table also compares the exact $E[D_k]$ with its estimate $\hat{E}[D_k]$ as obtained from the average of D_k from the same 100 realizations. Both $E[D_k]$ and $\hat{E}[D_k]$ are rounded to the nearest integer.

As seen from the table, an interesting feature of the square root of the variance approximation (6.5) is that as a function of k, $1 \le k \le 6$, it increases or decreases very much like the sample estimate $\sqrt{\hat{\mathrm{V}}\mathrm{ar}[D_k]}$, and that the two are rather close.

It should be noted that the *ARMA* models listed in Table 6.1 are all continuous spectrum models. It has, however, been observed that the Markov approximation (6.5) need not hold for discrete spectrum processes. In fact, experience shows that in many cases (6.5) tends to underestimate the true variance in discrete and mixed spectrum time series.

6.2.2 The Case of m-Dependence

Another way to obtain an approximation for the variance of simple HOC, is to investigate the limit of $\mathrm{Var}[D_k]$ as k increases, but holding N fixed. Under a certain assumption on the dependence structure in $\{Z_t\}$ it will be shown that, for each fixed N, $\mathrm{Var}[D_k]$ behaves asymptotically as $k \to \infty$ like $(N-1)\lambda_1^{(k)}(1 - \lambda_1^{(k)})$. This is in general the case when $\mathrm{Cov}(d_t^{(k)}, d_s^{(k)})$ decreases faster than $\lambda_1^{(k)}$.

TABLE 6.1 Comparison of $\sqrt{\text{Var}[D_k]}$ from (6.5) with $\sqrt{\hat{\text{V}}\text{ar}[D_k]}$ from 100 Realizations.

Series	k	$E[D_k]$	$\hat{E}[D_k]$	Markov Approx. $\sqrt{\text{Var}[D_k]}$	Sample Approx. $\sqrt{\hat{\text{V}}\text{ar}[D_k]}$
White noise	1	500	497	15.81	15.96
	2	666	666	13.15	13.63
	3	732	732	12.16	12.53
	4	769	770	11.57	11.49
	5	794	795	11.18	11.05
	6	813	814	10.82	10.00
$MA(1)$	1	369	371	14.17	14.59
$\theta = -0.5$	2	553	551	12.44	12.27
	3	640	639	11.91	12.48
	4	694	693	11.66	12.22
	5	732	731	11.44	11.40
	6	759	759	11.22	11.02
$AR(2)$	1	424	425	9.64	9.67
$\phi_1 = 0.4$	2	484	485	9.38	9.13
$\phi_2 = -0.7$	3	536	537	10.29	10.81
	4	594	594	11.27	12.72
	5	651	652	11.87	12.02
	6	702	701	12.04	11.34
$ARMA(2,2)$	1	884	883	10.04	10.51
$\phi_1 = -1.4$	2	897	897	9.20	9.53
$\phi_2 = -0.5$	3	903	903	8.84	9.01
$\theta_1 = 0.2$	4	908	908	8.60	8.50
$\theta_2 = 0.1$	5	911	911	8.43	8.47
	6	914	914	8.29	8.38

Note: $N = 1000$.

To clarify the last statement, recall that by the higher order crossings theorem (Theorem 5.3), when π is included in the spectral support, $X_t^{(k)}$ approaches the degenerate state where a 0 follows a 1 and vice versa, and hence $\lambda_1^{(k)} \to 0$, as $k \to \infty$. It follows that if we can show that[1] $\text{Cov}(d_t^{(k)}, d_s^{(k)}) = o(\lambda_1^{(k)})$, then by (6.1) $\text{Var}[D_k]$ is asymptotically equivalent to $(N-1)\lambda_1^{(k)}(1-\lambda_1^{(k)})$, as $k \to \infty$. This is the subject of the next result stated for m-dependent processes.

[1] $\text{Cov}(d_t^{(k)}, d_s^{(k)}) = o(\lambda_1^{(k)})$ means that $|\text{Cov}(d_t^{(k)}, d_s^{(k)})|/\lambda_1^{(k)} \to 0$, as $\lambda_1^{(k)} \to 0$.

Definition. A stationary process $\{Z_t\}$, $t = 0, \pm 1, \pm 2, \dots$, is called an *m-dependent* process if there exists a nonnegative integer m such that for any integer t, the sequence $\dots, Z_{t-1}, Z_t$, is independent of $Z_{t+m+1}, Z_{t+m+2} \dots$. □

By this definition, white noise and its finite differences are m-dependent processes. If $\{u_t\}$ is white Gaussian noise, then $\{u_t\}$ is 0-dependent, and $\{\nabla^k Z_t\}$ is k-dependent. More generally, the $MA(q)$ process, $Z_t = u_t - \theta_1 u_{t-1} - \dots - \theta_q u_{t-q}$, is m-dependent with $m = q$. Problem 7 describes a method for generating m-dependent processes from continuous spectrum Gaussian processes. The central limit theorem for m-dependent stationary processes is proved in [1, p. 427] . This means that partial sums from an m-dependent process are asymptotically normally distributed.

Theorem 6.3. [11] *Let $\{Z_t\}$, $t = 0, \pm 1, \pm 2, \dots$, be an m-dependent stationary Gaussian process with π in the spectral support. Let $\{D_k\}$, $k = 1, 2, \dots$, be the simple HOC from differences. Then for fixed N,*

$$\lim_{k \to \infty} \frac{\text{Var}[D_k]}{(N-1)\lambda_1^{(k)}(1 - \lambda_1^{(k)})} = 1$$

The proof of Theorem 6.3 is given in the appendix to this chapter. It is instructive to identify exactly what we need to establish. To do so, we require the following two lemmas concerning axis-crossings by Gaussian sequences (not necessarily m-dependent).

Lemma 6.1. *Let $\{Z_t\}$, $t = 0, \pm 1, \pm 2, \dots$, be a zero-mean stationary Gaussian process. Then uniformly in k, s, t,*

$$\text{Cov}(X_s^{(k)}, d_t^{(k)}) = 0$$

Proof. See Problem 4. □

Lemma 6.2. *Let $\{Z_t\}$, $t = 0, \pm 1, \pm 2, \dots$, be a zero-mean stationary Gaussian process. Then the following two-sided inequality holds,*

$$\max\left(\frac{\lambda_{t-1}^{(k)}}{4} - \frac{\lambda_{t-3}^{(k)}}{4} - \frac{\lambda_1^{(k)}}{2} + \frac{(\lambda_1^{(k)})^2}{2}, -\frac{\lambda_{t-1}^{(k)}}{4} + \frac{\lambda_{t-3}^{(k)}}{4} - \frac{\lambda_1^{(k)}}{2} + \frac{(\lambda_1^{(k)})^2}{2}\right)$$

$$\leq \text{Cov}(d_2^{(k)}, X_t^{(k)} X_{t-1}^{(k)}) \leq \frac{(\lambda_1^{(k)})^2}{2}$$

Proof. Because $\{Z_t\}$ is Gaussian, it is sufficient to consider $k = 1$ only, in which case it is convenient to suppress k. First note that

$$\mathrm{Cov}(d_2, X_t X_{t-1}) = E[d_2 X_t X_{t-1}] - E[d_2]E[X_t X_{t-1}]$$
$$\leq E[X_t X_{t-1}]\{1 - E[d_2]\} = \frac{\lambda_1^2}{2}$$

Observe that $d_2 = X_1^2 + X_2^2 - 2X_1X_2 = X_1 + X_2 - 2X_1X_2$. For the left-hand side we have four cases corresponding to the removal of either X_1 or X_2 or X_t or X_{t-1}. Suppose we remove X_1. Then by Theorem 4.3,

$$-2E[X_1X_2X_{t-1}X_t] \geq -2E[X_2X_{t-1}X_t]$$
$$= -2\left(-\frac{1}{4} + \frac{\lambda_{t-3}}{4} + \frac{\lambda_{t-2}}{4} + \frac{\lambda_1}{4}\right)$$

from which follows that,

$$\mathrm{Cov}(d_2, X_tX_{t-1}) \geq \left(-\frac{1}{4} + \frac{\lambda_{t-1}}{4} + \frac{\lambda_{t-2}}{4}\right) + \left(-\frac{1}{4} + \frac{\lambda_{t-2}}{4} + \frac{\lambda_{t-3}}{4}\right)$$
$$+ \left(\frac{1}{2} - \frac{\lambda_{t-3}}{2} - \frac{\lambda_{t-2}}{2} - \frac{\lambda_1}{2} + \frac{\lambda_1^2}{2}\right)$$
$$= \frac{\lambda_{t-1}}{4} - \frac{\lambda_{t-3}}{4} - \frac{\lambda_1}{2} + \frac{\lambda_1^2}{2}$$

and similarly for the other three cases.

□

From Lemma 6.1 $\mathrm{Cov}(d_s^{(k)}, d_t^{(k)}) = \mathrm{Cov}(d_s^{(k)}, X_t^{(k)}X_{t-1}^{(k)})$. Therefore, Theorem 6.3 is proved if we can show that $\mathrm{Cov}(d_s^{(k)}, X_t^{(k)}X_{t-1}^{(k)}) = o(\lambda_1^{(k)})$, for arbitrary $s \neq t$. For this, by Lemma 6.2, it suffices to show after dropping the quadratic term $\frac{1}{2}(\lambda_1^{(k)})^2$ that for $t \neq 2$,

$$\lim_{k\to\infty} \frac{\lambda_{t-1}^{(k)}/4 - \lambda_{t-3}^{(k)}/4 - \lambda_1^{(k)}/2}{\lambda_1^{(k)}} = 0 \tag{6.7}$$

or the same result with the signs of the first two terms changed. This is what is shown in the appendix to this chapter.

6.2.3 A White Noise Test

Suppose $\{Z_t\}$ is white Gaussian noise consisting of uncorrelated $\mathcal{N}(0, \sigma^2)$ random variables. For such a process we expect D_1 to fall within reasonable distance from $\frac{1}{2}N$, and that D_2 falls within reasonable distance from $\frac{2}{3}N$. In general, D_k admits values around its expected value $E[D_k] = (N-1)(1-\lambda_1^{(k)})$. But how close D_k and $E[D_k]$ are, is determined to a large extent by $\mathrm{Var}[D_k]$.

From (6.6) and the fact that (Problem 16 in Chapter 5, or the long formula)

$$\rho_1^{(k)} = -\frac{k}{k+1}$$

we obtain

$$\lambda_1^{(k)} = \frac{1}{2} - \frac{1}{\pi}\sin^{-1}(\frac{k-1}{k}) \tag{6.8}$$

and therefore,

$$E[D_k] = (N-1)\left\{\frac{1}{2} + \frac{1}{\pi}\sin^{-1}(\frac{k-1}{k})\right\} \tag{6.9}$$

for $k = 1, 2, 3, \ldots$. Also, from Theorem 6.3, for large k

$$\text{Var}[D_k] \sim (N-1)\lambda^{(k)}(1-\lambda^{(k)}) \tag{6.10}$$

Observe that (6.10) is exact for $k = 1$, because for Gaussian white noise $\text{Var}[D_1] = (N-1)/4$. With $N = 1000$, the values of $\{(N-1)\lambda^{(k)}(1-\lambda^{(k)})\}^{1/2}$ for $k = 1, \ldots, 6$, are 15.80, 14.90, 14.00, 13.30, 12.80, 12.31, respectively, and this is not out of line with the white noise results in Table 6.1. Thus for white noise the approximation (6.10) is of value also for small $k > 1$. This, and the fact that $\{\nabla^k Z_t\}$ is k-dependent yield approximate probability limits for D_k when N is large. By combining (6.8), (6.9), and (6.10), and appealing to the central limit theorem for m-dependent processes, the probability that D_k falls within the limits

$$(N-1)\left\{\frac{1}{2} + \frac{1}{\pi}\sin^{-1}(\frac{k-1}{k})\right\}$$
$$\pm 1.96(N-1)^{1/2}\left\{\frac{1}{4} - [\frac{1}{\pi}\sin^{-1}(\frac{k-1}{k})]^2\right\}^{1/2} \tag{6.11}$$

is approximately 95% for each k. This fact can be used for testing whether a time series oscillates as Gaussian white noise. The proposed test is to reject the hypothesis of white noise when at least one D_k, $k = 1, 2, \ldots, K$, for some K, falls outside the limits (6.11) [11].

In passing we note that under white noise, as $k \to \infty$, $\rho_1^{(k)} \to -1$, $\lambda_1^{(k)} \to 0$, and $E[D_k] \to N-1$, in agreement with the HOCT.

Example 6.1: Annual Mean Air Temperature [10], [11]

Figure 6.1 gives the annual mean air temperature from 1781 to 1980 at Hohenpeissenberg, Germany. The values for the years 1811 and 1812, were missing, and were replaced by the mean of neighboring observations. The actual data set is given in Table A.2 in Appendix A. With $N = 180$, the D_k, $k = 1, 2, \ldots, 8$, were obtained from the centered series, and are plotted (dashed curve) in Figure 6.2 together with the white noise probability limits (6.11) (solid curves). The D_k are well within the limits (6.11). From this point of view, the series oscillates pretty much like Gaussian white noise.

■

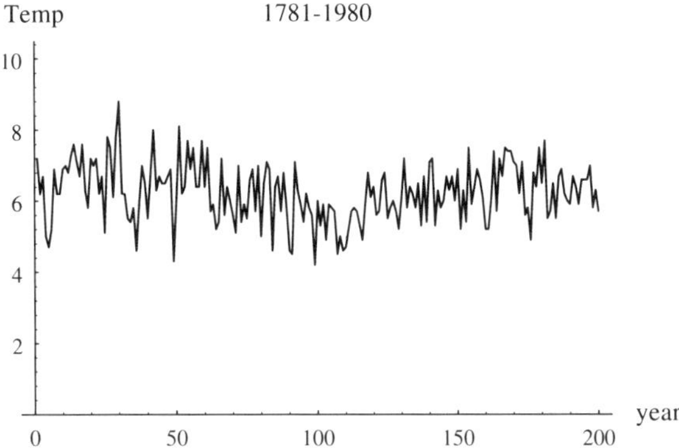

Figure 6.1: *Annual mean air temperature 1781–1980:* $Mean = 6.23deg$, $Variance = 0.69$. (*Source: Report 155 of the Deutschen Wetterdienstes, Germany, 1981.*)

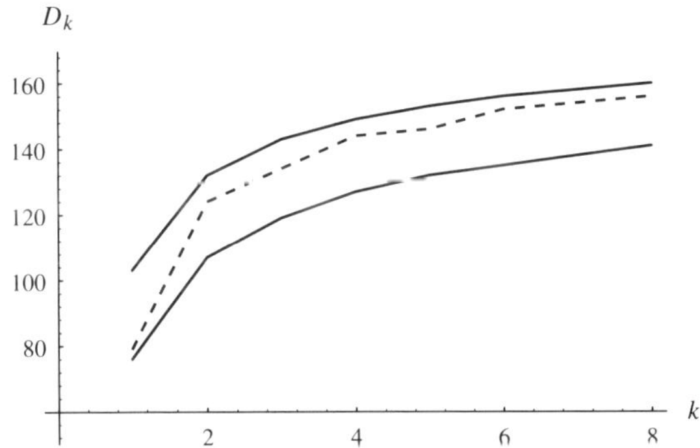

Figure 6.2: *Upper and lower probability limits under white noise* (*solid curves*) *and the HOC from the temperature series* (*dashed curve*). *The series oscillates as white noise.*

6.3 ASYMPTOTIC MIXED SPECTRUM RESULTS

Since we assume that $\{Z_t\}$ is strictly stationary, so is $\{X_t\}$, the latter being bounded, and therefore $D/(N-1)$ always converges to a random variable with probability one, regardless of the type of the spectrum ([5, p. 483]). Under what conditions does the sample zero-crossing rate $D/(N-1)$ converge to a *constant* as the series size N increases? That is, we want to know conditions under which the limiting random variable is a constant (a.s.). This basic question is intimately tied up with the asymptotic variance of $D/(N-1)$. As we saw above under certain conditions, in the continuous spectrum case $\text{Var}[D/(N-1)]$ decreases as $1/N$, and hence $D/(N-1)$ does converge in mean square (also almost surely by uniqueness of the limit) to its expected value (constant) $\arccos(\rho_1)/\pi$. The situation, however, is not that simple in the mixed spectrum case. As we shall see, under the Gaussian assumption, two or more jumps in the discrete spectral component prevent $\text{Var}[D/(N-1)]$ from vanishing as N grows. Things are not as decisive for a single jump, but we will study two important cases when the variance of the sample zero-crossing rate does vanish asymptotically.

The mixed spectrum predicament in the Gaussian case can be remedied if the process is transformed into a *random phases* process by *fixing* the random amplitudes of the discrete sinusoidal components. Conditional on the amplitudes, the process becomes *ergodic* and the empirical zero-crossing rate converges to a constant with probability one.

6.3.1 The Discrete Spectrum of d_t

For the material of this subsection, assume for simplicity and notational convenience that $\text{Var}(Z_t) = 1$, and let σ, σ_d, be the spectral measures corresponding to $\{Z_t\}$ and $\{d_t\}$, respectively. Then

$$\rho_k = \int_{-\pi}^{\pi} e^{ik\lambda}\sigma\,(d\lambda)$$

and we observe that σ becomes a probability measure on $[-\pi, \pi]$, and

$$\text{Cov}(d_t, d_0) = \int_{-\pi}^{\pi} e^{it\lambda}\sigma_d\,(d\lambda)$$

Note that σ_d is not normalized, and is therefore not a probability measure.

Pursuing the investigation of the variance of the number of zero-crossings, our next goal is to determine the asymptotic variance as $N \to \infty$ of the sample (i.e., observed or empirical) zero-crossing rate

$$\hat{\gamma} = \frac{1}{N-1}\sum_{t=2}^{N} d_t$$

in the general mixed spectrum case. That is, when the data contain sinusoidal components, endowing σ with a discrete spectral component. From the mean square ergodic theorem in Chapter 3, all that we need is to determine $\sigma_d(\{0\})$, because

$$\lim_{N\to\infty} \text{Var}[\hat{\gamma}] = \sigma_d(\{0\})$$

As it turns out, when the spectrum is of a mixed type, $\sigma_d(\{0\})$ is in general positive, and therefore the asymptotic variance does not vanish, except for some particular cases. This means that as an estimator of its expected value, $\hat{\gamma}$ need not be a consistent estimator under the Gaussian assumption.

Intuitively, there should be a connection between σ and σ_d, and the fact that the first is of a mixed type emboldens one to believe that so is the latter. The connection is rather subtle and is given in the following theorem.

Theorem 6.4. [12], [22] *Let* $\{Z_t;\ t = 0, \pm 1, \pm 2, \ldots\}$ *be a real-valued zero-mean stationary Gaussian process with normalized spectral measure* σ. *The discrete spectral masses corresponding to the points of jump in the spectrum of* $\{d_t\}$ *are given by*

$$\sigma_d(\{\omega\}) = \frac{1}{\pi^2} \sum_{m=1}^{\infty} \frac{1}{(2m)!} E\left\{ \left(\sum_{j=0}^{2m} S_{2m}^j C_{m,j}(\rho_1) \right)^2 I_{[\Lambda_1 + \cdots + \Lambda_{2m} = \omega]} \right\} \tag{6.12}$$

where the random variables Λ_j *are independent with distribution* σ *on* $[-\pi, \pi]$, *and addition is modulo* 2π; *where* $\{C_{m,j}(\cdot)\}$ *denotes a special family of hypergeometric functions determined for* $0 \le j \le 2m$ *by*

$$C_{m,j}(x) = C_{m,2m-j}(x)$$

$$C_{m+1,j}(x) = 2(m-j)C_{m,j}(x) + C_{m+1,j+2}(x)$$

$$C_{m,1}(x) = -\frac{(2m-2)!}{(m-1)!2^{m-1}}(1-x^2)^{-m+1/2}$$

$$C_{m,2}(x) = \frac{(2m-2)!}{(m-1)!2^{m-1}} x(1-x^2)^{-m+1/2} = -xC_{m,1}(x)$$

and where S_{2m}^j *is a function of* $\Lambda_1, \ldots, \Lambda_{2m}$, *defined by*

$$S_{2m}^0 = 1$$

$$S_{2m}^j = \sum \exp[i(\Lambda_{k_1} + \cdots + \Lambda_{k_j})], \qquad 1 \le j \le 2m$$

and the summation is over all distinct j-element subsets $\{k_1, \ldots, k_j\} \subset \{1, \ldots, 2m\}$.

The proof of this result is rather technical and is sketched briefly in the appendix to this chapter to satisfy the curiosity of the ardent reader. It may, however, be skipped quite safely upon first reading. We only mention that the proof depends on the remarkable fact that d_t admits the representation derived in [22],

$$d_t = \frac{1}{2} - \frac{1}{\pi} \sum_{m=0}^{\infty} (-1)^m I_{2m} \left(\exp[it(\lambda_1 + \cdots + \lambda_{2m})] \sum_{j=0}^{2m} S_{2m}^j C_{m,j}(\rho_1) \right) \tag{6.13}$$

where I_k are certain integral operators referred to as the Wiener–Itô integral operators (see [18] and also Problem 23 in Chapter 3). Now, Theorem 6.4 leads to the curious fact that $\sigma_d(\{0\})$ is positive whenever the data contain at least two sinusoidal components.

Theorem 6.5. [12] *Let* $\{Z_t;\ t = 0, \pm 1, \pm 2, \ldots\}$ *be a real-valued zero-mean stationary Gaussian process with normalized spectral measure* σ. *If* σ *has at least two* *points of jump (atoms) in* $[0, \pi]$, *then* $\sigma_d(\{0\}) > 0$. *That is, the spectrum of* $\{d_t\}$ *has a jump at* 0.

Proof. Let ω_1 and ω_2 be two distinct atoms of σ in $[0, \pi]$. It suffices to show that the summand for $m = 1$ in (6.12) with $\omega = 0$ is positive. Thus, we want to show

$$E\left\{\left(\sum_{j=0}^{2} S_2^j C_{1,j}(\rho_1)\right)^2 I_{[\Lambda_1+\Lambda_2=0]}\right\} > 0 \tag{6.14}$$

From Theorem 6.4, $C_{1,0}(\rho_1) = C_{1,2}(\rho_1) = \rho_1(1-\rho_1^2)^{-1/2}$, and $C_{1,1}(\rho_1) = -(1-\rho_1^2)^{-1/2}$. Moreover, on each of the events $\{|\Lambda_1| = \omega_k, \Lambda_1 + \Lambda_2 = 0\}$, $k = 1, 2$, both of which have positive probability given $\Lambda_1 + \Lambda_2 = 0$, we obtain $S_2^0 = S_2^2 = 1$, and $S_2^1 = 2\cos(\omega_k)$. Therefore on these events,

$$\sum_{j=0}^{2} S_2^j C_{1,j}(\rho_1) = 2\{\rho_1 - \cos(\omega_k)\}(1-\rho_1^2)^{-1/2}$$

which is nonzero for $k = 1$ or $k = 2$, since ω_1 and ω_2 are distinct. Therefore (6.14) holds and we are done.

□

Theorem 6.5 tells us that the variance of the empirical zero-crossing rate of a mixed spectrum Gaussian process does not tend to 0 as the series length grows when there are at least two jumps in $[0, \pi]$. This leaves for investigation the case of a single atom in $[0, \pi]$ with or without continuous spectrum noise. This is done in the next two subsections.

6.3.2 Case of a Sinusoid Plus Noise

Suppose $\{Z_t\}$ has the form,

$$Z_t = A\cos(\omega_0 t) + B\sin(\omega_0 t) + \epsilon_t$$

where A, B are independent $N(0, E(A^2))$ random variables, independent of the noise $\{\epsilon_t\}$, and where $\omega_0 \in (0, \pi]$. The noise is assumed to be a continuous spectrum Gaussian process with autocorrelation $\rho_\epsilon(k)$. Then $\{Z_t\}$ admits the representation,

$$Z_t = R\cos(\omega_0 t + \varphi) + \epsilon_t \tag{6.15}$$

where $R^2 = A^2 + B^2$ is proportional to a χ_2^2 random variable (the square root of a χ_2^2 is a Rayleigh random variable with density $re^{-r^2/2}$ on $[0, \infty)$), the random phase φ is uniformly distributed in $[-\pi, \pi]$ independently of R^2, and (R, φ) is independent of $\{\epsilon_t\}$. In this subsection we do not require $\text{Var}(Z_t)$ to be equal to 1.

For our next result it is convenient to define a signal-to-noise ratio by the quantity

$$\alpha \equiv \frac{E(R^2)}{E(R^2) + 2\text{Var}[\epsilon_t]}$$

where "signal" refers to the random sinusoid. Then

$$\rho_1 = \alpha \cos(\omega_0) + (1-\alpha)\rho_\epsilon(1)$$

The following result gives the exact asymptotic variance of the sample zero-crossing rate for a sinusoid plus continuous spectrum noise (see Section 3.4.3).

Theorem 6.6. [12] *Suppose that $\{Z_t\}$ is given by (6.15), with $\omega_0 \neq 0$ and first order autocorrelation ρ_1. Then the limiting large sample variance $\sigma_d(\{0\})$ for the sample zero-crossing rate $\hat{\gamma} = \bar{d} = \sum_{t=2}^{N} d_t/(N-1)$ is given by*

$$\sigma_d(\{0\}) = \frac{1}{\pi^2}\sum_{m=1}^{\infty}\frac{\alpha^{2m}}{(2m)!2^{2m}}\binom{2m}{m} \times \left\{\sum_{j=0}^{2m}\sum_{k=0}^{j}\binom{m}{k}\binom{m}{j-k}\cos[(2k-j)\omega_0]C_{m,j}(\rho_1)\right\}^2 \tag{6.16}$$

In particular, if $\rho_1 = \cos(\omega_0)$, then $\sigma_d(\{0\}) = 0$.

The proof depends on several combinatorial facts and is given in [12]. Instead of formal proof, we shall compare $\sigma_d(\{0\})$ as given in (6.16) with sample estimates. Before doing that, we have:

Corollary 6.1. *If both $R\cos(\omega_0 t + \varphi)$ and ϵ_t have the same first order autocorrelation, then $\sigma_d(\{0\}) = 0$, and $\cos(\pi\hat{\gamma})$ is a consistent estimator of ρ_1.*

Proof. If both $R\cos(\omega_0 t + \varphi)$ and ϵ_t have the same first order autocorrelation, then it must be $\cos(\omega_0)$, and hence also $\rho_1 = \cos(\omega_0)$.

□

Thus, when the sinusoid and the noise have the exact same first order autocorrelation (i.e., both equal $\cos(\omega_0)$), then $\lim_{N\to\infty} \mathrm{Var}[D/(N-1)] = 0$. In the next chapter we give an example in frequency estimation where in the limit a sinusoidal signal and a superimposed noise component have the same first order autocorrelation.

We next turn to some direct calculations of $\sigma_d(\{0\})$ using the formula (6.16). For small α, and ω_0 well removed from 0 and π, the convergence in (6.16) is very rapid. This can be seen from Table 6.2, which gives the partial sums from the first 15 terms in the infinite sum (6.16) (the first term, the sum of the first two terms, the sum of the first three terms, etc.). The very last row in Table 6.2 gives the values of $\hat{\mathrm{Var}}(\pi\hat{\gamma})$ obtained from 500 independent realizations, each of length $N = 10^4$. The simulation results agree rather well with the row just above them, which gives the sum of the first 15 terms in (6.16).

6.3.3 Zero-Crossing Rate of Band-Limited Processes

From the preceding development we now know that when $\{Z_t\}$ is a mixed spectrum process, the zero-crossing rate need not be a consistent estimate for its

TABLE 6.2 Partial Sums of up to 15 Terms in the Expansion (6.16) Representing $\pi^2\sigma_d(\{0\})$ for $\rho_\epsilon(1) = \cos(0.8)$.

Number of Terms	$\omega_0 = 1$, $\alpha = 0.5$, $\rho_z(1) = 0.6185$	$\omega_0 = 1$, $\alpha = 0.25$, $\rho_z(1) = 0.6576$	$\omega_0 = 2$, $\alpha = 0.25$, $\rho_z(1) = 0.4185$
1	2.47614×10^{-3}	1.51528×10^{-3}	5.27833×10^{-2}
2	3.14462×10^{-3}	1.62329×10^{-3}	5.76271×10^{-2}
3	3.32615×10^{-3}	1.63111×10^{-3}	5.81911×10^{-2}
4	3.37575×10^{-3}	1.63169×10^{-3}	5.82721×10^{-2}
5	3.38939×10^{-3}	1.63173×10^{-3}	5.82860×10^{-2}
6	3.39316×10^{-3}	1.63173×10^{-3}	5.82887×10^{-2}
7	3.39421×10^{-3}	1.63173×10^{-3}	5.82893×10^{-2}
8	3.39451×10^{-3}	1.63173×10^{-3}	5.82895×10^{-2}
9	3.39459×10^{-3}	1.63173×10^{-3}	5.82895×10^{-2}
10	3.39462×10^{-3}	1.63173×10^{-3}	5.82895×10^{-2}
11	3.39462×10^{-3}	1.63173×10^{-3}	5.82895×10^{-2}
12	3.39463×10^{-3}	1.63173×10^{-3}	5.82895×10^{-2}
13	3.39463×10^{-3}	1.63173×10^{-3}	5.82895×10^{-2}
14	3.39464×10^{-3}	1.63173×10^{-3}	5.82895×10^{-2}
15	3.39467×10^{-3}	1.63173×10^{-3}	5.82895×10^{-2}
$\hat{\mathrm{V}}\mathrm{ar}(\pi\hat{\gamma})$:	3.43333×10^{-3}	1.69203×10^{-3}	5.79279×10^{-2}

Note: Values of $\hat{\mathrm{V}}\mathrm{ar}(\pi\hat{\gamma})$ obtained by simulation appear in the last row.

expectation. However, for a band-limited spectrum, the normalized (i.e., multiplied by π) asymptotic observed zero-crossing rate always lies within the positive part of the spectral band with probability one *regardless* of the spectral type, and this is equally true for Gaussian as well as for non-Gaussian processes.

This fact casts light on the asymptotic variability of the observed zero-crossing rate in general.

To prove the last claim we resort to zero-crossing counts obtained from repeated differencing and repeated summation. Define the asymptotic sample zero-crossing rate corresponding to the $(j-1)$th difference by the almost sure limit,

$$\gamma_j \equiv \lim_{N\to\infty} \frac{D_j}{N-1}$$

Similarly, the asymptotic rate corresponding to the $(j-1)$th summation is

$${}_j\gamma \equiv \lim_{N\to\infty} \frac{{}_jD}{N-1}$$

Clearly ${}_1\gamma = \gamma_1 = \lim_{N\to\infty} \hat{\gamma}$.

Theorem 6.7. [12] *Assume that a stationary, zero-mean (not necessarily Gaussian) process $\{Z_t\}$ has spectral measure σ supported on $\mathbf{J} \cup (-\mathbf{J})$ where*

$$\mathbf{J} = [{}_*\omega, \omega^*] \subset [0, \pi]$$

Then, whether σ has atoms (jumps) or not, the limit of the normalized empirical zero-crossing rate, $\lim_{N\to\infty} \pi\hat{\gamma}$, lies in $\mathbf{J}$ with probability one.

Proof. Observe that

$$D_j \leq D_{j+1} + 1$$

and therefore by dividing both sides of the preceding inequality by N and then letting N tend to infinity,

$$\gamma_1 \leq \gamma_2 \leq \gamma_3 \leq \cdots$$

This is a monotone and bounded (by 1) sequence and therefore it converges, say to γ^*, with probability one. In symbols, as $j \to \infty$,

$$\gamma_j \to \gamma^* \quad \text{a.s.}$$

Let $\nu_j(d\omega)$, $j = 0, 1, 2, \ldots$, be the spectral measure of the jth difference as in the proof of Theorem 5.2. Then we know that

$$\nu_j \to \frac{1}{2}\delta_{-\omega^*} + \frac{1}{2}\delta_{\omega^*}, \qquad j \to \infty$$

This implies that the asymptotic normalized zero-crossing rate sequence $\{\pi\gamma_j\}$ converges in probability to the asymptotic zero-crossing rate of a sinusoid supported at ω^*, or

$$\pi\gamma_j \xrightarrow{P} \omega^*$$

But since γ_j also converges with probability one, and the limit is unique, we have

$$\pi\gamma_j \xrightarrow{\text{a.s.}} \omega^*$$

It follows that

$$\pi\gamma_1 \leq \pi\gamma_2 \leq \pi\gamma_3 \leq \cdots \leq \omega^*$$

and in particular

$$\pi\gamma_1 \leq \omega^*$$

Now use repeated summation to show in the same way that

$$_*\omega \leq {}_1\gamma\pi$$

□

Theorem 6.7 tells us that

$$0 \leq_* \omega \leq \lim_{N\to\infty} \pi\hat{\gamma} \leq \omega^* \leq \pi.$$

It follows that in the extreme case when ${}_*\omega = \omega^*$, $\pi\hat{\gamma}$ converges with probability one to the *constant* ω^*.

Corollary 6.2. *Suppose that* $\{Z_t\}$ *is a pure random sinusoid with frequency* $\omega_0 \in (0, \pi]$. *That is, a strictly stationary discrete time process with mean* 0 *and spectral measure* σ *supported on* $\{-\omega_0\} \cup \{\omega_0\}$. *Then,*

$$(a) \qquad \lim_{N\to\infty} \frac{\pi}{N-1} \sum_{t=2}^{N} d_t = \omega_0, \quad \text{a.s.}$$

Equivalently,

$$(b) \qquad \sigma_d(\{0\}) = 0$$

The statistical import of Theorem 6.7 is that after applying an approximate band-pass filter to an underlying process in such a way that the output process has a single discrete frequency within a narrow interval of spectral support, the empirical zero-crossing rate is an extremely simple and effective statistic for estimating that frequency. In particular, it is possible to construct a recursive sequence of zero-crossing rates that converges to the frequency within the band. This idea will be explored in detail in the next chapter.

From Theorem 6.7 we would expect the variance of $\pi\hat{\gamma}$ to be on the order of $\eta^2 = |\omega^* -_* \omega|^2$ or better for sufficiently large N. Some evidence supporting this claim is given in Table 6.3. The table gives the estimates of $\text{Var}(\pi\hat{\gamma})$ from 500 independent realizations, each of size $N = 10^4$, obtained from the superposition of 2 and 4 random Gaussian sinusoids.

As a check of the simulation results, Table 6.3 also gives the values of $\hat{E}(\pi\hat{\gamma})$, which estimate $E(\pi\hat{\gamma})$. The table shows that in all the considered cases, $\hat{E}(\pi\hat{\gamma})$ falls between the highest and lowest frequencies as expected.

6.3.4 Model of Random Phases

When the spectrum of a stationary Gaussian process $\{Z_t\}$ is of mixed type, σ assigns a positive measure to discrete frequencies. In this case, $\{Z_t\}$ is nonergodic [15]. However, by conditioning on the amplitudes of the discrete spectral components, we can obtain under a certain condition on the discrete frequencies a new process which is a sum of sinusoids, having fixed amplitudes but random phases, plus an ergodic Gaussian noise process. The new conditional process is a stationary ergodic process which is no longer Gaussian. If we denote the discrete frequencies by λ_i, the condition needed for ergodicity is that there does not exist a finite sequence $\{\lambda_1, \ldots, \lambda_m\}$ of frequencies for which $\sigma(\{\lambda_i\}) > 0$ and integers $\{n_i\}$ such that $\lambda_1 n_1 + \cdots + \lambda_m n_m = 0 \bmod 2\pi$. The λ_i are then said to be "linearly incommensurate with π." In this case the empirical zero-crossing rate converges almost surely to its expectation, conditional on the amplitudes.

TABLE 6.3 The Order of Magnitude of $\hat{\mathrm{V}}\mathrm{ar}(\pi\hat{\gamma})$ Compared with that of $\eta^2 = |\omega^* -_* \omega|^2$

ω_1	ω_2	ω_3	ω_4	$\hat{E}(\pi\hat{\gamma})$	$\hat{\mathrm{V}}\mathrm{ar}(\pi\hat{\gamma})$	η^2
0.50000	1.10000	—	—	0.86535	10^{-2}	10^{-1}
1.09000	1.10000	—	—	1.09534	10^{-5}	10^{-4}
1.10000	2.10000	—	—	1.59599	10^{-1}	10^{0}
1.10000	1.20000	—	—	1.15528	10^{-3}	10^{-2}
1.10000	1.10500	—	—	1.10255	10^{-6}	10^{-5}
0.01000	0.02000	0.03000	0.04000	0.02716	10^{-5}	10^{-4}
0.50000	0.80000	1.20000	2.00000	0.78430	10^{-2}	10^{0}
0.75000	0.76000	0.77000	0.78000	0.75679	10^{-5}	10^{-4}
0.77000	0.77040	0.77080	0.78000	0.77262	10^{-6}	10^{-4}
1.10000	1.10010	1.10020	1.10030	1.10015	10^{-8}	10^{-8}
1.20000	1.20100	1.20200	1.20300	1.20145	10^{-7}	10^{-6}
1.50000	1.50010	1.50020	1.50030	1.50014	10^{-8}	10^{-8}
2.00000	2.00010	2.00020	2.00030	2.00014	10^{-8}	10^{-8}
2.59000	2.59100	2.5920	2.59300	2.59145	10^{-7}	10^{-6}
2.59970	2.59980	2.59990	2.60000	2.59985	10^{-8}	10^{-8}
3.00000	3.00100	3.00200	3.00300	3.00143	10^{-7}	10^{-6}

Note: All estimates are obtained from 500 independent realizations of length $N = 10^4$.

A process of the type just described is the so-called *model of random phases*. It is a stationary non-Gaussian process of the form

$$Z_t = \sum_{k=1}^{p} \beta_k \cos(\omega_k t + \phi_k) + \epsilon_t \tag{6.17}$$

where the β_k are *fixed* amplitudes, $0 < \omega_1 < \omega_2 < \cdots < \omega_p < \pi$, and the ϕ_k are independent random variables uniformly distributed in $[-\pi, \pi]$ and independent of the continuous spectrum Gaussian noise $\{\epsilon_t\}$.

6.4 CONSISTENCY OF THE SAMPLE AUTOCORRELATION

The sample autocorrelation follows the same pattern as the number of zero-crossings. Under the Gaussian assumption, the sample autocorrelation is a consistent estimator of the theoretical autocorrelation in the continuous spectrum case, but is inconsistent when the spectrum is of a mixed type (e.g., see [14, p. 65]). To see this, it suffices to consider the sample covariance obtained from a series of length N,

$$\hat{C}(k) = \frac{1}{N}\sum_{t=1}^{N} Z_t Z_{t-k}$$

Then one can show [12],

$$\hat{C}(k) \overset{\text{q.m.}}{\rightarrow} E[Z_t Z_{t-k}] + (|\beta(\{0\})|^2 - \sigma(\{0\}))$$

$$+2\sum_{\lambda_j>0} \cos(k\lambda_j)\{|\beta(\{\lambda_j\})|^2 - \sigma(\{\lambda_j\})\}, \qquad N \rightarrow \infty$$

where the $\beta(\{\lambda_j\})$ are the random amplitudes of the discrete frequencies λ_j, and $\sigma(\{\lambda_j\}) = E|\beta(\{\lambda_j\})|^2$. When the discrete spectrum is missing, then $\hat{C}(k) \overset{\text{q.m.}}{\rightarrow} E[Z_t Z_{t-k}]$ (q.m. = quadratic mean), and we have consistency.

In the remainder of the chapter we focus briefly on the sample autocorrelation in filtered data (HOC from sample autocorrelation), given a process of random phases as in (6.17). We specialize to the case where $\{\epsilon_t\}$ is a linear process,

$$\epsilon_t = \sum_j \psi_j u_{t-j}$$

the u_t being independent $\mathcal{N}(0,1)$ random variables, and $\sum_j |\psi_j| < \infty$. Also we assume the frequencies ω_j are linearly incommensurable. The noise spectral distribution function is given by

$$F_\epsilon(\omega) = \frac{1}{2\pi}\int_{-\pi}^{\omega} |\sum_j \psi_j \exp(-ij\lambda)|^2 \, d\lambda$$

As discussed above, under these conditions the process is ergodic, and the sample autocorrelation converges almost surely to its theoretical counterpart. Furthermore, under some additional assumptions the convergence of sample HOC (from the sample autocorrelation), obtained from a parametric family of filters, is uniform in the parameter.

Let $\mathcal{L}_\alpha$, $\alpha \in (-1,1)$, be a family of causal time-invariant linear filters with impulse response $\{h_j(\alpha)\}_{j=0}^{\infty}$, and corresponding transfer function

$$H(\omega;\alpha) = \sum_{j=0}^{\infty} h_j(\alpha)\exp(-ij\omega)$$

Given a finite time series, $Z_0, Z_1, \ldots, Z_{N-1}$, define the filtered data by

$$\hat{Z}_t(\alpha) \equiv \sum_{j=0}^{t} h_j(\alpha) Z_{t-j}, \qquad t = 0, 1, 2, \ldots, N-1$$

Define the sample HOC by

$$\hat{\rho}_1(\alpha) = \frac{\sum_{t=1}^{N-1} \hat{Z}_t(\alpha)\hat{Z}_{t-1}(\alpha)}{\sum_{t=0}^{N-1} \hat{Z}_t^2(\alpha)}$$

Assume,

H1. There exists constants $a_j > 0$, such that $\sum_{j=0}^{\infty} ja_j < \infty$ and $|h_j(\alpha)| \leq a_j$ for all $j = 0, 1, 2, \ldots$, and all $\alpha \in A'$, where A' is a closed subset of $(-1, 1)$.

H2. $h'_j(\alpha)$ exists and there are constants $b_j > 0$, such that $\sum_{j=0}^{\infty} jb_j < \infty$ and $|h'_j(\alpha)| \leq b_j$ for all $j = 0, 1, 2, \ldots$, and all $\alpha \in A'$.

Clearly, the α-filter, $Y_t = \alpha Y_{t-1} + Z_t$, satisfies H1 and H2. We now have:

Theorem 6.8. [16] *Suppose $\{Z_t\}$, $t = 0, \pm 1, \pm 2, \ldots$, is the random phases model (6.17), where $\epsilon_t = \sum_j \psi_j u_{t-j}$. Then under H1, H2, as $N \to \infty$,*

$$\hat{\rho}_1(\alpha) \xrightarrow{\text{a.s.}} \rho_1(\alpha)$$

and

$$\frac{d}{d\alpha}\hat{\rho}_1(\alpha) \xrightarrow{\text{a.s.}} \frac{d}{d\alpha}\rho_1(\alpha)$$

uniformly in $\alpha \in A'$.

The proof of this theorem is long and tedious and is omitted. We only note that the proof depends critically on the fact that if both $\{h_j(\alpha)\}_{j=0}^{\infty}$ and $\{g_j(\alpha)\}_{j=0}^{\infty}$ satisfy assumption H1, and $H(\omega; \alpha), G(\omega; \alpha)$ are the corresponding respective transfer functions, then as $N \to \infty$,

$$\frac{1}{N}\sum_{t=0}^{N-\tau-1}\left(\sum_{j=0}^{t+\tau} h_j(\alpha) Z_{t-j+\tau}\right)\left(\sum_{j=0}^{t} g_j(\alpha) Z_{t-j}\right) \xrightarrow{\text{a.s.}}$$
$$\sum_{k=1}^{p}\frac{\beta_k^2}{2}\Re\left\{H(\omega_k;\alpha)\overline{G(\omega_k;\alpha)}\exp(i\tau\omega_k)\right\} + \int_{-\pi}^{\pi} H(\omega;\alpha)\overline{G(\omega;\alpha)}\, dF_\epsilon(\omega)$$

uniformly in $\alpha \in A'$.

6.5 APPENDIX TO CHAPTER 6

Proof of Theorem 6.1 Following the method of proof in [6], we define an M-dependent process $Z_t^{(M)}$ as in Problem 7. Let $d_t^{(M)}$ be related to $Z_t^{(M)}$ as d_t is to Z_t. Then for every $\epsilon > 0$ we can find M_1 and N_1 such that for all $M > M_1$ and for all $N > N_1$,

$$E\left\{\frac{\sum_{t=2}^{N} d_t^{(M)} - \sum_{t=2}^{N} E[d_t^{(M)}]}{\sqrt{N-1}} - \frac{\sum_{t=2}^{N} d_t - \sum_{t=2}^{N} E[d_t]}{\sqrt{N-1}}\right\}^2 < \epsilon$$

This establishes the mean square convergence,

$$\frac{\sum_{t=2}^{N} d_t^{(M)} - \sum_{t=2}^{N} E[d_t^{(M)}]}{\sqrt{N-1}} \xrightarrow{\text{q.m.}} \frac{\sum_{t=2}^{N} d_t - \sum_{t=2}^{N} E[d_t]}{\sqrt{N-1}}, \qquad M \to \infty \quad (6.18)$$

uniformly in $N > N_1$. But by the Central Limit Theorem for m-dependent processes, for each M the left-hand side of (6.18) is asymptotically (as N grows) normal with mean 0 and variance σ_M^2, where $\sigma_M^2 \to \sigma_1^2$. Also, from theorem 4.1 in [6], $\kappa_x(1, -j, 1-j)$ is absolutely summable.

□

Proof of Theorem 6.3 Using (6.6) we replace λ's by ρ's in (6.7). We want to show that

$$\lim_{k\to\infty} \frac{\sin^{-1}\rho_{j-1}^{(k)} - \sin^{-1}\rho_{j-3}^{(k)} - 2\sin^{-1}\rho_1^{(k)} - \pi}{\frac{1}{2} + \frac{1}{\pi}\sin^{-1}\rho_1^{(k)}} = 0 \tag{6.19}$$

It can be shown [11], [21],

$$a_1 \equiv \lim_{k\to\infty} \frac{d\rho_1^{(k+1)}}{d\rho_1^{(k)}} = 1 \tag{6.20}$$

See Problem 5. Define $a_0 \equiv 0$, and for $j \geq 2$,

$$a_j \equiv \lim_{k\to\infty} \frac{d\rho_j^{(k)}}{d\rho_1^{(k)}}$$

Then (6.20) together with the fact that

$$\rho_{j+1}^{(k)} = -\rho_{j-1}^{(k)} + 2\rho_j^{(k)} - 2\rho_j^{(k+1)}(1 - \rho_1^{(k)})$$

imply the difference equation

$$a_{j+1} + 2a_j + a_{j-1} = 2(-1)^j$$

whose solution, since $a_2 = -4, a_3 = 9$, is

$$a_j = (-1)^{j+1} j^2$$

Equipped with this, we can evaluate (6.19) using l'Hôpital's rule. Thus, suppressing k, we need to evaluate

$$\lim_{\rho_1\to -1} \frac{\{1-\rho_{j-1}^2\}^{-1/2}\left(\dfrac{d\rho_{j-1}}{d\rho_1}\right) - \{1-\rho_{j-3}^2\}^{-1/2}\left(\dfrac{d\rho_{j-3}}{d\rho_1}\right) - 2\{1-\rho_1^2\}^{-1/2}}{4\{1-\rho_1^2\}^{-1/2}} \tag{6.21}$$

since $\rho_1 \to -1$ when $k \to \infty$ by the higher order crossings theorem. But

$$\lim_{\rho_1\to -1} \frac{1-\rho_1^2}{1-\rho_j^2} = \lim_{\rho_1\to -1} \frac{2\rho_1}{2\rho_j(d\rho_j/d\rho_1)} = \frac{1}{|a_j|}$$

and (6.21) becomes

$$\frac{1}{4}\frac{a_{j-1}}{\sqrt{|a_{j-1}|}} - \frac{1}{4}\frac{a_{j-3}}{\sqrt{|a_{j-3}|}} - \frac{1}{2}$$

When $j \neq 2$ is even and $j \geq 4$, this gives

$$\frac{1}{4}(j-1) - \frac{1}{4}(j-3) - \frac{1}{2} = 0$$

and similarly for j odd, but using the counterpart of (6.7) with the signs of the first two terms changed. Thus, we have shown that for $s \neq t$, $\mathrm{Cov}(d_s^{(k)}, d_t^{(k)}) = o(\lambda_1^{(k)})$, and this completes our proof.

□

Proof of Theorem 6.4 Using the Wiener–Itô calculus ([18], [22]), the representation of d_t in (6.13) implies for continuous spectrum,

$$\pi^2 \,\mathrm{Cov}(d_0, d_t) = E\left\{\sum_{m\geq 1} \frac{1}{(2m)!} \exp[it(\Lambda_1 + \cdots + \Lambda_{2m})] \left(\sum_{j=0}^{2m} S_{2m}^j C_{m,j}(\rho_z(1))\right)^2\right\} \tag{6.22}$$

Therefore, by iterated expectation, conditioning on $\Upsilon_{2m} \equiv \Lambda_1 + \cdots + \Lambda_{2m}$, we can rewrite (6.22) as

$$\pi^2 \int_{-\pi}^{\pi} e^{it\lambda} \sigma_d\,(d\lambda) = \pi^2 \,\mathrm{Cov}(d_0, d_t) = E\left\{\sum_{m\geq 1} \frac{1}{(2m)!} \exp[it\Upsilon_{2m}] E\left\{\left(\sum_{j=0}^{2m} S_{2m}^j C_{m,j}(\rho_z(1))\right)^2 \middle| \Upsilon_{2m}\right\}\right\}$$

Write the outer expectation as an integral with respect to the distribution $\sigma^{\star 2m}$ of $\Upsilon_{2m} = \Lambda_1 + \cdots + \Lambda_{2m}$ on $[-\pi, \pi]$ to conclude that

$$\pi^2 \int_{-\pi}^{\pi} e^{it\lambda} \sigma_d\,(d\lambda) = \sum_{m=1}^{\infty} \frac{1}{(2m)!} \int_{-\pi}^{\pi} e^{it\lambda} E\left\{\left(\sum_{j=0}^{2m} S_{2m}^j C_{m,j}(\rho_z(1))\right)^2 \middle| \Upsilon_{2m} = \lambda\right\} \sigma^{\star 2m}\,(d\lambda) \tag{6.23}$$

Then by a sequence of approximations it follows that (6.23) also holds in the mixed spectrum case. By appealing to the uniqueness of inverse Fourier transforms, we find

$$\pi^2 \sigma_d(d\lambda) = \sum_{m=1}^{\infty} \frac{1}{(2m)!} E\left\{\left(\sum_{j=0}^{2m} S_{2m}^j C_{m,j}(\rho_z(1))\right)^2 \middle| \Upsilon_{2m} = \lambda\right\} \sigma^{\star 2m}\,(d\lambda)$$

Thus σ_d is absolutely continuous with respect to the finite measure

$$\sum_{m=1}^{\infty} \frac{1}{(2m)!} \sigma^{\star 2\alpha m}$$

and its atoms must form a subset of the atoms of $\sum_{m=1}^{\infty}[1/(2m)!]\sigma^{\star 2m}$. The equality of the last two displayed measures on singletons $\{\omega\}$ together with the definition of conditional expectation imply the assertion (6.12).

□

6.6 PROBLEMS AND COMPLEMENTS

1. *Markov approximation [9].* Consider the second order moving average

$$Z_t = u_t - \theta_1 u_{t-1} - \theta_2 u_{t-2}$$

where $\{u_t\}$ is white Gaussian noise with variance $\sigma_u^2 = 1$.

(a) Argue that the process is stationary for any θ_1, θ_2, and show that

$$\rho_1 = \frac{-\theta_1(1-\theta_2)}{1+\theta_1^2+\theta_2^2}$$

$$\rho_2 = \frac{-\theta_2}{1+\theta_1^2+\theta_2^2}$$

and $\rho_3 = \rho_4 = \rho_5 = \rho_6 = 0$.

(b) Let $\theta_1 = -0.5, \theta_2 = -0.8$. For $N = 1000$, follow Algorithm 6.1 in showing that the Markov approximation to $\sqrt{\text{Var}[D_k]}$, $k = 1, \ldots, 6$, is given to one decimal place by 18.3, 17.7, 13.8, 11.4, 10.2, 9.7, respectively.

(c) Assuming that D_k are approximately normally distributed, use the results in (b) to construct a 95% confidence interval for $E[D_k]$, $k = 1, \ldots, 6$.

2. *m-Dependence approximation.* Disregarding the fact that k is small, compare your results in Problem 1 with the approximation suggested by Theorem 6.3:

$$\{\text{Var}[D_k]\}^{1/2} \sim \{(N-1)\lambda_1^{(k)}(1-\lambda_1^{(k)})\}^{1/2}$$

3. *A connection between the sample covariance and D.* Let $\{Z_t\}$ be a zero-mean stationary Gaussian process, $t = 0, \pm 1, \pm 2, \ldots$. The sample autocovariance for $X_t = I_{[Z_t \geq 0]}$ is defined by

$$\hat{C}_x(k) \equiv \frac{1}{N-k}\sum_{t=2}^{N}(X_t - \frac{1}{2})(X_{t-k} - \frac{1}{2})$$

Show that

$$\hat{C}_x(1) = \frac{1}{4} - \frac{D}{2(N-1)}$$

4. Use the cosine formula and Theorem 4.3 to prove Lemma 6.1. Argue that the result implies $\text{Cov}(g(X_s), d_t) = 0$, for any (measurable) function g.

5. Verify (6.20) for $m = 1$ using the fact that

$$\rho_1^{(k)} = \frac{-\binom{2k}{k-1} + \rho_1\left[\binom{2k}{k} + \binom{2k}{k-2}\right]}{\binom{2k}{k} - 2\rho_1\binom{2k}{k-1}}$$

(Hint: Use the chain rule relative to ρ_1 [11].)

6. Show that if $\sum_j |\rho_j| < \infty$, then $\sum_j \rho_j^2 < \infty$.

7. *Construction of m-dependence [3], [19]*. Let $\{Z_t\}, t = 0, \pm 1, \pm 2, \ldots$, be a continuous spectrum zero-mean stationary Gaussian process with spectral density f. Then the process admits the representation

$$Z_t = \int_{-\pi}^{\pi} e^{it\omega}\sqrt{f(\omega)}\, dB(\omega)$$

where $E|dB(\omega)|^2 = d\omega$. Consider the Fejer Kernel

$$W_M(\lambda) = \frac{\sin^2 \frac{1}{2}\lambda M}{2\pi M \sin^2 \frac{1}{2}\lambda}, \qquad -\pi \leq \lambda \leq \pi$$

and define

$$Z_t^{(M)} \equiv \int_{-\pi}^{\pi} e^{it\omega}(f \otimes W_M)^{1/2}(\omega)\, dB(\omega)$$

where $\otimes$ denotes convolution.

(a) Show that $W_M(\lambda)$ is periodic with period 2π, that it has a global maximum $W_M(0) = M/(2\pi)$ at $\lambda = 0$, and that it integrates to 1 over $[-\pi, \pi]$.

(b) Show that $\{Z_t, Z_t^{(M)}\}$ are jointly stationary and Gaussian with mean zero.

(c) Show

$$E[Z_t^{(M)} Z_{t-k}^{(M)}] = w_M(k) E[Z_t Z_{t-k}]$$

where

$$\begin{aligned} w_M(k) &= \int_{-\pi}^{\pi} e^{ik\lambda} W_M(\lambda)\, d\lambda \\ &= \begin{cases} 1 - \frac{|k|}{M}, & k = 0, \pm 1, \ldots, \pm(M-1) \\ 0, & \text{otherwise} \end{cases} \end{aligned}$$

(d) Conclude that $\{Z_t^{(M)}\}$ is M-dependent, and $\text{Var}[Z_t^{(M)}] = \text{Var}[Z_t]$.

8. *Spectral atoms of clipped processes (C. Houdré in a private communication, [12])*. Let $\{Z_t\}$ be a mixed spectrum zero-mean stationary Gaussian sequence with spectral measure σ_z, and let σ_x be the spectral measure of $X_t = I_{[Z_t \geq 0]}$. Let $f_z(\omega)$, $f_x(\omega)$, be the correspondong generalized spectral densities (i.e., they may contain delta functions).

(a) Show that $f_x(\omega)$ is given by the convolution expansion

$$\frac{\pi}{2} f_x(\omega) =$$
$$f_z(\omega) + \frac{1}{6} f_z \otimes f_z \otimes f_z(\omega) + \frac{1 \cdot 3}{2 \cdot 4 \cdot 5} f_z \otimes f_z \otimes f_z \otimes f_z \otimes f_z(\omega) + \cdots$$

(Hint: Expand $\rho_x(k) = (2/\pi)\sin^{-1}(\rho_z(k))$ in Taylor series.)

(b) Assume that Z_t is a pure sinusoid with frequency $2\pi/3$. Show that $f_z \otimes f_z \otimes f_z(\omega)$ puts some discrete mass at 0. (Hint: Think about the meaning of convolution of probability densities, and note that $2\pi/3 + 2\pi/3 + 2\pi/3 = 2\pi$.)

(c) Conclude that for a pure sinusoid with frequency $2\pi/3, 2\pi/5, 2\pi/7$, etc, $F_x(\omega)$ has a jump at 0, and hence that the continuity condition, $\sigma_z(\{0\}) = 0$, does not imply that $\sigma_x(\{0\}) = 0$.

(d) More generally conclude that *given a process with a mixed spectrum, the continuity of the spectrum at* 0 *does not guarantee the continuity at* 0 *of the spectrum of a function of the process.*

9. [3] Let $\{Z_t\}$ be a stationary Gaussian process with autocorrelation ρ_k. Show that

$$\text{Corr}(|Z_t|, |Z_{t-k}|) \le \rho_k^2$$

10. *Analysis of variance [8].* Under the Gaussian assumption and stationarity for $\{Z_t\}$, let $X_t = I_{[Z_t \ge 0]}$, $S = \sum_{t=1}^N X_t$, $R = \sum_{t=2}^N X_t X_{t-1} - \frac{1}{2}(X_1 + X_N)$. Then $D = 2S - 2R$.

(a) Use Lemma 6.1 to show that

$$\text{Var}[D] = 4[\text{Var}(R) - \text{Var}(S)]$$

and

$$\tau^2 \equiv \text{Corr}^2(R, S) = \frac{\text{Var[S]}}{\text{Var[R]}}$$

so that

$$\text{Var}\left[\frac{D}{N}\right] = \left[\frac{1}{\tau^2} - 1\right] \text{Var}\left[\frac{S}{N}\right]$$

(b) Show that for white noise $\tau^2 \to 4/5$.

(c) Conclude that a sufficient condition for $\text{Var}[D/N] \to 0$ is that, as $N \to \infty$, τ^2 approaches a positive constant, and $\text{Var}[S/N]$ goes to 0.

(d) Along the same line, argue that for simple HOC

$$\lim_{N\to\infty} \lim_{j\to\infty} \text{Var}[\frac{D_j}{N}] = 0$$

so that $\pi D_j/(N-1)$ is a consistent estimator of ω^*, the highest frequency in the spectral support.

11. Let $Z_t = \beta \sin(\omega_1 t + \phi)$, $t = 0, 1, 2, \ldots, N$, where β is a constant, $\omega_1 \in (0, \pi)$, and ϕ is uniformly distributed in $(-\pi, \pi)$. Show directly that

$$\frac{1}{N}\sum_{t=1}^{N} Z_t Z_{t-1} \stackrel{\text{a.s.}}{\to} \frac{\beta^2}{2}\cos(\omega_1), \qquad N \to \infty$$

12. [6] Let $X_1, X_2, \ldots, X_N$ be a binary sequence taking the values 0 or 1. Denote by D the number of symbol changes in the sequence, and let $S = \sum_{t=1}^{N} X_t$, $H = X_1 + X_N$.

(a) Show that the number of binary sequences of size N for which $S = s, D = d, H = h$, is given by

$$\binom{2}{h}\binom{s-1}{\frac{d+h-2}{2}}\binom{N-s-1}{\frac{d-h}{2}}$$

(b) From this find the number of sequences for which $S = s, D = d$.

13. [13] *Stationary binary Markov chains.* Let $\{X_t\}$, $t = 0, \pm 1, \pm 2, \ldots$, be a stationary 0-1 Markov chain, with parameters $p = P(X_t = 1)$, and $\lambda = P(X_t = 1|X_{t-1} = 1)$.

(a) Show that for $k = 0, 1, 2, \ldots$, and $q = 1 - p$,

$$P(X_t = 1|X_{t-k} = 1) = p + q\left(\frac{\lambda - p}{q}\right)^k$$

(b) Hence show that

$$\gamma_x(k) \equiv \text{Cov}(X_t, X_{t-k}) = pq\left(\frac{\lambda - p}{q}\right)^k$$

and

$$\rho_x(k) \equiv \text{Corr}(X_t, X_{t-k}) = \left(\frac{\lambda - p}{q}\right)^k$$

(c) Since $\rho_x(1) = (\lambda - p)/q$, conclude that $|(\lambda - p)/q| \leq 1$, so that the autocorrelation decays exponentially fast.

(d) Obtain the maximum likelihood estimator of (λ, p).

14. *Bernoulli trials.* You are given a binary sequence of independent 0–1 random variables, $X_1, X_2, \ldots, X_N$, such that for each i, $P(X_i = 1) = p$, $P(X_i = 0) = q$. Suppose the number of 1s is $S = \sum_{t=1}^{N} X_t = s$. What is the probability of at least one "11"? That is, if $R_1 = \sum_{t=2}^{N} X_t X_{t-1}$, we want $P(R_1 > 0)$. Answer:

$$P(\text{At least one } 11) = 1 - \sum_{s=0}^{[N+1/2]} \binom{N-s+1}{s} p^s q^{N-s}$$

15. *Indistinguishable particles [7], [20].* Let $X_1, X_2, \ldots, X_N$, be a 0–1 sequence such that the number of 1s is s, number of 11s is n_{11}, number of 111s is n_{111}, and the number of 101s is n_{101}. Show that the number of binary sequences of size N with given $s, n_{11}, n_{111}, n_{101}$ is

$$\binom{n_{11}-1}{n_{111}}\binom{s-n_{11}}{n_{11}-n_{111}}\binom{s-n_{11}-1}{n_{101}}\binom{N-2s+n_{11}+2}{s-n_{11}-n_{101}}$$

What is the number if you add the restriction n_{1001}? (Note: In 0011100011110101 $s = 9, n_{11} = 5, n_{111} = 3, n_{101} = 2$.)

16. *Estimation of spectral moments [17] [23].* Let $\{Z(t)\}$, $-\infty < t < \infty$, be a continuous spectrum zero-mean stationary Gaussian process with $\text{Var}[Z(t)] = 1$, correlation function $\rho(t)$, and spectral density $f(\lambda)$. Let $D_z(u)$ be the observed level-crossing rate per unit time of level u, and define the estimator

$$\hat{\gamma}(u) \equiv \pi D_z(u) \exp(\frac{1}{2}u^2)$$

(a) Use formula (4.49) to argue that $\hat{\gamma}(0)$ estimates $\sqrt{-\rho''(0)}$, the square root of the second spectral moment ($-\rho''(0) = \int_{-\infty}^{\infty} \lambda^2 f(\lambda)\, d\lambda$).

(b) Discuss the alternative estimator for $\sqrt{-\rho''(0)}$, obtained by using the level-crossing rates at different levels, $\sum_{k=1}^{p} a_k \hat{\gamma}(u_k)$, where $\sum_{k=1}^{p} a_k = 1$.

(c) Use a simulation to compare $\hat{\gamma}(0)$ with the estimator

$$\gamma^*(-\frac{2}{3}, 0, \frac{2}{3}) \equiv \frac{1}{3}\left\{\hat{\gamma}(-\frac{2}{3}) + \hat{\gamma}(0) + \hat{\gamma}(\frac{2}{3})\right\}$$

when $\rho(t) = \exp(-\frac{1}{2}t^2)$, and when $\rho(t) = [\sin(t\sqrt{3})/t\sqrt{3}]$.

REFERENCES

[1] Anderson, T.W., *The Statistical Analysis of Time Series*, New York: Wiley, 1971.

[2] Cramér, H. and M.R. Leadbetter, *Stationary and Related Stochastic Processes*, New York: Wiley, 1967.

[3] Cuzick, J., "A central limit theorem for the number of zeros of a stationary Gaussian process," *Ann. Probab.*, Vol. 4, No. 4, pp. 547–556, 1976.

[4] Ho, H.-C. and T.-C. Sun, "A central limit theorem for noninstantaneous filters of a stationary Gaussian process," *J. Multivariate Anal.*, Vol. 22, pp. 144–155, 1987.

[5] Karlin, S. and H.M. Taylor, *A First Course in Stochastic Processes*, New York: Academic Press, 1975.

[6] Kedem, B., *Binary Time Series*, New York: Dekker, 1980.

[7] Kedem, B., "On nearest neighbor degeneracies of indistinguishable particles," *J. Math. Phys.*, Vol. 22, No. 3, pp. 456–461, 1981.

[8] Kedem, B., "Detection of periodicities by higher order crossings," *J. Time Ser. Anal.*, Vol. 8, No. 1, pp. 39–50, 1987.

[9] Kedem, B., "Higher-order crossings in time series model identification," *Technometrics*, Vol. 29, No. 2, pp. 193–204, 1987.

[10] Kedem, B., "A fast graphical goodness of fit test for time series models," in *Time Series and Econometric Modeling*, I.B. MacNeill and G.J. Umphrey, Eds., pp. 65–76, Dordrecht, the Netherlands: Reidel, 1987.

[11] Kedem, B. and G. Reed, "On the variance of higher order crossings with special reference to a fast white noise test," *Biometrika*, Vol. 73, No. 1, pp. 143–149, 1986.

[12] Kedem, B. and E. Slud, "On autocorrelation estimation in mixed-spectrum Gaussian processes," *Stochastic Processes and their Applications*, 1993.

[13] Klotz, J., "Statistical inference in Bernoulli trials with dependence," *Ann. Stat.*, Vol. 1, No. 2, pp. 373–379, 1973.

[14] Koopmans, L.H., *Spectral Analysis of Time Series*, New York: Academic Press, 1974.

[15] Kornfeld, I., Fomin, S., and Ya. Sinai, *Ergodic Theory*, New York: Springer-Verlag, 1984.

[16] Li, T. and B. Kedem, "Strong consistency of the contraction mapping method for frequency estimation," accepted for publication in *IEEE Trans. Inform. Theory*, 1993.

[17] Lindgren, G., "Spectral moment estimation by means of level crossings," *Biometrika*, Vol. 61, No. 3, pp. 401–418, 1974.

[18] Major, P., *Multiple Wiener Itô Integrals,* New York: Springer-Verlag, 1981.

[19] Malevich, T.L., "Asymptotic normality of the number of crossings of level zero by a Gaussian process," *Theory Probab. Appl.*, Vol. 14, pp. 287–295, 1969.

[20] McQuistan, R.B., "Exact next nearest neighbor degeneracy," *J. Math. Phys.*, Vol. 15, No. 11, pp. 1845–1848, 1974.

[21] Reed, G.W., *Some Properties and Applications of Higher Order Crossings*, Doctoral Dissertation, Department of Mathematics, Univ. of Maryland, College Park, 1983.

[22] Slud, E., "Multiple Wiener-Itô integral expansions for level-crossings-count functionals," *Probab. Theory Relat. Fields*, Vol. 87, pp. 349–364, 1991.

[23] Steinberg, H., P.M. Schultheiss, C.A. Wogrin, and F. Zweig, "Short-time frequency measurements of narrow-band random signals by means of a zero counting process," *J. Appl. Phys.*, Vol. 26, pp. 195–201, 1955.

7

Frequency Estimation by HOC

We are now in a position to make a meaningful use of the theory presented in the preceding three chapters. In this chapter HOC analysis is applied in the classical time series problem of discrete frequency estimation in the presence of noise. Our goal is to show how to construct normalized HOC sequences that approximate discrete frequencies in the presence of continuous spectrum noise.

The present chapter has two main parts. In the first, we illustrate a certain curious tendency of simple (observed normalized) HOC to "cling" to discrete frequencies. Evidence of this tendency will be given in terms of real and artificial data. This part should be viewed as data analytic (experimental). In the second part, we introduce a novel iterative method for frequency estimation based on HOC and referred to as the *contraction mapping* (*CM*) *method*. Here we will discuss rigorous convergence results. The CM method provides a mechanism whereby parametric filters are reparametrized (tuned) iteratively using HOC subsequences. The method yields very precise frequency estimates.

7.1 INTRODUCTION AND OVERVIEW

Consider the mixed spectrum stationary process,

$$Z_t = \sum_{j=1}^{p}(A_j \cos(\omega_j t) + B_j \sin(\omega_j t)) + \zeta_t \tag{7.1}$$

where, $t = 0, \pm1, \pm2, \ldots$, the A's and B's are all uncorrelated, $E(A_j) = E(B_j) = 0$, and $\text{Var}(A_j) = \text{Var}(B_j) = \sigma_j^2$. Further, suppose $\{\zeta_t\}$ is colored stationary noise with mean 0 and variance σ_ζ^2, independent of the A's and B's. The noise is assumed to possess an absolutely continuous spectral distribution function $F_\zeta(\omega)$ with spectral density $f_\zeta(\omega)$, $\omega \in [-\pi, \pi]$. We do not restrict the process $\{Z_t\}$ to be Gaussian, unless such a restriction is made explicitly. Without loss of generality assume that the frequencies are ordered fixed constants,

$$0 < \omega_1 < \omega_2 < \cdots < \omega_p < \pi$$

Our problem is to locate or estimate the frequencies, $\omega_1, \omega_2, \ldots, \omega_p$, from a time series, $Z_1, Z_2, \ldots, Z_N$, using HOC.

When the noise component $\{\zeta_t\}$ is absent, a highly idealized case, we can use Algorithm 5.1 to detect the frequencies. However, in the presence of noise, that algorithm is clearly not applicable (why?). More sophisticated filtering than simple differencing is needed.

The problem of estimating the frequencies of sinusoidal components in ambient noise is one of the oldest and best known problems in time series analysis (see the review articles [4], [15], and [32, chs. 6,8]). It arises in connection with man-made systems such as rotating machinery and communication systems, as well as in physical and economical systems due to the cyclical motion of our planet Earth. The celebrated solution advanced by Schuster [34] in 1898, almost one hundred years ago, in the form of periodogram analysis is still, to a large extent, the prevailing approach to this problem today, and especially so since the popularization of the "fast Fourier transform" (FFT) by Cooley and Tukey [6] in 1965. The periodogram, denoted by $I_N(\omega)$, is defined by

$$I_N(\omega) = \frac{2}{N}\left|\sum_{t=1}^{N} Z_t \exp(-i\omega t)\right|^2, \qquad \omega \in [-\pi, \pi] \tag{7.2}$$

Periodogram analysis consists of the search for significant peaks in $I_N(\omega)$ where the latter is evaluated usually at the set of discrete Fourier frequencies,

$$\frac{2\pi k}{N}, \qquad k = 0, 1, \ldots, \left[\frac{N}{2}\right]$$

known as Fourier frequencies. Here $[x]$ stands for the greatest integer less than or equal to x. Under some fairly general conditions it can be shown that, as N increases, the expected order of magnitude of a periodogram ordinate at a Fourier frequency that approaches one of the ω_j is $O(N)$;[1] otherwise—for Fourier frequencies that are sufficiently far away from the ω_j—the order of magnitude is $O(1)$ (e.g., see [32, p. 396]). This suggests that a frequency which gives rise to a large peak, approximates a true discrete frequency.

Incidentally, an important fact, yet mostly overlooked, is that it is useful to consider certain valleys in addition to peaks when examining the graph of $I_N(\omega)$. In this connection it is beneficial to replace the factor $2/N$ by $N^{-31/16}$.

[1]The "big O" notation $f(N) = O(N)$ means that as N increases $|f(N)| \leq K \times N$, where K is a positive constant.

The advantage of this modification is that the peaks increase and the valleys decrease at the same rate. See [10], [39], and to some extent also [11].

The FFT, whose use is widespread in science and engineering, enables the computation of the periodogram at Fourier frequencies from a data record of length N by requiring only $O(N \log N)$ computational complexity, a great improvement over $O(N^2)$ needed for direct computation. By resorting to HOC analysis, however, we can achieve $O(N)$ computational complexity. This reduction in computation is meaningful when facing very large data records. A good example is provided by the Search for Extra-Terrestrial Intelligence (SETI) project recently launched by the National Aeronautics and Space Administration (NASA). The project mission is to scan the sky continuously for a period of time using data from telescopes situated around the globe in the hope of detecting "intelligent signals" from outer space. The idea really is to search for some frequencies that can be attributed sensibly to advanced technologies (rather than beings) presumed to exist in the universe. The amount of data involved is indeed enormous, and any saving in computational effort can be very significant.

To gain some insight into the problem at hand, consider a single pure sinusoid in discrete time. It is clear, without getting into the problem of precision, that the number of observed cycles in a record of length N, or equivalently the number of zero-crossings, can provide an estimate of the frequency (or equivalently the period), and this arguably requires only $O(N)$ operations versus the $O(N \log N)$ operations needed for the FFT. One of the problems we face is how to execute a similar $O(N)$ complexity analysis in the multiple frequency case even in the presence of colored noise.

In regard to the two main parts of the chapter, in the early part, we introduce an experimental method for frequency detection by which the periodogram is applied only to a small number of points (frequencies). The points are suggested by the sequence of simple HOC obtained *after* the data are first filtered by a low-pass filter. This method provides a heuristic search that works surprisingly well in many cases. However, as remarked previously, this data analytic procedure is brought up only to point to the potential of HOC in frequency detection and estimation, and not as a recommended technique.

In the latter part of the chapter (Sections 7.3 and 7.4), we introduce a rigorous search method for frequency estimation by making use of certain *contraction mappings* explored in various forms in [8], [11], [14], [20], [21], [25], [29], and [41]. Let θ be a parameter taking values in $(-1, 1)$, and let $C(\theta)$ be a function such that $0 < C(\theta) < 1$. In their simplest form, our contraction mappings can be expressed as

$$\rho_1(\theta) = r^* + C(\theta)(\theta - r^*) \tag{7.3}$$

where r^* is the cosine of the frequency to be detected. Then there exists a sequence $\{\theta_j\}$ such that

$$\theta_{j+1} = \rho_1(\theta_j)$$

and $\theta_j \to r^*$, as $j \to \infty$. This is the *contraction mapping algorithm*. It is relatively simple and can easily be implemented in practice. From a computational point of view the algorithm is attractive, as each iteration of the recursion requires $O(N)$ operations, and usually only very few iterations are needed for a satisfactory level of precision. As defined earlier, $\rho_1(\theta)$ is a HOC family indexed by θ, and hence the sequence $\{\theta_j\}$ itself is a HOC sequence. It can be shown that this method, under certain conditions, leads to estimates which can be as precise as those obtained by nonlinear least squares, thus surpassing the precision of the FFT estimates. Furthermore, experimental results indicate that the iterative method is quite resistant to fairly high levels of noise.

Section 7.4 discusses a certain hybrid of the iterative (CM) and the well-known autoregressive (AR) methods. This brings to light an extension of the HOC idea.

A Note about Amplitude Estimation

In order to achieve stationarity in the model (7.1) the amplitudes are assumed to be random variables. This is nothing but a mathematical convenience. The methods of frequency estimation discussed in this chapter are just as applicable in fixed amplitude models. Clearly, when the amplitudes are fixed (not random) they are considered as parameters, and it is then meaningful to speak of their estimation. The problem of estimating the frequencies, as well as the amplitudes, is usually carried out by nonlinear least squares, which is equivalent to maximum likelihood estimation under normality of the noise. For example, with $p = 1$ and N observations, $Z_1, \ldots, Z_N$, the estimates for A_1, B_1, ω_1 are obtained by minimizing the sum of squares (see Problem 19)

$$\sum_{t=1}^{N}\{Z_t - A\cos(\omega t) - B\sin(\omega t)\}^2$$

with respect to A, B, ω. Starting from an initial guess, Newton–Raphson-type procedures are then used for the optimization problem. It turns out that this problem is very sensitive to the initial guess of the frequency. If the frequency is not resolved to order $o(N^{-1})$, the amplitudes estimates are very biased [33]. Fortunately, under some fairly general conditions, the CM method—with certain parametric filters—is capable of approaching this type of required precision in frequency.

7.2 HOC AND PERIODOGRAM

An early attempt to apply HOC in mixed spectrum estimation was carried out by evaluating the periodogram only at the sequence of normalized simple HOC

$$\frac{\pi D_k}{N-1}, \qquad k = 1, 2, \ldots, K$$

obtained from a low-pass filtered series[2] [16], [17], [18]. That is, prior to obtaining the D_k, the series was filtered by a low-pass filter so as to "push" D_1

[2]With some exceptions, to simplify the notation, no special symbol is used in conjunction with D_k to mark the prefiltering operation.

sufficiently close to 0. In this way, the sequence $\pi D_k/(N-1)$ is distributed more evenly across the spectral support $[0, \pi]$. The rationale behind this is predicated on the repeated observation using real data that, on their way toward the highest frequency (in the filtered process), the $\pi D_k/(N-1)$ tend to "visit" or enter small neighborhoods of true discrete frequencies as k increases.

We have already seen an example of this behavior in Figure 5.5 in the case of a sum of five sinusoids (without additive noise) whose frequencies and amplitudes are given in Table 5.1. Consider the same data again. The frequencies are

$$\underline{\omega} = (0.75, 1.25, 1.60, 2.00, 2.30)$$

After operating on the series with the low-pass filter $(1 + \mathcal{B})^2$, the sequence $\pi D_k/(N-1)$, $k = 1, 2, \ldots, 12$, was

$$\underline{1.245}, 1.490, \underline{1.630}, 1.686, 1.679, 1.784, \underline{1.952}, \underline{2.022}, 2.148, 2.259, 2.288, \underline{2.301}$$

where an underline indicates that a visit has occurred. When we first operated on the series with the more pronounced low-pass filter $(1 + \mathcal{B})^{10}$, the sequence $\pi D_k/(N-1)$ was

$$\underline{0.749}, \underline{0.749}, \underline{0.749}, \underline{0.749}, 0.903, 1.154, \underline{1.254}, 1.378, 1.546, \underline{1.595}, \underline{1.602}, \underline{1.602}$$

We can see that between these two low-pass operations, all the frequencies were visited at least once. Thus, the last 24 points contain excellent guesses of the true frequencies. We can find out which point is more likely to give a good approximation by obtaining its relative power. This, however, can be done by evaluating the periodogram only at these 24 points.

Had we not first low-pass filtered the data, the sequence $\pi D_k/(N-1)$ would have started off increasing from $1.595, 1.791, \ldots$, so that information regarding frequencies lower than 1.595 would have been completely lost.

The same tendency of simple HOC to "visit" discrete frequencies has been observed time and again using quite a few different real data sets, some of which are discussed in the next few examples. An intuitive explanation of this fact is furnished by appealing to the dominant frequency principle in the sense that when a certain true frequency becomes locally dominant, or becomes part of a dominant spectral band as a result of repeated filtering, it attracts "passing by" normalized zero-crossing counts. As a result, a periodogram ordinate corresponding to a "visit" tends to be inflated.

The above motivates the following *experimental* procedure [16], [18].

Experimental Procedure

(a) Apply one or more low-pass filters to the observed time series.

(b) Observe the simple HOC sequence $\pi D_k/(N-1)$, $k = 1, 2, 3, \ldots, K$, (e.g., K=10) in the filtered data.

(c) Determine the relative power associated with the $\pi D_k/(N-1)$ by evaluating the normalized periodogram $I_N^*(\pi D_k/(N-1))$, $k = 1, 2, 3, \ldots, K$, defined by,

$$I_N^*\left(\frac{\pi D_k}{N-1}\right) = \frac{I_N(\pi D_k/(N-1))}{\sum_{j=1}^{K} I_N(\pi D_j/(N-1))} \tag{7.4}$$

A fact useful in testing the significance of an I_N^* ordinate is that under the hypothesis that the series is white noise [16], [18], [32],

$$P\left\{\max_{1\le k\le K} I_N^*\left(\frac{\pi D_k}{N-1}\right) > \frac{x}{2K}\right\} \sim K\left(1-\frac{x}{2K}\right)^{K-1} \tag{7.5}$$

This approximation is asymptotic and requires large K, N such that $K << N$. However, the approximation is surprisingly good even for moderate K, N, as can be seen from Table 7.1. The table gives a comparison between an estimate $\hat{P}$ of the true probability in (7.5) obtained from 200 independent white noise time series, and the approximation $P' = K(1 - x/2K)^{K-1}$, for different choices of N. The table gives an idea of the magnitude of significant I_N^* ordinates.

TABLE 7.1 Comparison of $\hat{P}$, Obtained from 200 Independent Realizations of Size N, and $P' = K(1 - x/2K)^{K-1}$

N	K	$x/2K$	P'	$\hat{P}$
550	16	0.33	0.0390	0.0400
		0.66	0	0
148	16	0.40	0.0075	0.0050
		0.52	0.0003	0
105	8	0.77	0.0003	0

To illustrate how the preceding data analytic procedure works in practice, we next discuss several convincing real cases.

Variable Star Series

Whittaker and Robinson [40] supply the now well-known series of the magnitude (measure of brightness) of a variable star on 600 successive midnights. This set of data has been analyzed by many authors including Whittaker and Robinson, and by now it is well established that the series displays two significant periods of approximately 24 and 29 days [3]. It is interesting to see what a HOC analysis yields.

The data, whose graph is displayed in Figure 7.1, were prefiltered using two summations $(1 + \mathcal{B})^2$ followed by the $AR(1)$ filter

$$\mathcal{L}_\alpha \equiv 1 + \alpha\mathcal{B} + \alpha^2\mathcal{B}^2 + \cdots$$

with a parameter value $\alpha = 0.999999$ (close to 1). Figure 7.1 indicates that a model of the form (7.1) is quite adequate. The normalized simple HOC

$$\frac{\pi D_j}{549}, \qquad j = 1, 2, 3, \ldots, 16$$

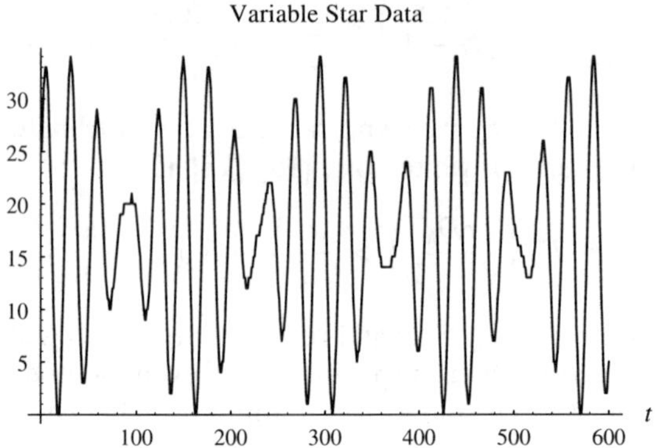

Figure 7.1: *Variable star series.*

were obtained from the filtered data with $N = 550$ (to take care of loss of observations due to differencing and prefiltering). The HOC sequence together with the corresponding I_N^* ordinates are given in Table 7.2 and Figure 7.2. Here we have resorted to a minor modification by evaluating I_N^* from the complete set of data with six hundred observations. For a rough significance test at level 0.05, use the value 0.32 corresponding to $N = 550$ and $K = 16$ in Table 7.1. That is, I_N^* ordinates greater than 0.32 are considered significant. When an ordinate is significant, the corresponding (frequency) period is considered a true period. Both Table 7.2 and Figure 7.2 show very clearly that the variable star series consists mainly of two periodic components with periods,

$$\frac{2\pi}{0.2174} = 28.9015 \text{ days}, \qquad \frac{2\pi}{0.2632} = 23.8723 \text{ days}$$

Thus, the two significant frequencies were visited by $\pi D_2/549$ and $\pi D_3/549$, respectively, and the periodogram was evaluated very economically only at 16 "promising" points instead of the 275 points or so required by the standard periodogram (harmonic) analysis. Recall that, in general, harmonic analysis calls for the evaluation of the periodogram at $[N/2]$ points, while our procedure calls for, by far, fewer ordinate evaluations, perhaps 16, regardless of N.

Diurnal Cycle in GATE

Our second case involves an interesting precipitation time series. We shall apply HOC analysis in the detection of a diurnal cycle (24 *hours*) in the so-called GATE data set.

GATE stands for Global Atmospheric Research Program, Atlantic Tropical Experiment. It is a field study conducted in 1974 in the eastern Atlantic off the coast of west Africa to study and collect rainfall data. Five ships equipped with precipitation radars that covered an area of about 400 *kilometers* in diameter, collected rainfall data in the form of huge snapshots of radar reflectivity taken every 15 minutes. The ships were also equipped with rain gauges, needed for calibration, and other relevant atmospheric probes ([35, p. 37]). The resulting data

TABLE 7.2 Normalized Simple HOC and Corresponding I_N^* Ordinates in the Filtered Variable Star Data

j	$\pi D_j/549$	$I_N^*(\pi D_j/549)$
1	0.1030	0.0000
2	0.2174	0.6521*
3	0.2632	0.3477*
4	0.6752	0.0000
5	1.5794	0.0000
.	.	.
.	.	.
.	.	.
16	2.3748	0.0000

Note: A "⋆" indicates significance.

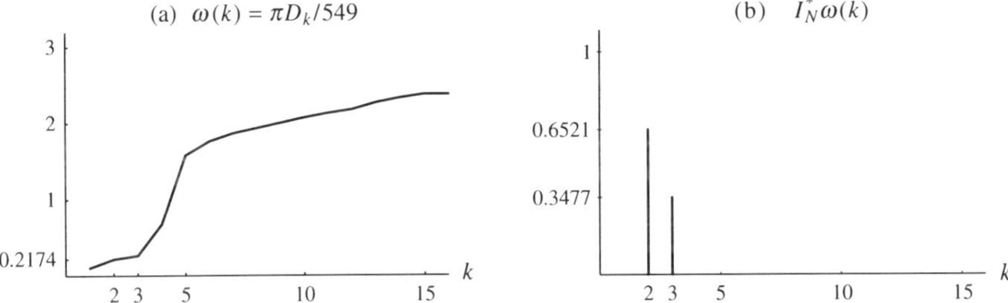

Figure 7.2: *Normalized HOC from the variable star data visit significant frequencies. (a) $\pi D_k/549$ obtained after applying $(1 - B)^2$, and the $AR(1)$ filter with parameter $\alpha = 0.999999$. (b) Normalized periodogram ordinates $I_N^*(\pi D_k/549)$, $k = 1, \ldots, 16$.*

set is known as the GATE data. See [35] for a detailed description of the GATE data set. There were several phases in the GATE study, the first two of which are commonly referred to as GATE-I and GATE-II, corresponding to data sets consisting of 1716 (18 days) and 1512 (15 days) snapshots, respectively. The radar reflectivity data were converted into rain rate data binned into 4×4 *kilometers*^{2} pixels, so that what we call a point in space is actually 4×4 *kilometers*^{2} pixels [13]. For technical reasons, some stretches of data are missing from the GATE data, so that observations in time are mostly, but not always, evenly spaced. This is not a serious problem if only very few zero-crossing counts are lost and N is large.

Although it is generally believed that a diurnal cycle exists in the GATE data, its detection is quite intriguing due to the fact that the cycle is not very strong uniformly throughout the different phases. A recent study in [2] raises the possibility that the cycle is significant in GATE-I but not in GATE-II.

HOC analysis was applied to two time series, centered at the mean, one from GATE-I (with $N = 1705$) and one from GATE-II (with $N = 1501$), of instantaneous rain rate averaged over an area of 28×28 *kilometers*^{2} from the "center" of the GATE snapshots. Note that 11 observations are lost due to differencing. The results are given in Table 7.3. By this analysis, a diurnal cycle of nearly 24 *hours* is detected in both phases of GATE. Prior to obtaining the HOC sequences, the data were filtered by the $AR(1)$ filter with $\alpha = 0.05$ (GATE-I; a very mild operation), and $\alpha = 0.46$ (GATE-II; a more pronounced operation). In both cases, I_N^* gave almost all its power to the normalized first zero-crossing count in the filtered data.

TABLE 7.3 Normalized Simple HOC and Corresponding I_N^* Ordinates in the Filtered GATE Data

	k	D_k	$\frac{\pi D_k}{N-1}$	Period	I_N^*
GATE-I	1	35	0.0645	24.3	0.9882
$N = 1705$	2	732	1.3496	1.2	0.0020
$\alpha = 0.05$	3	1068	1.9690	0.8	0.0003
	4	1204	2.2197	0.7	0.0008
	5	1270	2.3414	0.7	0.0048
	6	1320	2.4336	0.6	0.0011
	7	1358	2.5034	0.6	0.0005
	8	1400	2.5811	0.6	0.0012
	9	1426	2.6291	0.6	0.0010
	10	1442	2.6586	0.6	0.0001
GATE-II	1	31	0.0649	24.2	0.9962
$N = 1501$	2	454	0.9509	1.7	0.0003
$\alpha = 0.46$	3	829	1.7363	0.9	0.0008
	4	1010	2.1153	0.7	0.0014
	5	1100	2.3038	0.7	0.0001
	6	1153	2.4148	0.7	0.0003
	7	1191	2.494	0.6	0.0001
	8	1245	2.6075	0.6	0.0006
	9	1257	2.6327	0.6	0.0001
	10	1269	2.6578	0.6	0.0001

Note: The data are prefiltered with the $AR(1)$ filter with parameter α. Period is in hours. $\Delta t = 15$ min.

We shall return to this example again when applying the CM algorithm.

Chandler Wobble

In 1756, Euler determined that the pole of rotation of a rigid Earth could undergo a 305-day oscillation with respect to the crust [22]. A 305-day variation of latitude was sought by many astronomers prior to 1891, the year in which the American astronomer Chandler discovered an approximate 430-day (about 14 months) period which was subsequently shown to be the free Eulerian period [5], [22]. The difference is due to the fact that the Earth is not a perfectly rigid body, and the oscillation is called the Chandler wobble for the Earth. Today we know that the Chandler wobble has a period roughly between 430 to 440 days, but there remain some very controversial issues concerning the wobble variability, its source of excitation (earthquakes?), and whether it displays a single or perhaps two very close periods.

Whether the wobble period varies in time or not, whether the detected period is the result of two close periods or not, and regardless of its source of excitation, it is generally accepted that the Chandler wobble is associated with a significant period close to 14 months. We will apply our HOC procedure in the detection of this period.

A geographical coordinate system was attached to the body of the Earth early this century. The North Pole was defined as the average position of the rotation axis during the years 1900–1905. Relative to this coordinate system, the Bureau International de l'Heure (BIH) [3] has, since 1955, provided the position of the instantaneous rotation axis. Figure 7.3 shows the BIH pole path, or wobble, during 8 years starting from January 1, 1978, in 5-day increments. We denote these data by

$$(X_t, Y_t), \qquad t = 1, 2, \ldots, 585$$

Because there is no sense of "zero-crossings" for complex data, HOC analysis was applied independently to the univariate component time series $\{X_t\}$, $\{Y_t\}$, $t = 1, 2, \ldots, 585$. The univariate data are also shown in Figure 7.3.

With $N = 574$, $K - 10$, we counted in thc unfiltered raw X data $D_1 = 13$ zero-crossings so that

$$\frac{(N-1) \times 2 \times 5}{D_1} = 440.7692 \text{ days}$$

and, $I_N^*(440.7692) = 1.000$ (the significant ordinate out of of 10). Thus the Chandler wobble period is obtained just from the zero-crossing count in the raw X data.

To unearth the wobble period from the Y data, some prefiltering is needed. The Y data were prefiltered by the $AR(1)$ filter with parameter $\alpha = 0.99$. With $N = 574$, $K = 10$ we counted $D_1 = 13$, and again the period is 440.7692 days, but $I_N^*(440.7692) = 0.591$. In this case, the power was shared with a neighboring period of 409.2857 days obtained from D_2=14, and $I_N^*(409.2857) = 0.409$.

7.3 CONTRACTION MAPPINGS IN FREQUENCY ESTIMATION

Despite the fact that the preceding heuristic method seems to provide some useful insight, the results of that procedure should be viewed as preliminary and suggestive rather than conclusive. Yet, besides its experimental value, the method raises the possibility that there are HOC sequences, resulting from more clever

---

[3] Thanks are due to O.G. Jensen and L. Mansinha for the BIH data and clarifying discussions concerning the nature of the Earth's wobble.

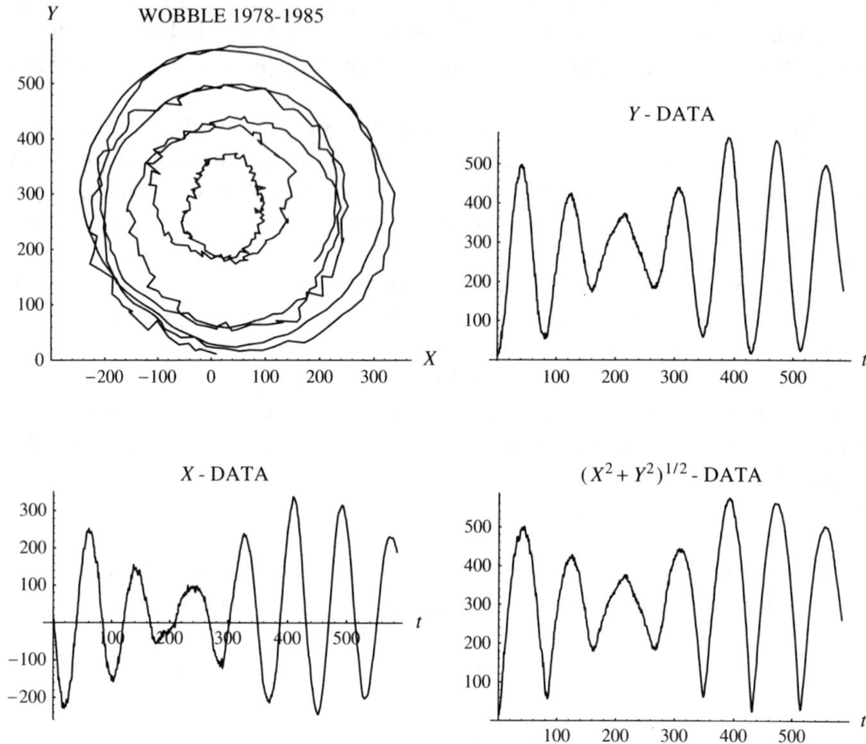

Figure 7.3: *The 5-day raw BIH data, 1978–1985, and their components.*

filtering than repeated differencing, whose convergence to the discrete frequencies in (7.1) can be proved with mathematical rigor. That this is indeed the case is the subject of the rest of this chapter.

7.3.1 Detection of a Single Frequency in Noise: The HK Algorithm

The iterative scheme to be described next is one of the highlights of the work on HOC. It epitomizes the notion of useful HOC sequences, and understanding it is important for subsequent extensions and generalizations.

Consider the process (7.1) with $p = 1$, and assume that $\{\zeta_t\}$ is white noise. The highest frequency is no longer ω_p but π. Consequently, by Theorem 5.2, $\pi E[D_j]/(N-1) \rightarrow \pi$, and the previous Algorithm 5.1 is no longer applicable. In order to produce a HOC sequence that converges to ω_1 we must therefore neutralize the noise in some way. The next result, from He and Kedem (HK) [11], does just that by making use of a HOC sequence from the $AR(1)$ filter. We note that very similar ideas have independently been arrived at by Dragošević and Stanković [8], [14], and Mataušek et al. [29].

Recall that the $AR(1)$ filter (α-filter) is defined by the operation,

$$Z_t(\alpha) = \mathcal{L}_\alpha(Z)_t = Z_t + \alpha Z_{t-1} + \alpha^2 Z_{t-2} + \cdots \tag{7.6}$$

whose squared gain $|H(\omega;\alpha)|^2$ is given by

$$|H(\omega;\alpha)|^2 = \frac{1}{1 - 2\alpha\cos(\omega) + \alpha^2}, \qquad \alpha \in (-1,1), \quad \omega \in [0,\pi] \tag{7.7}$$

Similarly define,

$$\zeta_t(\alpha) = \mathcal{L}_\alpha(\zeta)_t$$

and let

$$C(\alpha) = \frac{\mathrm{Var}(\zeta_t(\alpha))}{\mathrm{Var}(Z_t(\alpha))} \tag{7.8}$$

Then for $\alpha \in (-1,1)$,

$$0 < C(\alpha) < 1$$

Clearly $C(\alpha)$ also depends on ω_1, but this is not included to keep the notation simple.

Theorem 7.1. *[11] Suppose*

$$Z_t = A_1\cos(\omega_1 t) + B_1\sin(\omega_1 t) + \zeta_t, \qquad t = 0, \pm 1, \ldots$$

where $\omega_1 \in (0,\pi)$, A_1, B_1 are uncorrelated $N(0,\sigma_1^2)$ random variables, and $\{\zeta_t\}$ is Gaussian white noise with mean 0 and variance σ_ζ^2, independent of A_1, B_1. Let $\{D_\alpha\}$ be the HOC from the $AR(1)$ filter (7.6)*. Fix $\alpha_1 \in (-1,1)$, and define*

$$\alpha_{k+1} = \cos\left(\frac{\pi E[D_{\alpha_k}]}{N-1}\right), \qquad k = 1, 2, \ldots \tag{7.9}$$

Then, as $k \to \infty$,

$$\alpha_k \to \cos(\omega_1)$$

and

$$\frac{\pi E[D_{\alpha_k}]}{N-1} \to \omega_1 \tag{7.10}$$

Proof. Note that the special form (7.7) gives

$$\int_0^\pi |H(\omega;\alpha)|^2\, d\omega = \frac{\pi}{1-\alpha^2}$$

and

$$\int_0^\pi \cos(\omega)|H(\omega;\alpha)|^2\, d\omega = \frac{\pi}{1-\alpha^2} \times \alpha$$

Therefore, by symmetry, we obtain the *factorization*,

$$\int_{-\pi}^{\pi} \cos(\omega)|H(\omega;\alpha)|^2\,d\omega = \alpha \times \int_{-\pi}^{\pi} |H(\omega;\alpha)|^2\,d\omega \tag{7.11}$$

and so, from the zero-crossing spectral representation (4.17), and the cosine formula (5.6),

$$\rho_1(\alpha) = \cos\left(\frac{\pi E[D_\alpha]}{N-1}\right)$$

we have, with $dF_\zeta(\omega) = (1/2\pi)\sigma_\zeta^2\,d\omega$,

$$\rho_1(\alpha) = \frac{\sigma_1^2|H(\omega_1;\alpha)|^2 \times \cos(\omega_1) + \int_{-\pi}^{\pi} |H(\omega;\alpha)|^2\,dF_\zeta(\omega) \times \alpha}{\sigma_1^2|H(\omega_1;\alpha)|^2 + \int_{-\pi}^{\pi} |H(\omega;\alpha)|^2\,dF_\zeta(\omega)} \tag{7.12}$$

or, from the definition of $C(\alpha)$ in (7.8),

$$\rho_1(\alpha) = [1 - C(\alpha)] \times \cos(\omega_1) + C(\alpha) \times \alpha \tag{7.13}$$

We can see that $\rho_1(\alpha)$ is a convex combination of $\cos(\omega_1)$ and α, and that it also can be rewritten as a contraction mapping of the form (7.3),

$$\rho_1(\alpha) = \alpha^* + C(\alpha)(\alpha - \alpha^*) \tag{7.14}$$

where $\alpha^* = \cos(\omega_1)$. Invoke the cosine formula, and write the recursion (7.9) as,

$$\alpha_{k+1} = \rho_1(\alpha_k) \tag{7.15}$$

Starting with $k = 1$, substitute this in (7.14), iteratively, to obtain

$$\rho_1(\alpha_k) = \alpha^* + [\prod_{j=1}^{k} C(\alpha_j)](\alpha_1 - \alpha^*)$$

As $k \to \infty$, we have that $\prod_{j=1}^{k} C(\alpha_j) \to 0$, and this implies $\alpha_k \to \alpha^*$, and that α^* is a fixed point of $\rho_1(\cdot)$,

$$\alpha^* = \rho_1(\alpha^*)$$

or

$$\cos(\omega_1) = \cos\left(\frac{\pi E[D_{\alpha^*}]}{N-1}\right)$$

By the monotonicity of $\cos(x)$, $x \in [0, \pi]$,

$$\omega_1 = \frac{\pi E[D_{\alpha^*}]}{N-1}$$

□

An important fact in the preceding proof is the factorization equation (7.11) in which the parameter α is factorized outside the integral. As we shall see, this type of factorization is the fundamental fact needed for an extension of Theorem 7.1. The fact that the parameter α is "kicked out" in (7.11) is somewhat more apparent if we rewrite (7.11) as

$$\alpha = \rho_{1,\varsigma}(\alpha) = \frac{\int_{-\pi}^{\pi} \cos(\omega)|H(\omega;\alpha)|^2\, d\omega}{\int_{-\pi}^{\pi} |H(\omega;\alpha)|^2\, d\omega} \tag{7.16}$$

where $\rho_{1,\varsigma}(\alpha)$ is the first order autocorrelation of the filtered noise. The property (7.16) is what we call the *fundamental property* relative to a given family of filters and a noise spectrum. Thus, the fundamental property holds relative to the $AR(1)$ parametric filter and white noise. This, together with the correlation representation (7.13) lead to the contraction mapping (7.14), and eventually to the convergent HOC sequences α_k, and $(\pi E[D_{\alpha_k}])/(N-1)$. Fortunately, as we shall see very soon, factorizations of the form (7.11) are available at a surprisingly low cost to us.

The iteration (7.15) is called a *fixed point iteration* of the mapping (7.14). Its function is to locate an attracting fixed point of $\rho_1(\alpha)$, which in the present case is precisely given by $\alpha^* = \cos(\omega_1)$. The existence of an attracting fixed point of $\rho_1(\alpha)$ is illustrated in Figure 7.4. For more on this topic see [19], [25], [28], and Problem 17.

In terms of the *observed* zero-crossing rate, $\hat{\gamma}(\alpha) = D_\alpha/(N-1)$, the recursion (7.9) becomes

$$\alpha_{k+1} = \cos(\pi\hat{\gamma}(\alpha_k)) \tag{7.17}$$

Although in (7.9) we use expected HOC as opposed to (7.17) where observed HOC are used, it is convenient to refer to both (7.9) and (7.17) as the *HK algorithm*. It is a special case of CM. At a later section we will study the stochastic convergence of (7.17), as both N and k increase.

Convergence of Observed HOC from the *AR*(1) Filter

Table 7.4 shows some examples of the convergence (7.10) using simulated Gaussian data of a single sinusoid plus noise for different signal-to-noise ratios, and replacing the expected HOC by observed HOC. Recall that the signal-to-noise ratio (SNR) is defined as,

$$\text{SNR} \equiv 20\log_{10}\frac{\text{standard deviation of the signal}}{\text{standard deviation of the noise}}\ \text{dB}$$

and is given in dB. For the purpose of illustration, $N = 10,000$. However, N can be considerably smaller. Evidently, for a reasonable signal-to-noise ratio, the algorithm converges rather fast regardless of the initial value α_1. Note that no Fourier-type analysis has been used, and that ω_1 need not be a Fourier frequency of the form $2\pi k/N$. The fact that the sequence $\{\alpha_k\}$ is monotone (see Problem

7) is reflected very well by the observed normalized HOC $\pi D_{\alpha_j}/(N-1)$. The observed normalized HOC approach the frequency monotonically either from above or below the frequency, depending on the initial value $\alpha_1 \in (-1, 1)$.

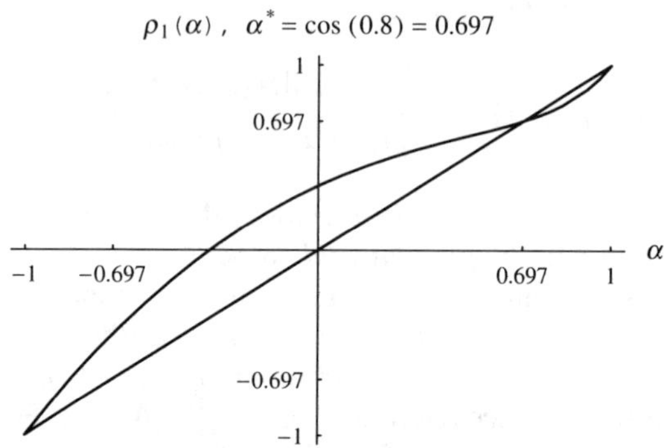

Figure 7.4: *An attracting fixed point* $\alpha^* = \cos(0.8)$ *of* $\rho_1(\alpha)$ *from the* $AR(1)$ *filter.* $p = 1, SNR = 0$.

TABLE 7.4 Illustration of the HK Algorithm

k	1 dB $\alpha_1 = -0.1$	0 dB $\alpha_1 = 0.9$	-1.94 dB $\alpha_1 = 0.2$	-6.02 dB $\alpha_1 = 0.5$
1	0.8848	0.5194	0.9127	0.9291
2	0.8006	0.5904	0.8222	0.8713
3	0.7987	0.6563	0.8015	0.8411
4	0.7987	0.7142	0.7965	0.8191
5	0.7987	0.7600	0.7952	0.8053
6	0.7987	0.7864	0.7952	0.8015
7	0.7987	0.8002	0.7952	0.7990
8	0.7987	0.8065	0.7952	0.7984
9	0.7987	0.8065	0.7952	0.7977
10	0.7987	0.8065	0.7952	0.7971
11	0.7987	0.8065	0.7952	0.7971
12	0.7987	0.8065	0.7952	0.7971
⋮	⋮	⋮	⋮	⋮

Note: Convergence of $\pi\hat{\gamma}(\alpha_k) = \pi D_{\alpha_k}/(N-1)$, $k \to \infty$, toward $\omega_1 = 0.8$ as a function of SNR. $N = 10,000$.

7.3.2 An Extension: The CM Algorithm

Looking back at α in (7.16), we can see that the parameter α is precisely the first order autocorrelation of the filtered white noise. In extending the work in [11], Yakowitz [41] has made the crucial observation that the same can be achieved without recourse to the Gaussian assumption and/or white noise. The key idea is to operate on the *noise* so that the parameter of the filter agrees with the first order autocorrelation of the filtered noise. This in fact is tantamount to the requirement that the fundamental property holds! In many cases this can be achieved by a *reparametrization* of the family of filters. Once a parametric family of filters possesses the fundamental property, it automatically generates the same contraction mapping as in (7.14).

Consider again the process (7.1) with $p = 1$, but this time the noise $\{\zeta_t\}$ is not necessarily white. Let r be a parameter that takes values in $(-1, 1)$, and let $\{H(\omega; r)\}$ be a family of transfer functions indexed by r. Assume that r is the first order autocorrelation of the filtered noise, where the filter corresponds to $H(\omega; r)$. Then the fundamental property takes on the form

$$r = \rho_{1,\zeta}(r) = \frac{\int_{-\pi}^{\pi} \cos(\omega)|H(\omega; r)|^2 \, dF_\zeta(\omega)}{\int_{-\pi}^{\pi} |H(\omega; r)|^2 \, dF_\zeta(\omega)} \tag{7.18}$$

From this we obtain a generalization of the factorization (7.11),

$$\int_{-\pi}^{\pi} \cos(\omega)|H(\omega; r)|^2 \, dF_\zeta(\omega) = r \times \int_{-\pi}^{\pi} |H(\omega; r)|^2 \, dF_\zeta(\omega) \tag{7.19}$$

As before, let $\rho_1(r)$ be the first order autocorrelation of the filtered *process* (HOC) using the same filter that was applied to the noise with transfer function $H(\omega; r)$. Then from the correlation representation in the mixed spectrum case, and (7.19),

$$\rho_1(r) = \frac{\sigma_1^2|H(\omega_1; r)|^2 \times \cos(\omega_1) + \int_{-\pi}^{\pi} |H(\omega; r)|^2 \, dF_\zeta(\omega) \times r}{\sigma_1^2|H(\omega_1; r)|^2 + \int_{-\pi}^{\pi} |H(\omega; r)|^2 \, dF_\zeta(\omega)} \tag{7.20}$$

which we recognize to be exactly the same as (7.12) with r replacing α. Everything now becomes entirely analogous to the previous case discussed in Theorem 7.1 where the $AR(1)$ filter was used. Thus, if as before we denote by $\{Z_t(r)\}$ and $\{\zeta_t(r)\}$ the filtered process and filtered noise, respectively, and let

$$C(r) = \frac{\text{Var}(\zeta_t(r))}{\text{Var}(Z_t(r))} \tag{7.21}$$

Then we obtain a general contraction mapping

$$\rho_1(r) = r^* + C(r)(r - r^*) \tag{7.22}$$

where $r^* = \cos(\omega_1)$. Assume that $C(r) < 1$ for all $r \in (-1, 1)$. It follows that for any initial point $r_1 \in (-1, 1)$, the sequence r_k defined by the *contraction mapping algorithm* (*CM*)

$$r_{k+1} = \rho_1(r_k) \tag{7.23}$$

converges to r^*,

$$r_k \to r^* \tag{7.24}$$

as $k \to \infty$. Thus r^* is a fixed point of $\rho_1(\cdot)$,

$$r^* = \rho_1(r^*) \tag{7.25}$$

and a limit point of the iterations (7.23). The desired frequency ω_1 is obtained from the inverse transformation $\omega_1 = \cos^{-1}(r^*)$. Clearly, under these conditions, Theorem 7.1 is now a special case, where the family of parametric filters is given by the $AR(1)$ filter with parameter $r = \alpha$, and the noise is white.

The reader should bear in mind that we do not really operate on the noise. After all, the noise component is never known to us in practice. Only the observed data, $Z_1, Z_2, \ldots, Z_N$, are operated on using $H(\omega; r)$. For the iterative method to work, we just need to know that (7.18), or equivalently the factorization (7.19), holds. Now, we will show shortly that there are families of complex band-pass filters which come arbitrarily close to satisfying (7.18), provided that the noise spectral density $f_\zeta(\omega)$ is sufficiently smooth. Thus in practice, knowing that $f_\zeta(\omega)$ is positive and sufficiently smooth, is sufficient for the application of the iterative method.

Another important point to bear in mind is that in the CM algorithm (7.23) we can either use higher order correlations, or higher order crossings. However, in order to use higher order crossings, we need a "cosine formula" that connects both types of HOC. In the Gaussian case this is clearly so, and we can use higher order crossings, in which case, as remarked earlier, we refer to the algorithm as the HK algorithm.

Example 7.1: The Critical *AR*(2) Filter

Let $\{Z_t\}$ be a zero-mean real-valued stationary Gaussian time series defined by the equation

$$Z_t = A\cos(\omega_1 t + \phi) + \zeta_t, \qquad t = 0, \pm 1, \pm 2, \ldots \tag{7.26}$$

where $\omega_1 \in (0, \pi)$ is a constant, A^2 is proportional to a χ_2^2, $\phi \sim U[-\pi, \pi]$ independently of A^2, and $\{\zeta_t\}$ is a zero-mean, continuous spectrum stationary Gaussian noise with a smooth spectral density $f_\zeta(\omega)$ which is positive in the neighborhood of ω_1. The pair (A, ϕ) is assumed to be independent of the noise.

Consider the (unstable) family $\{\mathcal{L}_r\}$ of $AR(2)$ filters indexed by a parameter r and defined by

$$\mathcal{L}_r \equiv \frac{1}{1 - r\mathcal{B} + \mathcal{B}^2}, \qquad r = 2\cos(\theta), \quad \theta \in (0, \pi) \tag{7.27}$$

or

$$Z_t(r) - rZ_{t-1}(r) + Z_{t-2}(r) = Z_t,$$

The filter has poles at $\exp(\pm i\theta)$ and is centered at $\theta = \cos^{-1}(r/2)$.

By truncating the impulse response appropriately, we obtain a modified stable filter which has a very narrow bandwidth in a neighborhood of θ. See Figure 7.5.

It follows from the fact that the gain approximates a delta function, that the fundamental property holds to a great degree, "but with $r/2$":

$$\rho_{1,\zeta}(r) \equiv \mathrm{Corr}(\mathcal{L}_r(\zeta)_t, \mathcal{L}_r(\zeta)_{t-1}) \doteq \frac{r}{2}$$

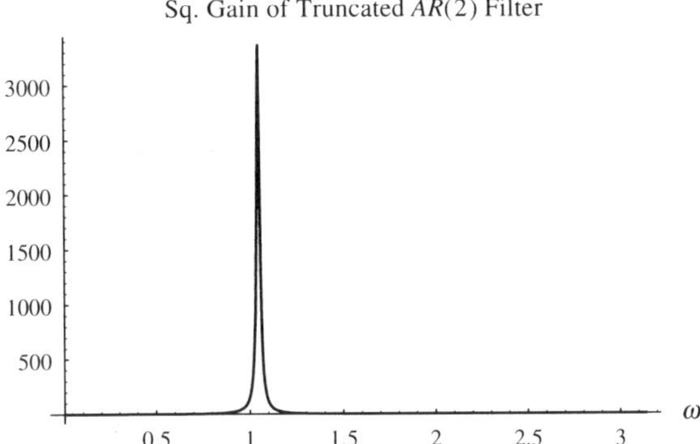

Figure 7.5: *Squared gain of the truncated $AR(2)$ filter with $r = 1$, centered at theta= $\pi/3$.*

The "$r/2$" is not a hindrance, and the contraction mapping takes the form,

$$\text{Corr}(\mathcal{L}_r(Z)_t, \mathcal{L}_r(Z)_{t-1}) \equiv \rho_1(r) = r^* + C(r)[\frac{r}{2} - r^*]$$

where $r^* = \cos(\omega_1)$, and

$$C(r) = \frac{E|\zeta_t(r)|^2}{E|Z_t(r)|^2}$$

Clearly, $0 \leq C(r) \leq 1$, and $C(r) = 1$ when the filter does not pass ω_1. When ω_1 passes, $0 < C(r) < 1$, and the CM algorithm becomes

$$r_{k+1} = 2\rho_1(r_k) \tag{7.28}$$

■

This example constitutes an important special case of the CM algorithm, and we shall consider it again—more rigorously—at a later stage by moving away from the unit circle, replacing $\mathcal{B}$ by $\eta\mathcal{B}$, $0 < \eta < 1$ ([8], [14], [26], [27], [29], [31], [38]).

7.3.3 Contractions from Band-Pass Filters

Consider again the case $p = 1$, and let us take a close look at the contraction mapping (7.22). In it, the contraction factor $C(r)$,

$$C(r) = \frac{\text{Var}(\zeta_t(r))}{\text{Var}(Z_t(r))}$$

plays a decisive role as it controls the rate of convergence of the HOC sequence $\rho_1(r_k)$ to $\cos(\omega_1)$. The smaller $C(r)$ is (that is closer to 0), the faster is the convergence, and so, an accelerated rate of convergence can be achieved if $C(r)$ is forced to become smaller and smaller with each iteration. This goal is within reach whenever the parametric family of filters is a band-pass family. By controlling the bandwidth according to a certain design, it is possible to obtain a greatly enhanced contraction, and hence, an accelerated convergence [8], [20], [21].

There is another good reason for resorting to band-pass filters: in the general multiple frequency case of p frequencies, with the help of band-pass filters the same procedure can be applied *separately* for each frequency. This means that in principle a general solution to our problem is in fact obtained by treating the case of a single frequency. Accordingly, in this section we treat the general case of p frequencies by focusing on a single frequency at the time.

We say that a frequency is "captured" by a band-pass filter when it falls within the bandwidth of the filter. Suppose now that a single frequency has been captured. Then it can be shown that by narrowing the bandwidth *judiciously*, the contraction factor $C(r)$ becomes smaller and the signal-to-noise ratio (SNR) corresponding to the captured frequency increases simultaneously. As a consequence, a speeded up convergence of the corresponding HOC sequence is achieved. We emphasize that the reduction in bandwidth must be carried out with care. Otherwise, an undue reduction may result in a lose or "release" of a "captured frequency" from the bandwidth, in which case no convergence occurs.

In what follows, it is convenient to let the family of band-pass filters be complex. Two of our best examples are in terms of complex band-pass filters.

Suppose we have a parametric family of band-pass filters indexed by $r \in (-1, 1)$, and by a bandwidth parameter M:

$$\{\mathcal{L}_{r,M}(\cdot),\ r \in (-1,1),\ M = 1, 2, \ldots\}$$

Let $h(n; r, M)$ and $H(\omega; r, M)$ be the corresponding complex impulse response and transfer function, respectively. It is required that as $M \to \infty$, $|H(\omega; r, M)|^2$ converges to a Dirac delta function centered at $\theta(r) \equiv \cos^{-1}(r)$. The filtered process and filtered noise are now denoted by $\{Z_t(r, M)\}$ and $\{\zeta_t(r, M)\}$, respectively. Assume further that the *filter passes only the* (*positive*) *discrete frequency to be detected*; suppose it is ω_1. Also, observe that

$$\Re\left\{\frac{E[\zeta_t(r,M)\overline{\zeta_{t-1}(r,M)}]}{E|\zeta_t(r,M)|^2}\right\} = \frac{\int_{-\pi}^{\pi} \cos(\omega)|H(\omega;r,M)|^2\, dF_\zeta(\omega)}{\int_{-\pi}^{\pi} |H(\omega;r,M)|^2\, dF_\zeta(\omega)}$$

where the overbar denotes "complex conjugate."

We now repeat the same procedure as before. Suppose that for *any* M the fundamental property takes the form

$$r = \Re\left\{\frac{E[\zeta_t(r,M)\overline{\zeta_{t-1}(r,M)}]}{E|\zeta_t(r,M)|^2}\right\} \tag{7.29}$$

and define

$$\rho_1(r,M) \equiv \Re\left\{\frac{E[Z_t(r,M)\overline{Z_{t-1}(r,M)}]}{E|Z_t(r,M)|^2}\right\} \tag{7.30}$$

Clearly, the full representation of $\rho_1(r, M)$ is given by

$$\rho_1(r, M) = \frac{\frac{1}{2}\sigma_1^2|H(\omega_1; r, M)|^2 \times \cos(\omega_1) + \int_{-\pi}^{\pi} |H(\omega; r, M)|^2 \, dF_\zeta(\omega) \times r}{\frac{1}{2}\sigma_1^2|H(\omega_1; r, M)|^2 + \int_{-\pi}^{\pi} |H(\omega; r, M)|^2 \, dF_\zeta(\omega)} \tag{7.31}$$

which is the same as (7.20) except for the $\frac{1}{2}$ factor. This factor is needed because now the gain is not symmetric and the complex filter passes only ω_1, but not $-\omega_1$. When $-\omega_1$ also passes, as is the case for symmetric gains, the exact same analysis still goes through. Now if we let

$$C(r, M) = \frac{E|\zeta_t(r, M)|^2}{E|Z_t(r, M)|^2}$$

then the basic contraction (7.22) has now an extra parameter M:

$$\rho_1(r, M) = r^* + C(r, M)(r - r^*) \tag{7.32}$$

where $r^* = \cos(\omega_1)$, and ω_1 is the frequency to be detected. The CM algorithm now takes the form

$$r_{k+1} = \rho_1(r_k, M_k) \tag{7.33}$$

We assume there is an initial search interval Ω in the spectral support containing ω_1:

$$\omega_1 \in \Omega \equiv [\omega_a, \omega_b] \subset (0, \pi)$$

For simplicity we assume that none of the other discrete frequencies is contained in Ω. *The general idea underlying our estimation method is to iterate the CM algorithm* (7.33) *while shrinking the bandwidth* (*increasing* M_k) *judiciously at each step, thus creating a sequence of shrinking* (*possibly nested*) *intervals that contain* ω_1 *until* r_k *converges to* $r^* = \cos(\omega_1)$. This procedure will be illustrated in Section 7.3.6, which is devoted to examples.

What happens if ω_1 is not captured in Ω? To answer this, observe that when the filter passes ω_1,

$$C(r, M) = \frac{E|\zeta_t(r, M)|^2}{(\sigma_1^2/2)|H(\omega_1; r, M)|^2 + E|\zeta_t(r, M)|^2} \tag{7.34}$$

It is important to note from (7.34) that $C(r, M)$ and the SNR, here relative to the first sinusoid, are inversely related, a point to be considered below. Now, it follows from (7.34) that as long as the filter passes ω_1, the same analysis as above, only slightly modified, still applies. However, when M is sufficiently large so that $|H(\omega; r, M)|^2$ is narrow enough and only passes an interval Ω that does not contain ω_1, the discrete spectral distribution does not have a jump at Ω, and this implies

$$C(r, M) = 1$$

or, from (7.32),

$$\rho_1(r, M) = r$$

Consequently, r does not change, the filter does not change its location, the signal-to-noise ratio does not increase, and no convergence occurs. We refer to this by saying that the "filter does not move."

Connection between *C(r, M)* and the SNR

The SNR and the contraction factor $C(r, M)$ are related by the relationship,

$$\text{SNR} = 10\log_{10}\left\{\frac{1}{C(r, M)} - 1\right\} \tag{7.35}$$

Thus, the SNR and $C(r, M)$ are inversely related:

$$\text{SNR} \to \infty \iff C(r, M) \to 0$$

$$\text{SNR} \to -\infty \iff C(r, M) \to 1$$

It follows that the smaller $C(r, M)$ is, the greater is the signal-to-noise ratio, and a faster speed of convergence is achieved. In essence, when a single frequency is captured, the SNR increases with each iteration of the CM algorithm, until detection occurs.

7.3.4 Two Parametric Families of Filters

This section discusses the CM plan relative to two useful parametric bandpass filters. In addition, it is shown that the contraction factor can in fact be controlled by the filter bandwidth.

The Complex Exponential Filter

Pertaining to the contraction mapping algorithm as given by (7.33), a useful parametric family is the exponential family of parametric filters defined by the parametric impulse response [20], [41],

$$h(n; r, M) = \begin{cases} \exp\left(in\theta(r)\right)/\sqrt{2M+1}, & |n| \le M \\ 0, & |n| > M \end{cases} \tag{7.36}$$

The corresponding family of squared gains is given by the Fejér kernel,

$$|H(\omega; r, M)|^2 = \frac{1}{2M+1}\frac{\sin^2[\frac{1}{2}(2M+1)(\omega - \theta(r))]}{\sin^2[\frac{1}{2}(\omega - \theta(r))]}, \qquad -\pi \le \omega \le \pi \tag{7.37}$$

Assume the noise process $\{\zeta_t\}$ is white noise. This assumption can be disposed of in practice, as we shall argue below. For

$$\theta(r) = \cos^{-1}\left(\frac{2M+1}{2M}r\right)$$

the fundamental property (7.29)

$$r = \Re\left\{\frac{E[\zeta_t(r, M)\overline{\zeta_{t-1}(r, M)}]}{E|\zeta_t(r, M)|^2}\right\} \tag{7.38}$$

holds for $|r| \leq 2M/(2M+1)$. Observe that $|H(\omega; r, M)|^2$ is now centered at $\theta(r)$, the value at the mode is $2M+1$, and that it approaches the Dirac delta function as $M \to \infty$. See Figure 7.6 and Problems 10–12. The complex exponential filter is essentially a band-pass filter with a dominant lobe supported over the interval

$$\mathcal{I} = (\theta(r) - 2\pi/(2M+1), \theta(r) + 2\pi/(2M+1))$$

For sufficiently large M, we can think of $\mathcal{I}$ as the effective bandwidth of the filter. When the filter passes ω_1 and M is sufficiently large, $-\omega_1$ is suppressed as we assumed above, and the contraction factor is given by,

$$C(r, M) \doteq \frac{\sigma_\zeta^2}{(\sigma_1^2/2)|H(\omega_1; r, M)|^2 + \sigma_\zeta^2} \tag{7.39}$$

The CM recursion, $r_{k+1} = \rho_1(r_k, M_k)$, yields the convergence $r_k \to r^*$, provided $\Pi_{j=1}^k C(r_j, M_j) \to 0$, as $k \to \infty$. It follows that the CM algorithm is executable under the assumption of white noise.

When $\{\zeta_t\}$ is not white noise but merely possesses a sufficiently smooth spectral density $f_\zeta(\omega) > 0$, $\omega \in (-\pi, \pi]$, we have asymptotically as M increases (the Fejér kernel then approximates a Dirac delta)

$$\begin{aligned}\Re\left\{\frac{E[\zeta_t(r, M)\overline{\zeta_{t-1}(r, M)}]}{E|\zeta_t(r, M)|^2}\right\} &= \frac{\int_{-\pi}^{\pi}\cos(\omega)|H(\omega; r, M)|^2 f_\zeta(\omega)\,d\omega}{\int_{-\pi}^{\pi}|H(\omega; r, M)|^2 f_\zeta(\omega)\,d\omega} \\ &\approx \cos(\cos^{-1}(r)) = r\end{aligned} \tag{7.40}$$

and so, the above procedure can be reasonably adapted to this case as well. In short, exact knowledge of the noise spectral density is not needed, a fact alluded to earlier.

Bounding the Contraction Factor by the Bandwidth

It was pointed out in the preceding discussion that an enhanced speed of convergence toward r^* can be obtained by decreasing the contraction factor $C(r_j, M_j)$ with each iteration of the CM recursion [20], [21]. This, however, can be achieved by controlling the effective bandwidth of the filter. The following result, which helps to motivate a certain assumption, renders this idea more precise.

Proposition 7.1. [20], [21]. *Under the assumption of white noise, let the squared gain $|H(\omega_1; r, M)|^2$ be as in (7.37), and assume,*

$$\omega_1 \in [\omega_a, \omega_b] \subset \mathcal{I}$$

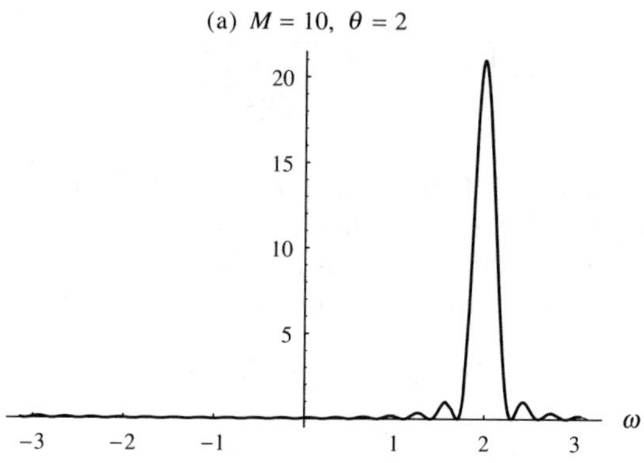

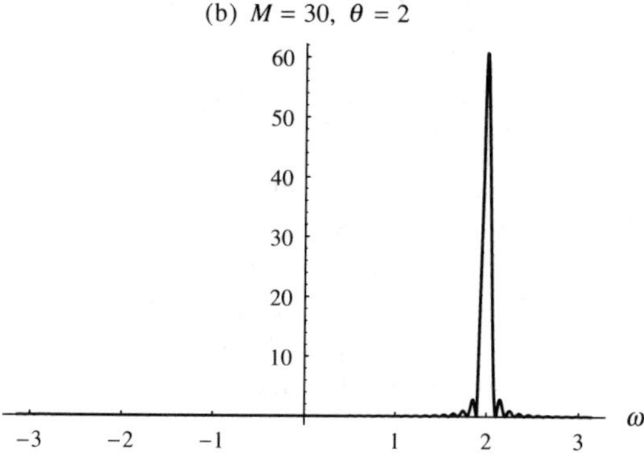

Figure 7.6: *Fejér kernel* $|H(\omega; r, M)|^2$ *centered at* $\theta(r) = 2$. *(a)* $M = 10$. *(b)* $M = 30$.

Define

$$\Delta \equiv \omega_b - \omega_a$$

Then there exists a $\overline{C}(M)$*, which does not depend on* r*, satisfying*

$$0 < C(r, M) \leq \overline{C}(M) < 1$$

and $\overline{C}(M)$ *is of order* $O(\Delta)$ *as* $\Delta \to 0$.

Proof. By centering the filter at 0,

$$|H(\omega_1; r, M)|^2 = |H(\omega_1 - \theta(r); \frac{2M}{2M+1}, M)|^2$$

Therefore, when $\omega_1, \theta(r) \in [\omega_a, \omega_b]$,

$$|H(\Delta; \frac{2M}{2M+1}, M)|^2 < |H(\omega_1 - \theta(r); \frac{2M}{2M+1}, M)|^2$$

Define

$$\overline{C}(M) = \frac{\sigma_\zeta^2}{(\sigma_1^2/2)|H(\Delta; \frac{2M}{2M+1}, M)|^2 + \sigma_\zeta^2} \tag{7.41}$$

Then in light of (7.39),

$$0 < C(r, M) \leq \overline{C}(M) < 1$$

As $\Delta \to 0$, choose $M = O(1/\Delta)$.

□

By minimizing (7.41) with respect to M, the optimal value of M is approximately $1.165/\Delta$.

Proposition 7.1 can be demonstrated graphically as well. Figure 7.7(a) shows the graph of $\rho_1(r, 20)$ when $\omega_1 = 0.8$ radians per unit time, and $\sigma_1^2 = \sigma_\zeta^2 = 1$. In the neighborhood of $r^* = \cos(0.8) = 0.697$, we can see from the figure that the derivative $\rho_1'(r, 20)$ is very close to 0, and this is already with $M = 20$. Therefore, in the neighborhood of r^*,

$$0 \approx \rho_1'(r^*, 20) \approx \frac{\rho_1(r, 20) - r^*}{r - r^*} = C(r, 20)$$

However, the effective bandwidth of the filter in the neighborhood of $\theta = 0.8$ is considerably greater than 0, as can be seen from Figure 7.7(b).

To summarize the preceding discussion, by shrinking the bandwidth we can force the contraction factor to vanish. This motivates assumption **A1** below.

A Band-Pass Filter

We mention briefly another band-pass family of filters that shares essentially the same properties as the previous exponential family, except that the gain does not have side lobes (see Figure 7.8). It is defined—recall Eq. (3.85)—by

$$Z_t(r; M) = \left(1 + e^{i\theta(r)}\mathcal{B}\right)^M Z_t \tag{7.42}$$

where M is a positive integer, $r \in (-1, 1)$, and $\theta(r) \in (-\pi, \pi)$. For sufficiently large M, for example, $M = 30$, this filter acts as a band-pass filter with impulse response and squared gain given by, respectively,

$$h(n; r, M) = \begin{cases} \binom{M}{n} e^{i\theta(r)n}, & n = 0, \ldots, M \\ 0, & \text{otherwise} \end{cases} \tag{7.43}$$

and

$$|H(\omega; r, M)|^2 = 4^M \cos^{2M}\left(\frac{\omega - \theta(r)}{2}\right) \tag{7.44}$$

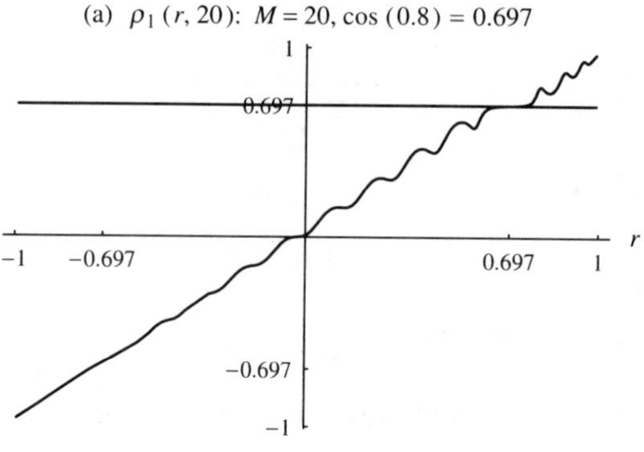

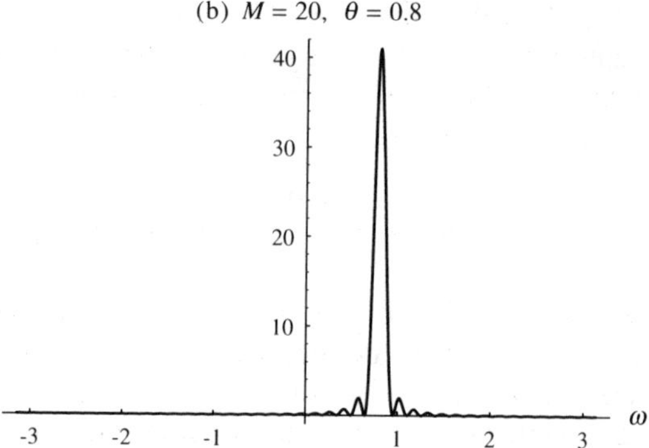

Figure 7.7: *(a) Plot of* $\rho_1(r, 20)$ *from the Fejér kernel* $|H(\omega; r, 20)|^2$ *centered at* $\theta = 0.8$. *The derivative of* $\rho_1(r, 20)$ *in the neighborhood of* $\cos(0.8)$ *is very close to* 0. *(b) The squared gain of the Fejér kernel* $|H(\omega; r, 20)|^2$ *centered at* $\theta = 0.8$. *The effective bandwidth is considerably larger than* 0.

The fundamental property is obtained by *defining*

$$r = \frac{\int_{-\pi}^{\pi} \cos^{2M}\left([\lambda - \theta(r)]/2\right) \cos(\lambda)\, dF_\zeta(\lambda)}{\int_{-\pi}^{\pi} \cos^{2M}\left([\lambda - \theta(r)]/2\right)\, dF_\zeta(\lambda)} \tag{7.45}$$

When $\{\zeta_t\}$ is white noise [11],

$$r \to \cos(\theta(r)),\ M \to \infty$$

and we have the approximation $\theta(r) \approx \cos^{-1}(r)$. As in the previous case, it is easy to see that this approximation holds true for any continuous spectrum noise, provided the spectral density is (positive) sufficiently smooth. For sufficiently large M, for example, $M = 100$, when the filter is centered at $\theta(r)$ near ω_1, it only passes a band of frequencies in the neighborhood of ω_1 (see Figure 7.8), and we obtain the contraction mapping (7.32). See Problem 14.

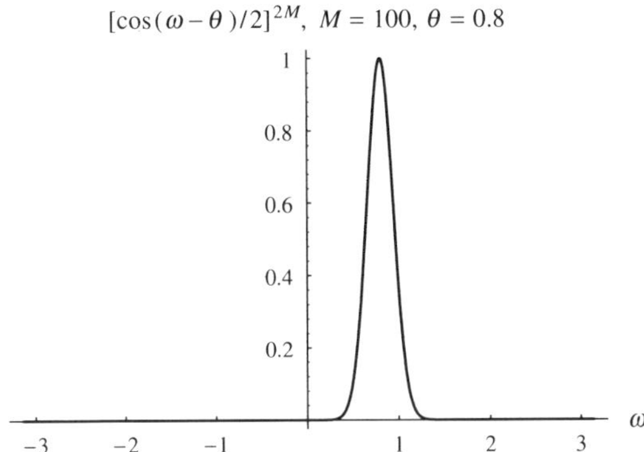

Figure 7.8: *The normalized squared gain of the band-pass filter* (7.42) *centered at* $\theta = 0.8$, *with* $M = 100$.

The disadvantage of the parametric filter (7.42) is that to achieve a sharp reduction of bandwidth, M must be rather large as compared with the complex exponential filter. This can create computational difficulties.

7.3.5 Stochastic Convergence of the HK and CM Algorithms

So far we dealt with the convergence of the CM algorithm using theoretical HOC sequences of the form $\{\rho_1(\alpha_k)\}$, $k = 1, 2, \ldots$. In this section we deal with the asymptotic stochastic convergence of the CM algorithm replacing $\rho_1(\alpha_k)$ by sample estimates. This is done mostly in the band-pass case in terms of HOC from zero-crossings and under the Gaussian assumption. Then the CM algorithm reduces to the HK algorithm. In pursuing this goal, it is important to keep in mind that we have at our disposal two variables whose growth can be controlled, the series size N and the number of iterations of the algorithm. As before, since for the most part we have in mind band-pass filters, we assume without loss of generality that $p = 1$.

Convergence of the HK Algorithm

Suppose the parametric family $\{\mathcal{L}_r\}$ defines a family of band-pass filters satisfying (7.18), where $\mathcal{L}_r$ has bandwidth of size W_r stretching from ${}_*\omega$ to ω^*, $[{}_*\omega, \omega^*] \subset [0, \pi]$. Let $\{D_r\}$ be the corresponding HOC family obtained from a time series of length N. Consider the asymptotic empirical zero-crossing rate,

$$\gamma_r \equiv \lim_{N\to\infty} \hat{\gamma}(r) = \lim_{N\to\infty} \frac{D_r}{N-1}$$

From the discussion in Chapter 6 we know that $\pi\gamma_r$ has a good chance of being a random variable (mixed spectrum in a Gaussian process), but from Theorem 6.7 we also know that it (and hence also its expected value) falls in the band-pass with probability one. Thus, by controlling W_r, we can control the variability of the normalized rate $\pi\gamma_r$, and make it arbitrarily small. In fact, it follows from Theorem 6.7 that for any number a,

$$\pi|\gamma_r - E(\gamma_r)||\sin(a)| \leq W_r \tag{7.46}$$

with probability one. This motivates assumption **A2** below.

To simplify our notation, we write the contraction mapping (7.32) in its basic form, suppressing the dependence on the bandwidth W_r,

$$\rho_1(r) = r^* + C(r)(r - r^*)$$

This and the cosine formula give

$$\rho_1(r) = \cos(E(\pi\gamma_r)) = r^* + C(r)(r - r^*)$$

The Mean Value Theorem implies

$$\cos(\pi\gamma_r) - \cos(E(\pi\gamma_r)) = -\pi(\gamma_r - E(\gamma_r))\sin(z_r)$$

where z_r is between $\pi\gamma_r$ and $E(\pi\gamma_r)$. If we define $y_r \equiv -\pi(\gamma_r - E(\gamma_r))\sin(z_r)$, then we obtain the *observed contraction mapping*

$$\cos(\pi\gamma_r) = r^* + C(r)(r - r^*) + y_r \tag{7.47}$$

This expression identifies the exact error incurred by replacing $\cos(E(\pi\gamma_r))$ by $\cos(\pi\gamma_r)$. Now consider the HK algorithm

$$r_{k+1} = \cos(\pi\gamma_{r_k}) \tag{7.48}$$

This and (7.47) give

$$r_{k+1} = r^* + [\prod_{j=1}^{r_k} C(r_j)](r_1 - r^*) + [\prod_{j=2}^{r_k} C(r_j)]y_{r_1} + \cdots + C(r_k)y_{r_{k-1}} + y_{r_k} \tag{7.49}$$

To achieve convergence, the bandwidth must go to 0 at a certain rate. To uncover this rate, we resort to the following assumptions.

A1. Motivated by Proposition 7.1 and Figure 7.7, we assume that in the neighborhood of r^*, $C(r_k)$ is smaller than the bandwidth times a constant. Formally we assume that there is a positive constant K such that

$$0 < C(r_k) < KW_k$$

A2. Motivated by (7.46), we assume that

$$|y_{r_k}| < KW_k$$

where, for simplicity, K is as in **A1**.

Remark. 7.1. Figure 7.6 indicates that there are cases where $C(r_k)$ is very close to 0 in the neighborhood of r^* in addition to being much smaller than the bandwidth. Thus, **A1** is very plausible. Note, however, that we stress the requirement of being in the neighborhood of r^*, for otherwise, when the frequency "is not captured," $C(r_k)$ is equal to 1 (case of ideal band-pass) and convergence toward the frequency does not occur.

Remark. 7.2. In light of (7.46), **A2** is very plausible also. However, it should be noted that in (7.46), the parameter r is fixed, while it is random in **A2**.

So, under these assumptions,

$$|r_{k+1} - r^*| \leq KW_k \{1 + KW_{k-1} + \cdots + (KW_2 \cdots KW_{k-1})\} + \left\{ \prod_{j=1}^{k} KW_j \right\} (r_1 - r^* + 1) \qquad (7.50)$$

We can now shrink the bandwidth W_k, as $k \to \infty$, at a rate (for example, at a rate of $1/k$) that guarantees the almost sure convergence of r_k to r^*. In this case,

$$\pi \gamma_{r_k} \to \omega_1$$

almost surely. In summary we have.

Theorem 7.2. [21]. *Suppose that $\{Z_t\}$ in (7.1) is Gaussian, and let $p = 1$. Let $\{\mathcal{L}_r\}$, $r \in (-1, 1)$, be a parametric family of time-invariant linear band-pass filters, with bandwidth W_r, for which the fundamental property (7.18) holds, and one that passes ω_1. Assume that the filtered process $\{Z_t(r)\}$ is real-valued for all $r \in (-1, 1)$. Consider the recursion*

$$r_{k+1} = \cos(\pi \gamma_{r_k})$$

and suppose W_{r_k} is of the order $O(1/k)$. If A1 and A2 hold, then

$$\pi \gamma_{r_k} \to \omega_1 \qquad (7.51)$$

with probability one.

There is an interesting way to support the almost sure convergence of the HK algorithm in Theorem 7.2 by making reference to Corollary 6.1. To see that, consider the process (7.1) with $p = 1$, and let

$$S_t \equiv A_1 \cos(\omega_1 t) + B_1 \sin(\omega_1 t)$$

Then

$$Z_t = S_t + \zeta_t$$

Let $\{\mathcal{L}_r\}$, $r \in (-1, 1)$, be as in Theorem 7.2, and put $r = r^* = \cos(\omega_1)$. Then

$$\mathcal{L}_{r^*}(Z)_t = \mathcal{L}_{r^*}(S)_t + \mathcal{L}_{r^*}(\zeta)_t$$

Since r^* is a fixed point of $\rho_1(r)$, $\mathcal{L}_{r^*}(S)_t$ and $\mathcal{L}_{r^*}(\zeta)_t$ have r^* as their first order autocorrelation, and so

$$\pi\hat{\gamma}(r^*) \rightarrow E(\pi\gamma_{r^*}) = \omega_1$$

almost surely. Therefore, for r in sufficiently small neighborhoods of r^*, the empirical zero-crossing rate $\hat{\gamma}(r)$ approaches a constant, and we can expect the HK algorithm (in fact, band-pass filters or not) to converge to r^* with a high probability.

Convergence of the General CM Algorithm

Similar results can be proved for the general CM procedure using observed higher order correlations rather than higher order crossings. Under appropriate conditions the CM estimator is strongly consistent and asymptotically normal. For rigorous proofs the reader is referred to [23], [25], [26], [27]. In this subsection we specialize to the case $p = 1$, real filters, and concentrate briefly on the asymptotic normality result.

For $p = 1$ rewrite (7.1) as

$$Z_t = \beta\cos(\omega_1 t + \phi) + \epsilon_t$$

where β is a positive constant, and

$$\epsilon_t = \sum_j \psi_j u_{t-j}$$

with $\sum_j |\psi_j| < \infty$, and the u_t are independent and identically distributed random variables with mean 0 and variance σ_u^2. The phase ϕ is uniformly distributed in $(-\pi, \pi]$, independently of $\{u_t\}$. As before, we let $\mathcal{L}_\alpha$, $\alpha \in (-1, 1)$, be a family of causal time-invariant linear filters (band-pass or not) with impulse response $\{h_j(\alpha)\}_{j=0}^{\infty}$, and corresponding transfer function

$$H(\omega; \alpha) = \sum_{j=0}^{\infty} h_j(\alpha)\exp(-ij\omega)$$

and assume the fundamental property

$$\text{Corr}(\mathcal{L}_\alpha(\epsilon)_t, \mathcal{L}_\alpha(\epsilon)_{t-1}) = \alpha$$

In the notation of Chapter 6, given a finite time series, $Z_0, Z_2, \ldots, Z_{N-1}$, define the filtered data by

$$\hat{Z}_t(\alpha) \equiv \sum_{j=0}^{t} h_j(\alpha) Z_{t-j}, \qquad t = 0, 1, 2, \ldots, N-1$$

The sample HOC is given by

$$\hat{\rho}_1(\alpha) = \frac{\sum_{t=1}^{N-1} \hat{Z}_t(\alpha)\hat{Z}_{t-1}(\alpha)}{\sum_{t=0}^{N-1} \hat{Z}_t^2(\alpha)}$$

and the CM algorithm becomes

$$\alpha_{k+1} = \hat{\rho}_1(\alpha_k), \qquad k = 0, 1, 2, \ldots$$

As in the original HK, we assume that $\alpha^* = \cos(\omega_1)$ is a fixed point of $\rho_1(\alpha)$. The contraction mapping is then

$$\rho_1(\alpha) = \alpha^* + C(\alpha)(\alpha - \alpha^*)$$

It is convenient to introduce the quantities $G(\alpha) = 1 - C(\alpha)$ and

$$\text{snr}(\alpha) \equiv \frac{\text{Variance of the filtered signal}}{\text{Variance of the filtered noise}}$$

Then

$$C(\alpha) = \frac{1}{1 + \text{snr}(\alpha)}$$

and

$$G(\alpha) = \frac{\text{snr}(\alpha)}{1 + \text{snr}(\alpha)}$$

It has been shown in [25], under the conditions H1,H2 in Chapter 6, that if $C(\alpha^*) < 1$, then $\hat{\rho}_1(\alpha)$ has almost surely a unique fixed point, denoted by $\hat{\alpha}$, and that, as $N \to \infty$, $\hat{\alpha}$ converges almost surely to α^*, provided the initial guess α_1 is not too far from $\cos(\omega_1)$. It follows that the estimator

$$\hat{\omega}_1 \equiv \cos^{-1}(\hat{\rho}_1(\hat{\alpha}))$$

converges with probability one to ω_1. We refer to $\hat{\omega}_1$ and $\hat{\alpha}$ as the *CM estimators*. It turns out that both $\hat{\omega}_1$ and $\hat{\alpha}$ are asymptotically normally distributed.

Let $\rho_k^\epsilon(\alpha)$ be the autocorrelation of the filtered noise $\{\mathcal{L}_\alpha(\epsilon)_t\}$. Then we first have:

Lemma 7.1. [27] *Assume that $E[u^4] < \infty$, and that $\sum_j j|h_j(\alpha^*)| < \infty$. Then, as $N \to \infty$,*

$$\sqrt{N}(\hat{\rho}_1(\alpha^*) - \rho_1(\alpha^*)) \xrightarrow{\mathcal{L}} \mathcal{N}(0, v_\rho C^2(\alpha^*))$$

where

$$v_\rho = \sum_{k=-\infty}^{\infty} \{(1 + 2\alpha^{*2})(\rho_k^\epsilon(\alpha^*))^2 - 4\alpha^* \rho_k^\epsilon(\alpha^*)\rho_{k+1}^\epsilon(\alpha^*) + \rho_{k+1}^\epsilon(\alpha^*)\rho_{k-1}^\epsilon(\alpha^*)\}$$

From this we have:

Theorem 7.1. [27] *Assume that $E[u^4] < \infty$. Suppose H1 and H2 are satisfied and that $G(\alpha) \geq g > 0$ for all α in a neighborhood of α^*. Then, as $N \to \infty$,*

$$\sqrt{N}(\hat{\alpha} - \alpha^*) \xrightarrow{\mathcal{L}} \mathcal{N}(0, \sigma_\alpha^2)$$

and

$$\sqrt{N}(\hat{\omega} - \omega_1) \xrightarrow{\mathcal{L}} \mathcal{N}(0, \sigma_\omega^2)$$

where

$$\sigma_\alpha^2 = \frac{\upsilon_\rho}{snr^2(\alpha^*)} \qquad \textit{and} \qquad \sigma_\omega^2 = \frac{\upsilon_\rho}{snr^2(\alpha^*)(1 - \alpha^{*2})}$$

and υ_ρ *is as in Lemma 7.1.*

The proofs of these results are too technical and are omitted.

The *AR*(2) Filter

Regarding the asymptotic normality result of the previous subsection, we now consider an important special case studied recently in [31] and [38]. Assume $\epsilon_t \equiv u_t$ is white noise consisting of independently and identically distributed (IID) random variables with mean 0 and variance σ_ϵ^2, and consider the stable $AR(2)$ filter [8], [12], [27],

$$\mathcal{L}_\alpha \equiv \frac{1}{1 + \theta(\alpha)\eta\mathcal{B} + \eta^2\mathcal{B}^2} \tag{7.52}$$

where $0 < \eta < 1$, and

$$\theta(\alpha) \equiv -\frac{1 + \eta^2}{\eta}\alpha \tag{7.53}$$

The squared gain is given by

$$|H(\omega; \alpha)|^2 = \frac{1}{a + b\cos(\omega) + c\cos(2\omega)}$$

where the dependence on η is suppressed for simplicity, and

$$a = 1 + (1 + \eta^2)^2\alpha^2 + \eta^4, \qquad b = -2(1 + \eta^2)^2\alpha, \qquad c = 2\eta^2$$

Assume that $|\alpha| < 2\eta/(1 + \eta^2)$. Then we can always write

$$\theta(\alpha) = -2\cos(\omega(\alpha))$$

for some $\omega(\alpha) \in [0, \pi]$. With this it is easy to verify that the poles of the filter are $\eta\exp(\pm i\omega(\alpha))$. The parameter η controls the bandwidth—smaller η implies a wider bandwidth—and as long as $\eta < 1$, the bandwidth is wider than in the limiting $AR(2)$ filter (7.27). For η close to 1, the filter center is approximately at $\cos^{-1}(\alpha)$. The dependence of the filter on η is illustrated in Figure 7.9 for two cases corresponding to $(\alpha = 0.5, \eta = 0.96)$, and $(\alpha = 0.5, \eta = 0.98)$

It can be checked that the fundamental property $\alpha = \rho_{1,\epsilon}(\alpha)$ holds relative to white noise [12], and some algebra shows that [27]

$$\sigma_\alpha^2 = \left(\frac{1 - \eta^2}{1 + \eta^2}\right)^3 \frac{4\sigma_\epsilon^4(1 - \alpha^{*2})}{\beta^4} \tag{7.54}$$

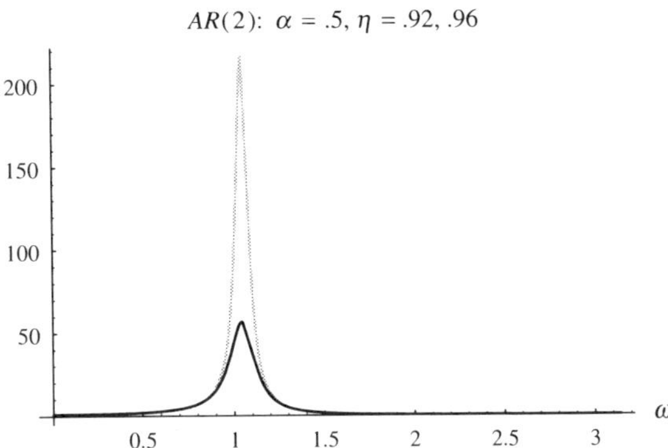

Figure 7.9: *Squared gain* $|H(\omega;\alpha)|^2$, $\omega \in [0,\pi]$, *of the stable AR(2) filter with* $(\alpha = 0.5, \eta = 0.92)$ *(dark line), and* $(\alpha = 0.5, \eta = 0.96)$ *(gray line). The approximate center is at* $\cos^{-1}(\alpha) = \pi/3$.

$$\sigma_\omega^2 = \left(\frac{1-\eta^2}{1+\eta^2}\right)^3 \frac{4\sigma_\epsilon^4}{\beta^4} = O((1-\eta)^3) \tag{7.55}$$

We can see from (7.55) that in order to achieve a small asymptotic variance η should be close to 1; however, the case $\eta = 1$ is of a different nature and requires special considerations. We cannot just go ahead and substitute $\eta = 1$ in (7.55), for then we get a nonsensical result of zero variance. In the extreme case when $\eta = 1$, the filter reduces to the special case (7.27), except for a factor of 2. In this case it can be shown that [31], [38],

$$N^{3/2}(\hat{\omega} - \omega_1) \xrightarrow{\mathcal{L}} \mathcal{N}(0, \sigma_{\text{NLS}}^2) \tag{7.56}$$

where $\sigma_{\text{NLS}}^2 = 24\sigma_\epsilon^2/\beta^2$, is the asymptotic (normalized) variance of the nonlinear least squares estimator of ω_1.

The fact that, as $\eta \to 1$ and $N \to \infty$, the CM estimator approaches the asymptotic efficiency of the nonlinear least squares estimator has been verified by an extensive simulation summarized for $\omega_1 = 0.42\pi$ and SNR$=0$ in Figure 7.10 [27]. The figure shows estimates of the $-\log$ (normalized) mean square error, MSE $\equiv E(\hat{\omega} - \omega_1)^2$, of the CM estimator (solid thin lines) together with the corresponding normalized asymptotic variance of the nonlinear least squares estimator (thick solid line)—it may be interpreted as an approximate (normalized) MSE—for various values of η and N. The MSE estimates were obtained from 100 independent realizations corresponding to (from the bottom up) $\eta = 0.85, 0.90, 0.95, 0.98, 0.99, 0.999$. To recover N, multiply the horizontal figures by 100.

The figure also displays the asymptotic variance (7.55) (dashed lines). for (from the bottom up) $\eta = 0.85, 0.90, 0.95, 0.98, 0.99$.

The figure shows that as $\eta \to 1$ and $N \to \infty$, the MSE of the CM estimator corresponding to the $AR(2)$ filter (7.52) approaches that of the nonlinear least

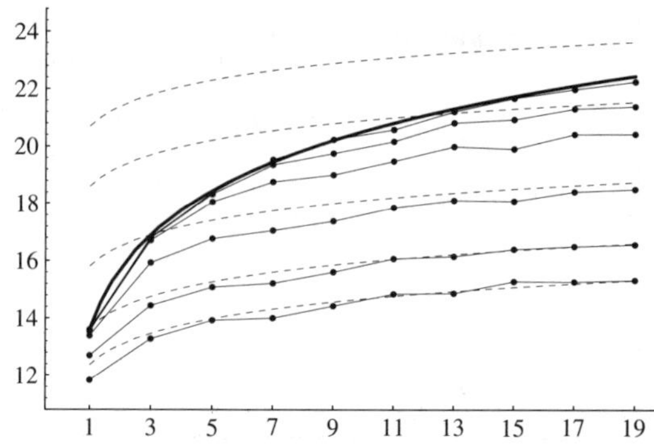

Figure 7.10: *−Log plot of (normalized) estimated MSE of the CM estimator (thin solid lines) corresponding to various values of η as a function of sample size ($\times 100$). The thick solid line is the approximate MSE of the nonlinear least squares estimator. The dashed lines are plots of the variance formula (7.55).* $\omega_1 = 0.42\pi, SNR = 0$.

squares frequency estimator. *More generally, this tells us that there are filters for which the CM estimator can achieve the precision of nonlinear least squares, but with much less computation.*

It should be pointed out, however, that the critical case $\eta = 1$ requires a rather close initial estimate (guess) of ω_1 to begin the CM procedure. A certain "accelerated" version discussed in [31] requires the initial guess to be accurate to order $O(N^{-1})$. This is so, because the bandwidth of the $AR(2)$ filter with $\eta = 1$ is very narrow and it can easily lose a potential frequency if the initial guess is not sufficiently precise (see Example 7.6). This problem can be rectified if the CM algorithm is used with $\eta < 1$, and η is allowed to approach 1 sequentially [8], [23]. In general, it is preferable to use filters whose bandwidths can be narrowed down gradually.

7.3.6 Examples

We shall now illustrate by several examples the CM method through its special case—the HK algorithm—using HOC from zero-crossings.

The HK algorithm, $r_{k+1} = \cos(\pi\hat{\gamma}(r_k))$, was applied (with shrinking bandwidth) to time series of length[4] $N = 2000$ from the process (7.1), with Gaussian white noise ζ_t, unless stated otherwise, using the real counterpart of the complex filter (7.36),

---

[4]More precisely, for a given M_k, the zero-crossing count is computed from time series of length $2000 + 2M_k$.

$$h(n;r,M) = \begin{cases} \cos(n\theta(r))/\sqrt{2M+1}, & |n| \leq M \\ 0, & |n| > M \end{cases} \tag{7.57}$$

We refer to this filter as the "cosine filter." The corresponding squared gain, $|\tilde{H}(\omega;r,M)|^2$, is closely related to $|H(\omega;r,M)|^2$ in (7.37) for $\omega > 0$,

$$|\tilde{H}(\omega;r,M)|^2 = \frac{1}{4}|H(\omega;r,M)|^2 + O(\frac{1}{M}) \tag{7.58}$$

Figure 7.11 shows the graph of $|\tilde{H}(\omega;r,M)|^2$, $\omega > 0$, with $(M_1 = 20, \theta(r_1) = 1)$ and $(M_2 = 40, \theta(r_2) = 1.1)$. The figure illustrates a shift in location accompanied by a reduced bandwidth, the idea behind the speeded up convergence.

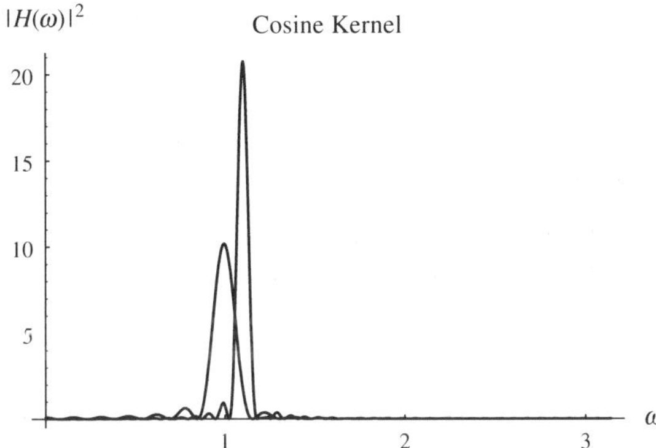

Figure 7.11: *The squared gain of the cosine filter centered at $\theta = 1$ with $M = 20$, and centered at $\theta = 1.1$ with $M = 40$.*

The convergence of the algorithm was enhanced by shrinking the bandwidth (linearly) at each iteration. As can be seen from the tables in the following examples, the convergence is very fast, and the frequency is never lost when M_k increases at a linear rate such as $\{5, 15, 25, 35, \ldots\}$. Roughly, this is equivalent to a bandwidth that shrinks at the order of $O(1/k)$. Also note that the sequence of observed zero-crossing rates $\{\hat{\gamma}(r_k)\}$ is monotone, with a possible exception at the very beginning.

Example 7.2: A Simple Case. $p = 1$

The results of this example are typical. In comparison with the results of Table 7.4, we can see that shrinking the bandwidth increases the precision of the frequency estimate, despite a sharp reduction in the series size.

TABLE 7.5 $\omega_1 = 0.8, \theta(r_0) = 0.4, A_1 = -0.91333, B_1 = 0.06975$, SNR = 0 dB

k	M_k	r_k	$\pi\hat{\gamma}(r_k)$
0	5	0.92106	0.65221
1	15	0.79475	0.75279
2	25	0.72979	0.79994
3	35	0.69675	0.79994
4	45	0.69675	0.79994
5	55	0.69675	0.79994
.	.	.	.
.	.	.	.
.	.	.	.
			$\hat{\omega}_1 = 0.79994$

■

Example 7.3: Non-Gaussian and Colored Noise. $p = 1$

In the band-pass case, when the fundamental property (7.18) holds for white noise, it also holds approximately for slowly varying continuous spectral densities, as long as the bandwidth is sufficiently narrow. In this respect, it is of interest to apply the HK algorithm with the cosine filter and colored Gaussian noise. It is also interesting to run the algorithm when the noise is white but non-Gaussian. Accordingly, we have applied the algorithm when ζ_t is Gaussian white noise (GWN), lognormal white noise (LNWN) (with parameters 0,1 , and properly centered and scaled), and a first order moving average (MA(1)), $u_t = \zeta_t - 0.8\zeta_{t-1}$, where ζ_t is Gaussian white noise. In all three cases the sinusoidal component was identical. The results, given in Table 7.6, show identical convergence. Evidently, in the band-pass case, the algorithm is quite robust.

TABLE 7.6 $\omega_1 = 0.71, \theta(r_0) = 1.1, A_1 = -0.954817, B_1 = 1.00687$, SNR = 0 dB

		GWN	LNWN	MA(1)
k	M_k	$\pi\hat{\gamma}(r_k)$	$\pi\hat{\gamma}(r_k)$	$\pi\hat{\gamma}(r_k)$
0	5	0.993240	0.913089	0.997955
1	15	0.920947	0.897373	0.928805
2	25	0.812508	0.766932	0.848654
3	35	0.766932	0.710355	0.748073
4	45	0.719785	0.710355	0.710355
5	55	0.710355	0.710355	0.710355
6	65	0.710355	0.710355	0.710355
.	.	.	.	.
.	.	.	.	.
.	.	.	.	.
		$\hat{\omega}_1 = 0.710355$	$\hat{\omega}_1 = 0.710355$	$\hat{\omega}_1 = 0.710355$

■

Example 7.4: Low SNR. $p = 1$

TABLE 7.7 $\omega_1 = 1.95$, $\theta(r_0) = 1.8$, $A_1 = 0.174937$, $B_1 = -0.263656$, SNR $= -13.9794$ dB

k	M_k	r_k	$\pi\hat{\gamma}(r_k)$
0	5	-0.22720	1.87647
1	15	-0.30094	1.89690
2	25	-0.32035	1.91576
3	35	-0.33816	1.93776
4	45	-0.35878	1.95191
5	55	-0.37195	1.94876
6	65	-0.36903	1.94876
7	75	-0.36903	1.94876
8	85	-0.36903	1.94876
.	.	.	.
.	.	.	.
.	.	.	.
			$\hat{\omega}_1 = 1.94876$

■

Example 7.5: Detection of a Diurnal Cycle

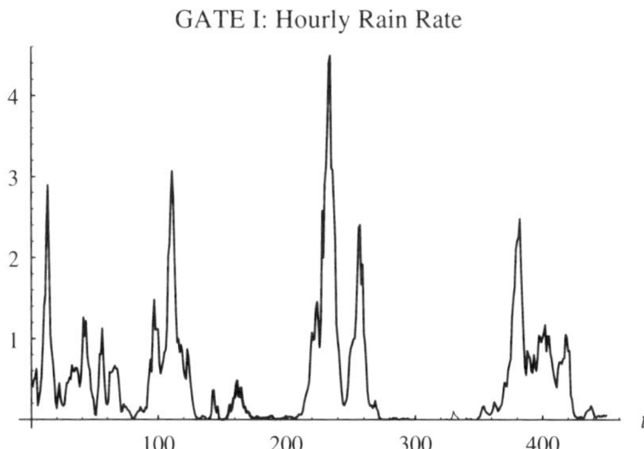

Figure 7.12: *GATE I data: interpolated hourly rain rate averaged over a* 280×280 *km*^{2} *area in the eastern Atlantic.* $N = 450$.

Recall the GATE I data set discussed early in the chapter. In search of the diurnal cycle, we have applied the HK algorithm, using the cosine filter, to a centered (i.e., the mean was subtracted) time series of length $N = 450$ of hourly rain rate[5] from GATE I, averaged over a region of 280×280 km^2. The data are recorded in Table A.1 in Appendix A and plotted in Figure 7.12. As can be seen from the following table, starting from various

[5]Thanks are due to Dr. Tom Bell, NASA, Goddard Space Flight Center, for kindly supplying the hourly GATE data.

points, the convergence is toward the value 0.262675 radians per hour. This leads to a period of

$$\frac{2\pi}{0.262675} = 23.92 \text{ hours}$$

which gives more credence to the the existence of a diurnal cycle in GATE I [2].

TABLE 7.8 Application of the HK Algorithm in the Detection of the Diurnal Cycle in GATE I

		$\theta(r_0) = 0.50$	$\theta(r_0) = 0.45$	$\theta(r_0) = 0.30$
k	M_k	$\pi\hat{\gamma}(r_k)$	$\pi\hat{\gamma}(r_k)$	$\pi\hat{\gamma}(r_k)$
0	55	0.483322	0.441294	0.273182
1	56	0.420280	0.441294	0.262675
2	57	0.367745	0.399266	0.262675
3	58	0.346731	0.346731	0.262675
4	59	0.304703	0.304703	0.262675
5	60	0.294196	0.294196	0.262675
6	61	0.283689	0.283689	0.262675
7	62	0.262675	0.262675	0.262675
8	63	0.262675	0.262675	0.262675
9	64	0.262675	0.262675	0.262675
.	.	.	.	.
.	.	.	.	.
.	.	.	.	.
		$\hat{\omega}_1 = 0.262675$	$\hat{\omega}_1 = 0.262675$	$\hat{\omega}_1 = 0.262675$

■

Example 7.6: Comparison with an *AR*(2) Filter

To apply the CM algorithm in conjunction with the (truncated) $AR(2)$ filter (7.27) in Example 7.1, the algorithm must be initiated with a guess close to the true frequency. The filter bandwidth is just too narrow so that by starting "a little too far" from the true frequency, the algorithm converges to a false frequency. The cosine filter with a variable bandwidth is more flexible in this respect, and can accommodate rough initial guesses. A comparison between the $AR(2)$ and cosine filters is given in Tables 7.9 and 7.10.

Suppose that $N = 2000$, $\omega_1 = 0.8$, and

$$Z_t = 0.733825\cos(0.8t) + 1.34406\sin(0.8t) + \zeta_t$$

where $\zeta_t \sim \mathcal{N}(0, 1)$. The HK algorithm was applied using the $AR(2)$ filter with a fixed bandwidth, and the cosine filter with a variable bandwidth. When the initial center of the filter is $\theta(r_0) = 0.82$, both filters perform equally well and give a precise estimate (Table 7.9). When $\theta(r_0) = 0.6$, the $AR(2)$ filter performs poorly, but the cosine filter still gives a precise estimate of $\omega_1 = 0.8$ (Table 7.10).

TABLE 7.9 "Starting Close:" $\theta(r_0) = 0.82$

k	M_k	Cosine Filter $\pi\hat{\gamma}(r_k)$	$AR(2)$ Filter $\pi\hat{\gamma}(r_k)$
0	25	0.79994	0.80937
1	35	0.79994	0.80465
2	45	0.79994	0.79994
3	55	0.79994	0.79994
4	65	0.79994	0.79994
5	75	0.79994	0.79994
.	.	.	.
.	.	.	.
.	.	.	.
		$\hat{\omega}_1 = 0.79994$	$\hat{\omega}_1 = 0.79994$

Note: Comparison between the $AR(2)$ and cosine filters. The M_k column pertains to the cosine filter only.

TABLE 7.10 "Starting Far:" $\theta(r_0) = 0.6$

k	M_k	Cosine Filter $\pi\hat{\gamma}(r_k)$	$AR(2)$ Filter $\pi\hat{\gamma}(r_k)$
0	25	0.77951	0.60506
1	35	0.79994	0.60977
2	45	0.79994	0.61292
3	55	0.79994	0.61763
4	65	0.79994	0.62392
5	75	0.79994	0.65692
.	.	.	0.69464
.	.	.	0.71664
.	.	.	0.72136
.	.	.	0.72293
.	.	.	0.72293
.	.	.	0.72293
		$\hat{\omega}_1 = 0.79994$	$\hat{\omega}_1 = 0.72293$

Note: Comparison between the $AR(2)$ and cosine filters. The M_k column pertains to the cosine filter only.

■

Example 7.7: Detection of a Weak Component

Here we have two sinusoidal components where the first one is relatively weak. The parameters are $p = 2$, $\omega_1 = 0.97$, $\omega_2 = 2.1$, $A_1 = -0.699457, B_1 = -0.116106$, $A_2 = -1.15000$, $B_2 = 2.45159$, $\sigma_\zeta^2 = 2$. With $\theta(r_0) = 0.85$, the normalized HOC sequence $\{\pi\hat{\gamma}(r_k)\}$ converges to 0.96967. For $\theta(r_0) = 1.8$, the convergence is toward 2.10121. The results are given in Table 7.11.

TABLE 7.11 Estimation of two Frequencies

k	M_k	$\omega_1 = 0.97, \theta(r_0) = 0.85$ $\pi\hat{\gamma}(r_k)$	$\omega_2 = 2.1, \theta(r_0) = 1.8$ $\pi\hat{\gamma}(r_k)$
0	10	0.95709	1.83718
1	20	0.97281	2.02891
2	30	0.96967	2.09963
3	40	0.96967	2.10121
4	50	0.96967	2.10121
5	60	0.96967	2.10121
		$\hat{\omega}_1 = 0.96967$	$\hat{\omega}_2 = 2.10121$

■

7.4 A MULTIVARIATE CM METHOD

7.4.1 Extending the HOC Idea: Parametrized Least Squares

In a strict sense, the underlying idea behind HOC is to filter the data and then count (zero-crossings), filter again and count, and so on, or filter and observe the (first-order) correlation, filter and observe the correlation, and so on. More generally, HOC families are created by parameterizing the zero-crossing count or the first order correlation through parametric filtering. In this way, the zero-crossing count or the observed correlation are endowed with a parameter. This idea of *parametrization of observables*,[6] however, can be extended in many different ways. Most notably, using parametric filters we can *parameterize the least squares estimator* (derived from the filtered data) of a certain quantity to gain precision, and hence improve the overall performance of the estimator. This ramification of the HOC idea will be exploited in this section in the simultaneous estimation of several frequencies in noise.

We might add in passing that parametrization, as we see it, is actually a widespread practice. For example, any transform, including the Fourier and wavelet transforms, can be thought of as a parametrization mechanism.

As remarked earlier, the AR method for multiple frequency estimation in the presence of ambient noise leads to biased estimates. However, this predicament of the AR method can be remedied by combining the method with a parametric filter which satisfies a certain "fundamental property" analogous to (7.18). When the fundamental property in question holds, in addition to some other conditions, we obtain a contraction mapping in terms of the least squares estimator (of the AR coefficients) parameterized by the filter parameter. As in

[6]Actually, these are stochastic processes.

the single frequency case, the mapping is then iterated by fixed point iterations until convergence to the AR coefficients is achieved. The frequency estimates are obtained from the AR coefficients. Experience shows that this hybrid is resistant to noise levels as low as −4 dB signal-to-noise ratio. Moreover, computer simulations reveal that when the signal-to-noise ratio reaches +3 dB or more, the method is capable of resolving successfully two closely spaced frequencies within a Fourier bin.

Thus, the combination of the CM and AR methods—which turns a least squares estimator into a (stochastic) multivariate contraction mapping—sets the stage for a significant improvement of the AR method. Special cases of this have also been discussed in [8], [14], [29], [31], [38]. We follow the more general development in [23], [26].

7.4.2 The *AR* Method and Its Unbiased Modification

Consider the model of random phases (6.17) with constant amplitudes,

$$Z_t = \sum_{k=1}^{p} \beta_k \cos(\omega_k t + \phi_k) + \epsilon_t \tag{7.59}$$

where p is known,

$$\epsilon_t = \sum_j \psi_j u_{t-j}, \qquad \sum_j |\psi_j| < \infty$$

and the u_t are independent $\mathcal{N}(0, \sigma_u^2)$ random variables, and are also independent of the ϕ_k. Given a time series of length N from (7.59), the nonlinear least squares approach to the frequency estimation problem is to fix the phases and then minimize the sum of squares $\sum_{t=1}^{N} \epsilon_t^2$ with respect to the resulting amplitudes and frequencies. On the other hand, the bulk of the AR approach is linear, a fact which makes it attractive in applications.

The AR method derives its name from the fact that (7.59) admits an autoregressive representation as we now explain (the method is also known as Prony's spectral line estimator [15]). Let $\mathcal{B}$ be the shift operator, $\mathcal{B}X_t = X_{t-1}$, and consider the polynomial

$$Q(\mathcal{B}) = \prod_{k=1}^{p} (\mathcal{B} - e^{i\omega_k})(\mathcal{B} - e^{-i\omega_k})$$

Then

$$Q(\mathcal{B})Z_t = Q(\mathcal{B}) \sum_{k=1}^{p} \beta_k \cos(\omega_k t + \phi_k) + Q(\mathcal{B})\epsilon_t$$

But since $Q(\mathcal{B}) \sum_{k=1}^{p} \beta_k \cos(\omega_k t + \phi_k) = 0$, we deduce

$$Q(\mathcal{B})Z_t = Q(\mathcal{B})\epsilon_t$$

Therefore, $\{Z_t\}$ follows an $AR(2p)$ model,[7]

$$\sum_{j=0}^{2p} a_j Z_{t-j} = e_t \tag{7.60}$$

where $e_t = \sum_{j=0}^{2p} a_j \epsilon_{t-j}$, and where the AR parameters a_j are real and identifiable in terms of the frequencies form the polynomial equation

$$Q(z) = \sum_{j=0}^{2p} a_j z^{2p-j} = \prod_{k=1}^{p} (z - e^{i\omega_k})(z - e^{-i\omega_k})$$

In addition, the a_j satisfy the symmetry relation

$$a_0 = 1, \qquad a_{2p-j} = a_j, \qquad j = 0, \ldots, p-1 \tag{7.61}$$

The key fact to observe is that the a_j are completely equivalent to the ω_k. That is, the parameters a_j determine and are determined by the ω_k. For example, in the single frequency case of $p = 1$, $a_0 = a_2 = 1$, and when—a_1 is known—ω_1 is obtained from the equation $a_1 = -2\cos(\omega_1)$. For $p = 2$, the frequencies can be obtained from the equations $a_1 = -2\cos(\omega_1) - 2\cos(\omega_2)$, $a_2 = 2 + 4\cos(\omega_1)\cos(\omega_2)$. In general, once the a_j become available, the frequencies are obtained from the zeros of $Q(z)$. It follows that the frequency estimation problem is equivalent to the estimation of the $p \times 1$ vector of AR parameters

$$\mathbf{a} = (a_1, a_2, \ldots, a_p)'$$

The latter can be estimated by several methods, of which the method of least squares is perhaps the first which comes to mind. However, surprisingly enough, the least squares estimator does not live up to expectations. This is discussed next.

Estimation of the *AR* Parameters

Given a time series $Z_1, Z_2, \ldots, Z_N$, from (7.60) with $N > 2p$, we wish to estimate the vector of AR parameters $\mathbf{a}$ by the method of least squares. With

$$\mathbf{z} \equiv \begin{bmatrix} Z_{2p+1} + Z_1 \\ \vdots \\ Z_N + Z_{N-2p} \end{bmatrix} \quad \mathbf{Z} \equiv \begin{bmatrix} Z_{2p} & \cdots & Z_2 \\ \vdots & & \vdots \\ Z_{N-1} & \cdots & Z_{N-2p+1} \end{bmatrix} \quad \mathbf{e} \equiv \begin{bmatrix} e_{2p+1} \\ \vdots \\ e_N \end{bmatrix}$$

and a $(2p-1) \times p$ matrix $\mathbf{Q}$ (similar but not quite the same as in[8]),

$$\mathbf{Q} \equiv \begin{bmatrix} \mathbf{I} & \mathbf{0} \\ \mathbf{0}' & 1 \\ \tilde{\mathbf{I}} & \mathbf{0} \end{bmatrix}$$

[7]Strictly speaking it is an $ARMA(2p, 2p)$ model.

where $\mathbf{I}$ is a $(p-1) \times (p-1)$ identity matrix, $\tilde{\mathbf{I}}$ is a $(p-1) \times (p-1)$ reverse permutation matrix with 1s on the antidiagonal and 0s elswhere, and $\mathbf{0}$ a $(p-1) \times 1$ zero vector, (7.61) together with (7.60) imply the linear model

$$\mathbf{z} = -\mathbf{ZQa} + \mathbf{e} \tag{7.62}$$

The least squares estimator of $\mathbf{a}$ is obtained by minimizing the sum of squares

$$\|\mathbf{e}\|^2 = \|\mathbf{z} + \mathbf{ZQa}\|^2$$

and is given by

$$\mathbf{a}_{\mathrm{LS}} = -(\mathbf{Q}'\mathbf{Z}'\mathbf{ZQ})^{-1}\mathbf{Q}'\mathbf{Z}'\mathbf{z} \tag{7.63}$$

We now show that the least squares estimator $\mathbf{a}_{\mathrm{LS}}$ is *biased* and hence *inconsistent*.

By the same arguments that lead to Theorem 6.8,

$$\frac{1}{N}\mathbf{Z}'\mathbf{Z} \overset{\text{a.s.}}{\rightarrow} \overline{\mathbf{R}}_z, \qquad \frac{1}{N}\mathbf{Z}'\mathbf{z} \overset{\text{a.s.}}{\rightarrow} \overline{\mathbf{r}}_z \tag{7.64}$$

where

$$\overline{\mathbf{R}}_z \equiv \begin{bmatrix} R_z(0) & R_z(-1) & \cdots & R_z(-2p+2) \\ R_z(1) & R_z(0) & \cdots & R_z(-2p+3) \\ \vdots & \vdots & \ddots & \vdots \\ R_z(2p-2) & R_z(2p-3) & \cdots & R_z(0) \end{bmatrix}$$

and

$$\overline{\mathbf{r}}_z \equiv \begin{bmatrix} R_z(1) \\ R_z(2) \\ \vdots \\ R_z(2p-1) \end{bmatrix} + \begin{bmatrix} R_z(2p-1) \\ R_z(2p-2) \\ \vdots \\ R_z(1) \end{bmatrix}$$

Clearly, by reversing the role of the Z's and ϵ's we also have, similarly to (7.62),

$$\boldsymbol{\epsilon} = -\mathbf{EQa} + \mathbf{e} \tag{7.65}$$

or

$$\mathbf{e} = \boldsymbol{\epsilon} + \mathbf{EQa} \tag{7.66}$$

(here $\mathbf{E}$ stands for uppercase $\boldsymbol{\epsilon}$) and we obtain the respective quantities $\overline{\mathbf{R}}_\epsilon$ and $\overline{\mathbf{r}}_\epsilon$. Therefore, by the independence of the ϕ_k and $\{\epsilon_t\}$, and from (7.62), (7.63), (7.64), and (7.66),

$$\begin{aligned} \mathbf{a}_{\mathrm{LS}} &\overset{\text{a.s.}}{\rightarrow} -(\mathbf{Q}'\overline{\mathbf{R}}_z\mathbf{Q})^{-1}\mathbf{Q}'\{-\overline{\mathbf{R}}_z\mathbf{Qa} + \overline{\mathbf{r}}_\epsilon + \overline{\mathbf{R}}_\epsilon\mathbf{Qa}\} \\ &= \mathbf{a} - \mathbf{R}_z^{-1}\{\mathbf{r}_\epsilon + \mathbf{R}_\epsilon\mathbf{a}\} \end{aligned} \tag{7.67}$$

where $\mathbf{R}_z \equiv \mathbf{Q}'\overline{\mathbf{R}}_z\mathbf{Q}$, $\mathbf{r}_z \equiv \mathbf{Q}'\overline{\mathbf{r}}_z$, and $\mathbf{R}_\epsilon, \mathbf{r}_\epsilon$ are defined in the same way from $\{\epsilon_t\}$. And so, *the least squares estimate is biased* [8]. We observe from (7.67) that as the signal-to-noise ratio tends to infinity, the bias disappears. This explains why the AR method requires a high signal-to-noise ratio (e.g., 30 dB) for successful implementation [9], [41]. But in general, even for a relatively high signal-to-noise ratio (e.g., 5 to 10 dB) the bias persists and distorts the frequency estimates. To get rid of the bias we turn to parametric filtering.

Unbiased *AR* Method

There are parametric filters which parameterize $\mathbf{a}_{\mathrm{LS}}$ and eliminate the bias iteratively. This is where the *fundamental property* enters into the picture. The end result is a modified least squares estimator which itself is a *contraction mapping in terms of the filter parameter* whose fixed point is a consistent estimator for $\mathbf{a}$ [23], [26].

Let $\mathcal{L}_{\boldsymbol{\alpha}}$ be a parametric family of linear filters which pass all the frequencies ω_k for each fixed $\boldsymbol{\alpha}$. Then the filtered process as well as the filtered signal and noise are endowed with an extra parameter $\boldsymbol{\alpha}$,[8]

$$Z_t(\boldsymbol{\alpha}) = S_t(\boldsymbol{\alpha}) + \epsilon_t(\boldsymbol{\alpha})$$

where

$$S_t(\boldsymbol{\alpha}) = \sum_{k=1}^{p} \beta_k(\boldsymbol{\alpha}) \cos(\omega_k t + \phi_k(\boldsymbol{\alpha}))$$

Except for the extra parameter, nothing has really been changed. The corresponding matrices and vectors are now parametrized by $\boldsymbol{\alpha}$, and we define

$$\mathbf{R}_s(\boldsymbol{\alpha}) = \mathbf{R}_z(\boldsymbol{\alpha}) - \mathbf{R}_\epsilon(\boldsymbol{\alpha}) \tag{7.68}$$

To simplify the exposition and cut down on technicalities, we shall assume that the family of filters $\mathcal{L}_{\boldsymbol{\alpha}}$ is such that all the relevant parametrized matrices are nonsingular, and that all the relevant parametrized limits exist. Let $\mathbf{a}(\boldsymbol{\alpha})$ be the almost sure limit of the least squares estimator from the filtered data, $a_{\mathrm{LS}}(\boldsymbol{\alpha})$,

$$a_{\mathrm{LS}}(\boldsymbol{\alpha}) \stackrel{\text{a.s.}}{\rightarrow} \mathbf{a}(\boldsymbol{\alpha}), \qquad N \rightarrow \infty$$

Then from (7.67)

$$\mathbf{a}(\boldsymbol{\alpha}) = \mathbf{a} - \mathbf{R}_z^{-1}(\boldsymbol{\alpha})\{\mathbf{r}_\epsilon(\boldsymbol{\alpha}) + \mathbf{R}_\epsilon(\boldsymbol{\alpha})\mathbf{a}\} \tag{7.69}$$

or because of (7.68),

$$\mathbf{a}(\boldsymbol{\alpha}) - \mathbf{a} = -\mathbf{R}_s^{-1}(\boldsymbol{\alpha})\{\mathbf{r}_\epsilon(\boldsymbol{\alpha}) + \mathbf{R}_\epsilon(\boldsymbol{\alpha})\mathbf{a}(\boldsymbol{\alpha})\} \tag{7.70}$$

In direct analogy with the fundamental property (7.18), $r = \rho_{1,\zeta}(r)$, we require that the filter be parametrized such that

$$\boldsymbol{\alpha} = -\mathbf{R}_\epsilon^{-1}(\boldsymbol{\alpha})\mathbf{r}_\epsilon(\boldsymbol{\alpha}) \tag{7.71}$$

[8] Technically, α takes values in an open set $\mathcal{A}$.

The rationale for this is the observation that *when* $\mathbf{a}(\boldsymbol{\alpha})$ *has a fixed point* $\boldsymbol{\alpha}^*$, *the bias vanishes*.

Equipped with the fundamental property (7.71), we obtain by substitution in (7.69),

$$\mathbf{a}(\boldsymbol{\alpha}) - \mathbf{a} = \mathbf{R}_z^{-1}(\boldsymbol{\alpha})\mathbf{R}_\epsilon(\boldsymbol{\alpha})(\boldsymbol{\alpha} - \mathbf{a}) \tag{7.72}$$

Now define

$$\mathbf{C}(\boldsymbol{\alpha}) = \mathbf{R}_z^{-1}(\boldsymbol{\alpha})\mathbf{R}_\epsilon(\boldsymbol{\alpha})$$

Then all the eigenvalues of $\mathbf{C}(\boldsymbol{\alpha})$ are strictly between 0 and 1, and hence there exsits a norm $\|\cdot\|$ such that $\|\mathbf{C}(\boldsymbol{\alpha})\| < 1$ [26]. The punch line is that (7.72) is a multivariate contraction at $\boldsymbol{\alpha}^* = \mathbf{a}$, with $\mathbf{C}(\boldsymbol{\alpha})$ as its contraction factor,

$$\mathbf{a}(\boldsymbol{\alpha}) - \boldsymbol{\alpha}^* = \mathbf{C}(\boldsymbol{\alpha})(\boldsymbol{\alpha} - \boldsymbol{\alpha}^*) \tag{7.73}$$

Therefore, starting with an initial guess $\boldsymbol{\alpha}_0$, the fixed point iterations

$$\boldsymbol{\alpha}_{k+1} = \mathbf{a}(\boldsymbol{\alpha}_k) \tag{7.74}$$

produce a convergent sequence $\{\boldsymbol{\alpha}_k\}$, $k = 1, 2, \ldots,$

$$\boldsymbol{\alpha}_k \overset{\text{a.s.}}{\to} \boldsymbol{\alpha}^* = \mathbf{a} \tag{7.75}$$

We conclude that, under some fairly general conditions on the parametric family $\mathcal{L}_{\boldsymbol{\alpha}}$, as long as the fundamental property (7.71) holds, and the p frequencies ω_k pass, *the fixed point of* $\mathbf{a}(\boldsymbol{\alpha})$, *denoted by* $\boldsymbol{\alpha}^*$, *coincides with the vector of AR parameters* $\mathbf{a}$. Thus, under these conditions, the CM algorithm yields an (a.s. consistent) estimate of $\mathbf{a}$. The asymptotic normality of the parameterized least squares estimator is discussed in [26].

We refer to the modified AR method as a multivariate CM method. Special cases of this method were discussed in terms of the $AR(2p)$ filter in [8], [14].

The *AR*(2*p*) Filter

As an illustration of the preceding discussion, assume that the noise $\{\epsilon_t\}$ is white noise, and consider the $AR(2p)$ parametric filter [8], [14], [26],

$$\mathcal{L}_{\boldsymbol{\alpha}} \equiv \frac{1}{1 + \theta_1(\boldsymbol{\alpha})\eta\mathcal{B} + \cdots + \theta_{2p}(\boldsymbol{\alpha})\eta^{2p}\mathcal{B}^{2p}}$$

or

$$y_t(\boldsymbol{\alpha}) + \theta_1(\boldsymbol{\alpha})\eta y_{t-1}(\boldsymbol{\alpha}) + \cdots + \theta_{2p}(\boldsymbol{\alpha})\eta^{2p} y_{t-2p}(\boldsymbol{\alpha}) = y_t, \tag{7.76}$$

where $0 < \eta \le 1$ controls the bandwidth, and

$$\theta_0(\boldsymbol{\alpha}) = 1, \qquad \theta_{2p-k}(\boldsymbol{\alpha}) = \theta_k(\boldsymbol{\alpha}), \qquad k = 0, 1, \ldots, p$$

For the fundamental property (7.71) to hold, we must choose the $\theta_j(\boldsymbol{\alpha})$ appropriately.

Define the column vector

$$\boldsymbol{\theta}(\boldsymbol{\alpha}) \equiv (\theta_1(\boldsymbol{\alpha}), \theta_2(\boldsymbol{\alpha}), \ldots, \theta_p(\boldsymbol{\alpha}))'$$

and a $p \times p$ diagonal matrix

$$\mathbf{T}_\eta \equiv \text{Diagonal}\left(\frac{1+\eta^{2p}}{\eta+\eta^{2p-1}}, \frac{1+\eta^{2p}}{\eta^2+\eta^{2p-2}}, \ldots, \frac{1+\eta^{2p}}{\eta^p+\eta^p}\right)$$

Then, relative to white noise, the fundamental property takes on the neat form [26],

$$\boldsymbol{\theta}(\boldsymbol{\alpha}) = \mathbf{T}_\eta \boldsymbol{\alpha} \tag{7.77}$$

For $p = 1$, (7.77) reduces to (7.53), replacing α by -2α, as it should.

Example 7.8: Spectral Analysis within a Fourier Bin

Fourier frequencies of the form $2\pi k/N$ are separated by an interval of length $2\pi/N$, the size of a Fourier bin, and hence are not resolvable by FFTs. Likewise, the periodogram $I_N(\omega)$, as a continuous function of ω, but with small N, has difficulties in resolving frequencies which are less than $2\pi/N$ apart. Such closely spaced frequencies tend to produce a single peak in $I_N(\omega)$, and hence remain unnoticeable by periodogram analysis. The multivariate CM method can help in these situations up tp a point. Experimental results indicate that, for reasonable SNR, the method is quite reliable when the separation between frequencies exceeds 25% of a Fourier bin size. However, the method becomes less reliable for a smaller separation, and seems to break down at or below the 10% mark as the following illustration indicates.

Table 7.12 gives the results of a computer simulation with $N = 100$, $p = 2$, fixed $\eta = 0.988$, and 3-dB SNR for each sinusoid. The separation between ω_1 and ω_2 is measured as a percent of the width 0.02π of a Fourier bin. At 10% separation the method fails—it gives a single frequency situated between the two true frequencies. In all the different runs, the CM procedure was initiated by Prony's estimator. In this as well as in many other simulations, convergence took place after about ten iterations.

TABLE 7.12 Results of an Application of the Multivariate CM Method using the Stable $AR(p)$ Filter

Separation %	ω_1	ω_2	$\hat{\omega}_1$	$\hat{\omega}_2$
50%	0.400π	0.410π	0.401689π	0.410990π
40%	0.400π	0.408π	0.401868π	0.408146π
25%	0.400π	0.405π	0.402064π	0.404194π
10%	0.400π	0.402π	0.401549π	0.401549π

Note: $p = 2, \eta = 0.988, N = 100$. (7.76).

■

7.5 SUMMARY

For the most part, we have explored the CM algorithm and some of its ramifications. In particular, the algorithm was exemplified through its special case, the HK algorithm that uses normalized higher order crossings. When combined with band-pass filters, and for a reasonable signal-to-noise ratio (e.g., SNR ≥ -6 dB), the method is fast, performs quite remarkably, and can be easily implemented.

We emphasized the single frequency case. However, as we saw in Example 7.7, an extension to the multiple frequency case is rather straightforward for well-separated frequencies because the CM method, when employing *band-pass filters*, can be applied independently to nonoverlapping frequency bands, and in this way detect or estimate all the frequencies. Therefore, the method is *amenable to parallel computation*.

A multivariate extension of the CM algorithm was obtained by a parametrization of Prony's AR estimator. The resulting hybrid is a modification of the AR method which provides *consistent* frequency estimates—improves as the series size increases—whereas the latter method is inconsistent. Experience indicates that the modified method performs well even for relatively small sample sizes provided the signal-to-noise ratio is not too low. For example, $N = 100$ for the case $p = 2$, and the SNR is 3 dB for each sinusoid. For both methods, the number of frequencies p is assumed known. In general, however, p is not known, and the rule of thumb is to use a number $q > p$.

From the works of Quinn and Fernandes [31] and Truong-Van [38] we know that when using the sample higher order correlation, the CM method with the (truncated) $AR(2)$ filter (7.27) in Example 7.1 is capable of producing estimates that are accurate to order $N^{-3/2}$, the precision of the nonlinear least squares frequency estimator, while the FFT, being confined to discrete Fourier frequencies only, has accuracy to order N^{-1}. However, the $AR(2)$ filter (7.27) has a very narrow bandwidth, and hence requires a rather precise initial guess. As shown in Example 7.6, if the initial guess is not sufficiently close to the true frequency, the algorithm may result in a biased estimate. By contrast, the cosine filter with a variable and initially wider bandwidth, and the more general stable $AR(2)$ filter (7.52), again with a somewhat wider bandwidth, are more flexible so that the CM algorithm may be initiated with a less precise guess, but still yielding very precise frequency estimates. This gives an indication that the class of filters for which optimal efficiency in frequency estimation is achieved by the CM method, with a variable bandwidth, may include filters whose initial bandwidth is somewhat wider than that of the (truncated) $AR(2)$ (7.27).

A few clarifying words concerning the relative performance of nonlinear least squares, periodogram analysis, and CM in frequency estimation. In the case of a single frequency, maximizing the periodogram $I_N(\omega)$ as a continuous function of ω is asymptotically equivalent to the nonlinear least squares procedure. Both methods give best estimates, but at a price: the respective optimizations must start roughly within $o(N^{-1})$ of the true frequency. The CM method, on the other hand, is computationally simpler and at the same time can approach the same level of precision starting from an initial guess precise to order $O(1)$ only. Similar remarks hold for p well-separated frequencies with the understanding that $I_N(\omega)$ is maximized locally. Things are not that simple for very closely spaced frequencies within a Fourier bin. In this case, $I_N(\omega)$ tends to "hover" above the frequencies within a bin with a single peak, while the nonlinear least squares method requires an extremely precise (multivariate) initial guess. As was demonstrated, the multivariate CM initiated with Prony's estimates can still resolve frequencies within a Fourier bin, provided they are sufficiently separated there.

The statistical and engineering literature on frequency estimation is rather wide and contains many more important references than included in this chapter. Many of these references, however, can be found in the works that we have cited. In general, the problem of frequency estimation is viewed as a regression problem where the periodic components are parts of the mean function. Our viewpoint, as expressed by the model (7.1), is that the periodic components are parts of the autocorrelation. From a practical point of view, the difference between the two approaches is rather philosophical and not substantive, for the end result is in many respects identical.

We mention in conclusion that modifications and extensions of the CM algorithm are possible, as illustrated by Problems 8 and 16.

7.6 PROBLEMS AND COMPLEMENTS

1. Consider the lynx data in Table A.3 in Appendix A ($N = 114$). Center the data (subtract the series average), apply a low-pass filter, and then obtain the simple HOC $D_1, \ldots, D_8$. Apply I_N^* to the normalized simple HOC in reference to Table 7.1 to determine the dominant period (about 9.5–10 years).
2. Suppose in the model (7.1), $p = 1$, A_1, B_1 are normally distributed, and the noise is Gaussian white noise. Denote the autocorrelation of $\{Z_t\}$ by ρ_k. Show that

$$\frac{\rho_2}{\rho_1} = \frac{2\cos^2(\omega_1) - 1}{\cos(\omega_1)}$$

3. (cont.) Suppose a time series of length N from the model with $p = 1$ is observed. Let D_1, D_2 be the first two simple HOC. Use the preceding equation to show that ω_1 can be determined from the equation

$$\left[-1 + 2\cos\left(\frac{\pi E[D_1]}{N-1}\right) - 2\cos\left(\frac{\pi E[D_2]}{N-1}\right) \times \left(1 - \cos\left(\frac{\pi E[D_1]}{N-1}\right)\right)\right]$$
$$\times \left[\cos\left(\frac{\pi E[D_1]}{N-1}\right)\right]^{-1} = \frac{2\cos^2(\omega_1) - 1}{\cos(\omega_1)}$$

4. (cont.) Show that the solution for ω_1 in terms of $E[D_1], E[D_2]$ is highly sensitive, even to minute changes in the expected HOC. Conclude that solving the preceding equation for ω_1 is impractical [7], [17].

5. (cont.) [17]. Assume that $\{Z_t\}$ is filtered with a linear filter whose squared gain is given by

$$\frac{(1-\alpha)^2}{1-2\alpha\cos(\omega)+\alpha^2}$$

for $0 < \alpha < 1$. Denote by D_α the corresponding HOC. Suppose we know that $0 < \omega_1 < \pi/2$. Show that for α sufficiently close to 0

$$\omega_1 \leq \frac{\pi E[D_\alpha]}{N-1} \leq \cos^{-1}(\alpha)$$

and that the inequalities are reversed for α sufficiently close to 1.

6. (cont.) [17]. Generate a time series of length $N = 2000$ from the model such that the SNR is -20 dB, and $\omega_1 = 0.75$. Apply to the time series two summations $(1+\mathcal{B})^2$ followed by the $AR(1)$ filter with parameter α. With 10 values of α such that,

$$\alpha_1 < \alpha_2 < \cdots < \alpha_{10}$$

compute I^* at the corresponding normalized observed HOC in the search for ω_1.

7. Show that the sequence α_k in (7.9) is a monotone sequence and that it must be a convergent sequence. (Hint: Deduce from (7.13) that $\rho_1(\alpha)$ must be between α and $\cos(\omega_1)$. Assume without loss of generality that $\alpha_0 \leq \cos(\omega_1)$. Then $\alpha_0 \leq \rho_1(\alpha_0) \leq \cos(\omega_1)$. Then by (7.15) $\alpha_0 \leq \alpha_1 \leq \cos(\omega_1)$. Now do the same for $\alpha_1, \rho_1(\alpha_1)$. Continue in this way and conclude that $\alpha_0 \leq \alpha_1 \leq \cdots \leq \alpha_j \leq \cdots \leq \cos(\omega_1)$.)

8. *A ramification of CM* [24]. Consider again the case $p = 1$ of a single frequency, and let $L(F_\zeta)$ be a family of filters. Suppose $\mathcal{L}(\cdot) \in L(F_\zeta)$ has a transfer function $H(\omega)$, and denote by $\rho(H)$ the first order autocorrelation of the filtered process $\{\mathcal{L}(Z)_t\}$. Assume the following.

$$|H(\omega)| = |H(-\omega)|$$
$$|H(\omega_1)| > 0$$
$$\int_{-\pi}^{\pi} |H(\omega)|^2 dF_\zeta(\omega) = 1$$

For every filter $\mathcal{L}_{k-1}(\cdot) \in L(F_\zeta)$, there exists a filter $\mathcal{L}_k(\cdot) \in L(F_\zeta)$ such that

$$\int_{-\pi}^{\pi} |H_k(\omega)|^2 \cos(\omega)\, dF_\zeta(\omega) = \rho(H_{k-1})$$

Show that starting with any filter $\mathcal{L}_0(\cdot) \in L(F_\zeta)$, we obtain the contraction

$$\rho(H_k) = \cos(\omega_1) + C(H_k)[\rho(H_{k-1}) - \cos(\omega_1)]$$

$k = 1, 2, \ldots,$ where

$$C(H_k) \equiv \frac{1}{1+\sigma_1^2|H_k(\omega_1)|^2}$$

is the contraction factor corresponding to $\mathcal{L}_k(\cdot)$. Conclude that as $k \to \infty$,

$$\rho(H_k) \to \cos(\omega_1)$$

9. [20]. Show that, when using a band-pass filter, the contraction mapping obtained under the model

$$Z_t = \sum_{j=1}^{p} A_j \cos(\omega_j t + \phi_j) + \zeta_t$$

where the A_j are fixed constants, and the ϕ_j are independently uniformly distributed in $(-\pi, \pi)$, has the same form as that obtained under (7.1).

10. Use the formula for geometric progression to conclude that for $\lambda \in [-\pi, \pi]$, and $M = 1, 2, \ldots,$

$$\sum_{t=-M}^{M} \exp(-it\lambda) = \frac{\sin[(2M+1)\frac{1}{2}\lambda]}{\sin(\frac{1}{2}\lambda)}$$

11. Show that for $\lambda \in [-\pi, \pi]$, $M = 1, 2, \ldots,$

$$\sum_{r=-(M-1)}^{M-1} \left(1 - \frac{|r|}{M}\right) \exp(ir\lambda) = \frac{1}{M} \left| \sum_{t=1}^{M} \exp(it\lambda) \right|^2 = \frac{1}{M} \left[\frac{\sin(\frac{1}{2}\lambda M)}{\sin(\frac{1}{2}\lambda)} \right]^2$$

Integrate term by term to conclude that

$$\int_{-\pi}^{\pi} \frac{1}{M} \left[\frac{\sin(\frac{1}{2}\lambda M)}{\sin(\frac{1}{2}\lambda)} \right]^2 d\lambda = 2\pi$$

12. Similarly, show

$$\int_{-\pi}^{\pi} \frac{1}{M} \left[\frac{\sin(\frac{1}{2}M(\omega-\theta))}{\sin(\frac{1}{2}(\omega-\theta))} \right]^2 \cos(\omega)\, d\omega = 2\pi \left(1 - \frac{1}{M}\right) \cos(\theta)$$

Thus with $|H(\omega; r, M)|^2$ as in (7.37), show that

$$\frac{\int_{-\pi}^{\pi} \cos(\omega) |H(\omega; r, M)|^2\, d\omega}{\int_{-\pi}^{\pi} |H(\omega; r, M)|^2\, d\omega} = \left(1 - \frac{1}{2M+1}\right) \cos(\theta(r)) \to \cos(\theta(r))$$

as $M \to \infty$.

13. Let $|H(\omega; r, M)|^2$ be given by (7.37), and suppose that $f_\zeta(\omega)$ is positive and continuous. Verify (7.40). (Hint: The normalized Fejér kernel acts as a Dirac delta as M increases; e.g., see [1, pp. 461–462].)

14. [11]. Consider the family of complex filters defined for each fixed θ by

$$Y_t = (1 + e^{i\theta}\mathcal{B})^n Z_t$$

where $\{Z_t\}$ is as in (7.1), the noise is white with variance σ_ζ^2, and $\theta \in [0, \pi]$.

(a) Show that the transfer function is given by

$$H(\lambda) = (1 + e^{i(\theta-\lambda)})^n$$

(b) Show that

$$\Re\left\{\frac{E[Y_t\overline{Y}_{t-1}]}{E[|Y_t|^2]}\right\} =$$

$$\frac{\sum_{j=1}^{p}\sigma_j^2\left[\cos^{2n}\left(\frac{\theta+\omega_j}{2}\right)+\cos^{2n}\left(\frac{\theta-\omega_j}{2}\right)\right]\cos(\omega_j)+\left(\frac{\sigma_\zeta^2}{\pi}\right)\int_{-\pi}^{\pi}\cos^{2n}\left(\frac{\lambda-\theta}{2}\right)\cos(\lambda)d\lambda}{\sum_{j=1}^{p}\sigma_j^2[\cos^{2n}\left(\frac{\theta+\omega_j}{2}\right)+\cos^{2n}\left(\frac{\theta-\omega_j}{2}\right)]+\left(\frac{\sigma_\zeta^2}{\pi}\right)\int_{-\pi}^{\pi}\cos^{2n}\left(\frac{\lambda-\theta}{2}\right)d\lambda}$$

(c) Suppose the spectrum is purely discrete, $\sigma_\zeta^2 = 0$ (no noise), and from all the frequencies, θ is closest to ω_r. Show that as $n \to \infty$,

$$\Re\left\{\frac{E[Y_t\overline{Y}_{t-1}]}{E[|Y_t|^2]}\right\} \to \cos(\omega_r)$$

(d) Show that in the presence of noise, and in fact regardless of its magnitude (could be very small),

$$\Re\left\{\frac{E[Y_t\overline{Y}_{t-1}]}{E[|Y_t|^2]}\right\} \to \cos(\theta)$$

and the preceding result in (c) breaks down. Historically, this shortcoming led to the HK algorithm.

(Hint: The sequence of probability density functions $f_n(\lambda;\theta)$, $\lambda \in (-\pi,\pi]$, defined by

$$f_n(\lambda;\theta) = \frac{\cos^{2n}\left(\frac{\theta-\lambda}{2}\right)}{\int_{-\pi}^{\pi}\cos^{2n}\left(\frac{\theta-\omega}{2}\right)\,d\omega}$$

satisfies, as $n \to \infty$,

$$\int_{-\pi}^{\pi} f_n(\lambda,\theta)\cos(\lambda)\,d\lambda \to \cos(\theta)$$

$0 \le \theta \le \pi$.)

15. Consider the "cosine binomial" filter

$$h(n;r,M) = \begin{cases} \binom{M}{n}\cos(\theta(r)n), & n = 0,\ldots,M \\ 0, & \text{otherwise} \end{cases}$$

Apply the HK algorithm in conjunction with the cosine binomial filter to simulated data from (7.1) with $p = 1$. Compare the performance of the cosine binomial filter with that of the cosine filter (7.57).

16. *A ramification of CM* [37], [42]. Suppose in (7.1), $\{Z_t\}$ is Gaussian and that the noise $\{\zeta_t\}$ is white Gaussian noise. We apply to $\{Z_t\}$ the ideal band-pass filter defined by the squared gain

$$|H(\omega;\theta)|^2 = \begin{cases} 1, & \omega \in [\theta-\delta,\theta+\delta] \\ 0, & \omega \in [0,\theta-\delta)\cup(\theta+\delta,\pi] \end{cases}$$

where $\delta < C/2$ and

$$C = \min\{|a - b| : a \text{ and } b \text{ are among } \omega_1, \omega_2, \ldots, \omega_p, 0, \pi\}$$

We say that the filter *captures* ω_j if ω_j lies in the interval $[\theta - \delta, \theta + \delta]$.

(a) Show that if for some θ_0, ω_i is captured by $|H(\omega;\theta_0)|^2$, and $\{\theta_j\}$ is determined from

$$\theta_{j+1} = \cos^{-1}\left[\frac{\delta\cos(\omega_{\theta_{j+1}})}{\sin(\delta)}\right]$$

then

$$\omega_{\theta_{j+1}} \equiv \frac{\pi E[D_{\theta_j}]}{N-1} \rightarrow \omega_i, \qquad j \rightarrow \infty$$

(b) Show that the convergence of ω_{θ_j} is monotone.

(c) Show that once the filter captures a frequency, it is never lost throughout the iterative procedure.

(d) Show that if no frequency is captured, the filter does not "move."

(e) Show that $E[D_{\theta_1}] \leq E[D_{\theta_2}]$ for $\theta_1 \leq \theta_2$.

17. *Fixed points interpretation of CM* [19], [28]. Consider the band-pass filter defined by (7.36), and assume the case $p = 2$ with ω_1, ω_2. Provide a formal proof for the fact that the essence of the CM method relative to this filter is to locate attracting fixed points of $\rho_1(r, M)$, and that, as $M \rightarrow \infty$, the relevant fixed points coincide with $\cos(\omega_1), \cos(\omega_2)$. See Figure 7.13.

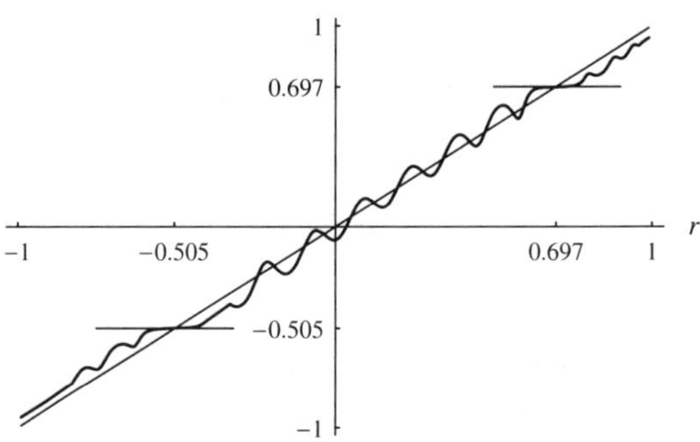

Figure 7.13: *Attracting fixed points* $\cos(0.8) = 0.697, \cos(2.1) = -0.505$ *of* $\rho_1(r, 20)$ *from the complex exponential filter (Fejér).* $p = 2, M = 20, \omega_1 = 0.8, \omega_2 = 2.1, SNR = 0$. *For the relevant fixed points of* $\rho_1(r, M)$ *to coincide with* $\cos(0.8)$ *and* $\cos(2.1)$, M *must increase.*

18. *Formant tracking* [30], [36]. Suppose a signal containing one formant can be modeled by the damped sinusoid

$$Z_t = 3\exp(-0.01t)\cos(0.8t) + \epsilon_t, \qquad t = 1, 2, \ldots, 300$$

where ϵ_t is white Gaussian noise with mean 0 and variance 0.08. Apply to the data the cosine filter (7.57). Conduct an experiment to illustrate the fact that when the the frequency is "captured," the sample standard deviation of the interval between zero-crossings tends to decrease. That is, the zero-crossings become "consistent."

19. *Nonlinear least squares estimates* [33]. Consider the model

$$Z_t = A\cos(\omega_1 t) + B\sin(\omega_1 t) + \epsilon_t$$

where ϵ_t is white Gaussian noise with mean 0, A, B are fixed amplitudes, $\omega_1 \in (0, \pi)$, and let $Z_1, \ldots, Z_N$ be a time series from the model. Argue that for sufficiently large N the least squares estimates of A, B, ω_1, are given by

$$\hat{A} = \frac{2}{N}\sum_{t=1}^{N} Z_t \cos(\hat{\omega}_1 t), \qquad \hat{B} = \frac{2}{N}\sum_{t=1}^{N} Z_t \sin(\hat{\omega}_1 t)$$

with $\hat{\omega}_1$ such that

$$I_N(\hat{\omega}_1) = \max_{0 \le \omega \le \pi} \{I_N(\omega)\}$$

REFERENCES

[1] Anderson, T.W., *The Statistical Analysis of Time Series*, New York: Wiley, 1971.

[2] Bell, T. L. and N. Reid. "Detecting of the diurnal cycle of rainfall using satellite observations." *J. Appl. Meteorol.*, Vol. 32, No. 2, pp. 311–322, 1993.

[3] Bloomfield, P., *Fourier Analysis of Time Series: An Introduction*, New York: Wiley, 1976.

[4] Brillinger, D.R., "Fitting cosines: Some procedures and some physical examples," in *Applied Probability, Stochastic Processes, and Sampling Theory*, I. B. MacNeill and G. J. Umphry, Eds., Dordrecht, The Netherlands: Reidel, pp. 75–100, 1987.

[5] Chandler, S.C., "On the variation of latitude," *Astron. J.*, Vol. 11, pp. 65–70, 1891.

[6] Cooley, J. W. and J. W. Tukey, "An algorithm for the machine calculation of complex Fourier series," *Math. Comput.*, Vol. 19, pp. 297–301, April 1965.

[7] Dechambre, M. and J. Lavergnat, "Frequency determination of a noisy signal by zero-crossing counting," *Signal Process.*, Vol. 8, pp. 93–105, 1985.

[8] Dragošević, M.V. and S.S. Stanković, "A generalized least squares method for frequency estimation," *IEEE Trans. Acoust., Speech, Signal Processing*, Vol. 37, No. 6, pp. 805–819, June 1989.

[9] Hannan, E.J. and B.G. Quinn, "The resolution of closely adjacent spectral lines," *J. Time Ser. Anal.*, Vol. 10, No. 1, pp. 13–31, 1989.

[10] He, S., "On estimating the hidden periodicities in linear time series models," *China-Japan Symposium on Statistics* (Beijing, China), pp. 94–97, 1984.

[11] He, S. and B. Kedem, "Higher order crossings of an almost periodic random sequence in noise," *IEEE Trans. Inform. Theory*, Vol. IT-35, No. 2, pp. 360–370, March 1989.

[12] He, S. and B. Kedem, "The zero-crossing rate of autoregressive processes and its link to unit roots," *J. Time Ser. Anal.*, Vol. 11, No. 3, pp. 201–213, 1990.

[13] Hudlow, M. D. and V. L. Patterson, *GATE Radar Rainfall Atlas*, NOAA Special Report, available from U.S. Government Printing Office, Washington DC 20402, 158 pp. 1979.

[14] Kay, S. M., "Accurate frequency estimation at low signal-to-noise ratio," *IEEE Trans. Acoust., Speech, Signal Processing*, Vol. ASSP-32, No. 3, pp. 540–547, June 1984.

[15] Kay, S. M. and S. W. Marple, "Spectrum analysis—A modern perspective," *Proc. IEEE*, Vol. 69, No. 11, pp. 1380–1419, Nov. 1981.

[16] Kedem, B., "Search for periodicities by axis-crossings of filtered time series," *Signal Processing*, Vol. 10, No. 2, pp. 129–144, March 1986.

[17] Kedem, B., "On frequency detection by zero-crossings," *Signal Processing*, Vol. 10, No. 3, pp. 303–306, April 1986.

[18] Kedem, B., "Detection of periodicities by higher order crossings," *J. Time Ser. Anal.*, Vol. 8, No. 1, pp. 39–50, 1987.

[19] Kedem, B. and S. Lopes, "Fixed points in mixed spectrum analysis," in *Probabilistic and Stochastic Methods in Analysis With Applications*, J. S. Byrnes et al., Eds., pp. 573–591, Norwell, Mass.: Kluwer, 1992.

[20] Kedem, B. and S. Yakowitz, "Practical aspects of a fast algorithm for frequency detection," accepted for publication in *IEEE Trans. Commun.*

[21] Kedem, B. and S. Yakowitz, "On the contraction mapping method for frequency detection." Tech. Rep. TR-92-45, Systems Research Center, Univ. of Maryland, College Park, 1992.

[22] Lambeck, K., *The Earth's Variable Rotation: Geophysical Causes and Consequences,* London: Cambridge Univ. Press, 1980.

[23] Li, T., *Multiple Frequency Estimation in Mixed Spectrum Time Series by Parametric Filtering*, Doctoral Dissertation, Department of Mathematics, Univ. of Maryland, College Park, 1992.

[24] Li, T. and B. Kedem, "Adaptive frequency tracking by zero-crossing counts." Tech. Rep. TR91-26, Department of Mathematics, Univ. of Maryland, College Park, 1991.

[25] Li, T. and B. Kedem, "Strong consistency of the contraction mapping method for frequency estimation." Accepted for publication in *IEEE Trans. Info. Theory*.

[26] Li, T. and B. Kedem, "Estimation of multiple sinusoids by parametric filtering," Tech. Rep. TR-92-51, Systems Research Center, Univ. of Maryland, College Park, 1992.

[27] Li, T., B. Kedem, and S. Yakowitz, "Asymptotic normality of the contraction mapping estimator for frequency estimation," Tech. Rep. TR-92-22, Systems Research Center, Univ. of Maryland, College Park, 1992.

[28] Lopes, S., *Spectral Analysis in Frequency Modulated Models*, Doctoral Dissertation, Department of Mathematics, Univ. of Maryland, College Park, 1991.

[29] Mataušek, M.R., S.S. Stanković, and D.V. Radović, "Iterative inverse filtering approach to the estimation of frequencies of noisy sinusoids," *IEEE Trans. Acoust., Speech, Signal Processing*, Vol. ASSP-31, No. 6, pp. 1456–1463, 1983.

[30] Niederjohn, R.J. and M. Lahat, "A zero-crossing consistency method for formant tracking of voiced speech in high noise levels," *IEEE Trans. Acoust., Speech, Signal Processing*, Vol. ASSP-33, No. 2, pp. 349–355, April 1985.

[31] Quinn, B.G., and J.M. Fernandes, "A fast efficient technique for the estimation of frequency," *Biometrika*, Vol. 78, No. 3, pp. 489–497, 1991.

[32] Priestley, M. B., *Spectral Analysis and Time Series*, Vol. 1, London: Academic Press, 1981.

[33] Rice, J. A. and M. Rosenblatt, "On frequency estimation," *Biometrika*, Vol. 75, No. 3, pp. 477–484, 1988.

[34] Schuster, A., "On the investigation of hidden periodicities with application to a supposed 26-day period of meteorological phenomena," *Terr. Magn. Atmos. Electro.*, Vol. 3, pp. 13–41, 1898.

[35] Simpson, J., "TRMM, A satellite mission to measure tropical rainfall," Report to the Science Steering Group, National Aeronautics and Space Administration, 1988.

[36] Sreenivas, T. V. and R. J. Niederjohn, "Zero-crossing based spectral analysis and SVD spectral analysis for formant frequency estimation in noise, *IEEE Trans. Signal Process.*, Vol. 40, No. 2, pp. 282–293, Feb. 1992.

[37] Troendle, J. F., *An Iterative Filtering Method of Frequency Detection in a Mixed Spectrum Model*, Doctoral Dissertation, Department of Mathematics, Univ. of Maryland, College Park, 1991.

[38] Truong-Van, B., "A new approach to frequency analysis with amplified harmonics," *J. R. Stat. Soc.* B, Vol. 52, No. 1, pp. 203–221, 1990.

[39] Wang, J., "Some estimation problems for linear models of time series," (in Chinese), *Chin. Ann. Math.*, Vol. 4, Ser. A, pp. 19–35, 1983.

[40] Whittaker, E.T. and G. Robinson, *The Calculus of Observation*, London: Blackie, 1924.

[41] Yakowitz, S., "Some contributions to a frequency location method due to He and Kedem," *IEEE Trans. Inform. Theory*, Vol. 37, No. 4, pp. 1177–1182, July 1991.

[42] Kedem, B. and J. F. Troendle, "An iterative filtering algorithm for non-Fourier frequency estimation," accepted for publication in *J. Time Ser. Analysis*.

8

Signal Discrimination by HOC

In addition to the fact that higher order crossings contain useful spectral information, evidence of which we saw in the previous chapters, they serve as a simple and effective means for achieving drastic data reduction important for discrimination and classification purposes. Moreover, viewed as discrimination features, ordinarily very few D_{θ_j} are needed for successful discrimination.

In this chapter we shall describe, mainly through examples, some distinctive aspects of higher order crossings (HOC) that were found useful in applications. These include in particular the so-called ψ^2 statistic and certain graphical similarity measures. However, it is important to note from the outset that the methods described in what follows comprise only a very partial list of possible zero-crossing-based methods and techniques.

Terms such as discrimination, classification, pattern recognition, goodness of fit, and signature analysis, are at times used somewhat loosely and interchangeably to signify the same thing. By discrimination between signals and time series, we shall mean the process of determining from HOC sequences $\{D_{\theta_j}\}$ the degree of similarity between two time series, or between a given time series and a hypothesized one, or between different sections of the same signal. For example, given the simple HOC $\{D_j\}$ of an engine vibration signature, one formulation of the discrimination problem is to determine how close the observed D_j are to the expected simple higher order crossings of a reliable engine. But we can also examine the D_j of an engine and compare them with the simple HOC of another engine, or with the simple HOC obtained from the same engine a year later, in order to detect changes.

As we have reiterated before, since D_θ is a function of *both* the parameter θ and the record size N, HOC lend themselves to manipulation through these two quantities. One possibility is to fix N and use the random vector of features $(D_{\theta_1}, D_{\theta_2}, \ldots, D_{\theta_K})$ to measure differences between time series. Another possibility is to fix the filter parameter, and look for changes (or convergence under stationarity) in the zero-crossing rate of the filtered data as the series size N grows. Obviously, the same can be repeated for different values of the parameter. Yet a third possibility—for each fixed θ—is to consider an adaptive version of the zero-crossing rate, treating it as a function of time while giving more weight to more current data. From such a procedure we can learn, for example, about the evolution of spectral changes in nonstationary data for tracking purposes.

8.1 THE ψ^2 STATISTIC

Given a vector of HOC $\mathbf{D} = (D_{\theta_1}, D_{\theta_2}, \ldots, D_{\theta_K})$, and a $K \times K$ matrix $\mathbf{C}$, we can always consider similarity or distance measures of the form

$$(\mathbf{D} - E[\mathbf{D}])'\mathbf{C}^{-1}(\mathbf{D} - E[\mathbf{D}])$$

to measure deviations from expected values [1], [20]. A related idea is to construct quadratic forms in terms of functions of the D_{θ_k}. A particular example of this is a quadratic form in terms of the increments of simple HOC.

Consider the simple observed HOC $\{D_k\}$. In previous chapters we saw ample evidence of the fact that for reasonably large records the simple HOC tend to increase monotonically with a high probability,

$$D_1 < D_2 < D_3 < \cdots < D_k < \cdots$$

and that the initial *rate* of increase may serve as a potent discriminator. However, as the difference order increases, so does the similarity between the resulting simple HOC from different processes, and their discrimination power tapers off. Figures 5.2–5.4 provide an illustration of this fact. What we need then is a way to quantify the initial monotone rate of increase in simple HOC. A useful statistic that does that is the so-called ψ^2 statistic which is defined next.

To capture the rate of increase in the first few D_k, we consider the *increments*

$$\Delta_k \equiv \begin{cases} D_1, & \text{if } k = 1 \\ D_k - D_{k-1}, & \text{if } k = 2, \ldots, K-1 \\ (N-1) - D_{K-1}, & \text{if } k = K \end{cases}$$

When N is sufficiently large (e.g., $N \geq 200$), then with a high probability $0 \leq \Delta_k \leq N - 1$ (see Problem 1), and

$$\sum_{k=1}^{K} \Delta_k = N - 1$$

Let $m_k = E[\Delta_k]$. Viewing the observed Δ_k as "frequencies" in a multinomial experiment, we define a general similarity measure by [13], [14],

$$\psi^2 \equiv \sum_{k=1}^{K} \frac{(\Delta_k - m_k)^2}{m_k} \tag{8.1}$$

where it is always assumed that the "expected" parameters m_k satisfy $m_k > 0$. As in multinomial experiments, the parameter K refers to the number of "classes" or "categories." The ψ^2 statistic can be used in testing the hypothesis $H_0 : m_k = m_k^{(0)}, k = 1, 2, \ldots, K$, or as a distance from a process with prescribed m_k. Its other usage is simply as a discrimination feature.

The ψ^2 statistic is a very simple quadratic form inspired by the celebrated chi square goodness of fit statistic. It involves only a trivial matrix inversion (see Problem 2), and yet its probability distribution is still an open question at present [19]. Under some regularity conditions when the m_k and Δ_k come from the same process [12], [14], the asymptotic distribution of ψ^2 as $N \to \infty$ approaches the distribution of the sum $\sum_{k=1}^{K-1} \alpha_k \eta_k^2$, where the α_k are nonnegative constants, and the η_k^2 are independent χ_1^2 random variables. Unfortunately, this by itself does not provide much about the true distribution of ψ^2. However, it has been observed that when the Δ_k come from stationary autoregressive moving average ($ARMA$) processes, certain tail probabilities of ψ^2, practically speaking, do not vary much from process to process, an issue to be discussed below.

From now on we set $K = 9$ unless otherwise is noted. This arbitrary choice of K is motivated from the empirical fact that only the first few D_k are useful for discrimination purposes, and $K = 9$ seems to work out quite well in several applications. But clearly, other small values of K are just as plausible and should be tried out in applications, as is indeed the case in our nondestructive evaluation (Examples 8.2 and 8.3) below.

In pursuing the probability distribution of ψ^2, we have resorted to a certain approximation which capitalizes on the method of moments. Figure 8.1 shows the histogram of ψ^2, with $K = 9, N = 1000$, obtained from 100 independent realizations of a stationary Gaussian (autoregressive) $AR(1)$ process with parameter 0.5. The figure, and other similar figures in [19], suggest that the distribution of ψ^2 can be approximated by a weighted or scaled chi square random variable. Recalling that the mean and variance of a χ_k^2 random variable with k degrees of freedom are k and $2k$, respectively, we are looking for a multiple of ψ^2 so that these two moment constraints are satisfied [2]. Accordingly, let $a = E[\psi^2]$ and $b = \frac{1}{2}\,\mathrm{Var}[\psi^2]$. Then the weighted statistic

$$\frac{a\psi^2}{b}$$

has mean a^2/b, and variance $2a^2/b$. One approximates the distribution of $a\psi^2/b$ by a chi square distribution with a^2/b degrees of freedom. Using this method, we list in Table 8.1 approximations to some tail probabilities of ψ^2. For each given

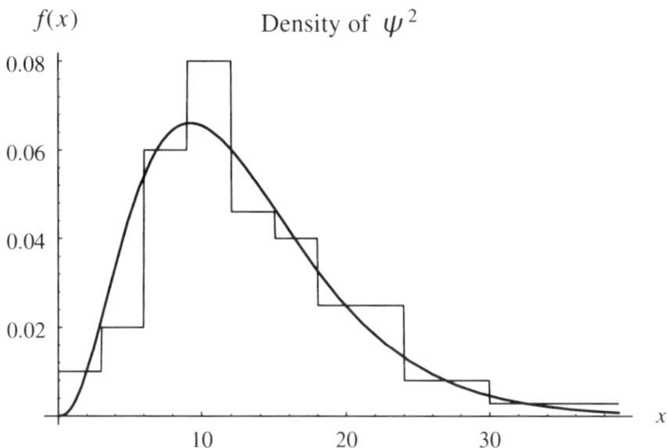

Figure 8.1: *The density of 1.847χ_7^2 versus an empirical distribution of ψ^2 ($K = 9, N = 1000$) obtained from 100 ψ^2 values. In 100 ψ^2 values, two exceeded 30, three exceeded 29, four exceeded 28.*

process, the mean and variance of ψ^2, with $K = 9, N = 1000$, were estimated from 100 independent realizations.

The results in Table 8.1 indicate a 5% critical value of 27–30, and a critical value 31–32 at a level anywhere between 1% and 2.5%. These results indicate roughly that for *ARMA* processes the critical ψ^2-value 30 may be used in testing the hypothesis $H_0 : m_k = m_k^{(0)}, k = 1, 2, \ldots, K$, at a significance level of less than 0.05. Very similar results were obtained by a direct simulation of the whole distribution of ψ^2, using stationary *ARMA* time series. However, serious departures from this rough rule of thumb can occur for non-*ARMA* processes.

TABLE 8.1 Method of Moments Approximation of $G(c) = P(\psi^2 > c), c = 30, 32$, $K = 9, N = 1000$

Process	a^2/b	$\hat{G}(30)$	$\hat{G}(32)$
$Z_t = u_t$	7.6	0.025	0.017
$Z_t = u_t + 0.4u_{t-1}$	8.0	0.020	0.012
$Z_t = u_t - 0.7u_{t-1}$	7.7	0.036	0.017
$Z_t = 0.25Z_{t-1} + u_t$	9.1	0.025	0.015
$Z_t = 0.8Z_{t-1} + u_t$	9.3	0.030	0.020
$Z_t = -0.36Z_{t-1} + u_t$	6.5	0.028	0.020
$Z_t = 2u_t + 85u_{t-1}$	8.9	0.027	0.018
$Z_t = \sin(2u_t + 85u_{t-1})$	8.9	0.016	0.010
$Z_t = 0.9u_t + 0.8u_{t-1} - 6.2u_{t-2} - 9.3u_{t-3} + 0.4u_{t-4}$	8.0	0.020	0.013
$Z_t = 20u_t - 5u_{t-1} + 30u_{t-2} - 8u_{t-3} + 75u_{t-4}$	5.9	0.045	0.035

Note: u_t is Gaussian white noise with mean 0 and variance 1.

The ensuing example illustrates one use of ψ^2 in practice.

Example 8.1: Illustration of the Use of ψ^2 [12], [14]

Consider the construction of the following non-Gaussian stationary process $\{Z_t\}$. Let $\{\epsilon_n\}_{n=-\infty}^{\infty}$ be a (Bernoulli) sequence of independent binomial $(1,p)$ random variables, and define

$$\{T_i\}_{i=-\infty}^{\infty} \equiv \{n \in (\ldots,-2,-1,0,1,2,\ldots) : \epsilon_n = 1\}$$

If $\{M_k\}_{k=-\infty}^{\infty}$ is a sequence of independent standard normal random variables $\mathcal{N}(0,1)$, independent of $\{\epsilon_n\}$, put

$$Z_t = M_k \quad \text{whenever} \quad T_k \le t < T_{k+1}$$

Then $\{Z_t\}$ is called a Bernoulli(p) point process with adjoined normal random variables (see Problem 3). Realizations from the process are shown in Figure 8.2. Clearly, as p increases we expect to see an increasing number of isolated points, whereas small p should give rise to longer runs ("lines") of consecutive points having the same value.

For each value of p, $p = 0.25, 0.5, 0.75$, Table 8.2 presents observed values[1] of D_k obtained from, $Z_1, Z_2, \ldots, Z_{1000}$, as well as averages of D_k obtained from 100 independent realizations of length $N = 1000$ each. The degrees of freedom a^2/b, and $\hat{G}(30), \hat{G}(32)$ were obtained as before. These tail probabilities resemble closely those of the *ARMA* processes in Table 8.1.

TABLE 8.2 Simple HOC (First Column) and Their Estimated Expected Values (Second Column) for Three Bernoulli(p) Processes with Adjoined Normal Random Variables

	$p = 0.25$	$p = 0.50$	$p = 0.75$
D_1	125 123.7	258 248.4	329 325.5
D_2	220 225.4	419 416.4	476 492.0
D_3	376 381.8	593 592.2	590 600.2
D_4	503 512.7	725 715.0	663 672.9
D_5	599 602.7	773 770.2	687 700.0
D_6	677 678.7	805 806.5	715 719.6
D_7	733 731.4	821 826.6	722 731.9
D_8	775 776.5	828 841.3	732 744.2
$(\text{Ave}(\psi^2), \hat{\text{Var}}(\psi^2))$	(10.1,45.4)	(12.7,47.6)	(13.0,47.2)
df $= a^2/b$	4.5	6.8	7.2
$(\hat{G}(30), \hat{G}(32))$	(0.014,0.010)	(0.022,0.015)	(0.023,0.015)

Note: $N = 1000$.

Consider the m_k and Δ_k from the time series with $p = 0.25$. Then

$$\begin{aligned}\psi^2 &= \frac{(125-123.7)^2}{123.7} + \frac{(95-101.7)^2}{101.7} + \frac{(156-156.4)^2}{156.4} + \cdots + \frac{(224-222.5)^2}{222.5} \\ &= 0.0137 + 0.441 + 0.001 + \cdots + 0.010 = 1.455 << 30\end{aligned}$$

[1] The values for $p = 0.75$ are curiously lower than expected.

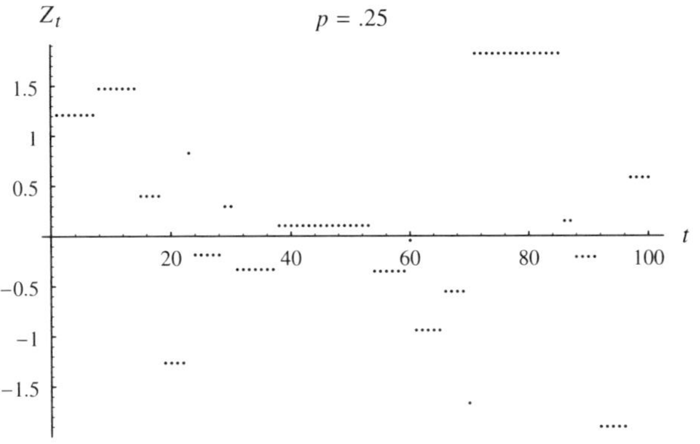

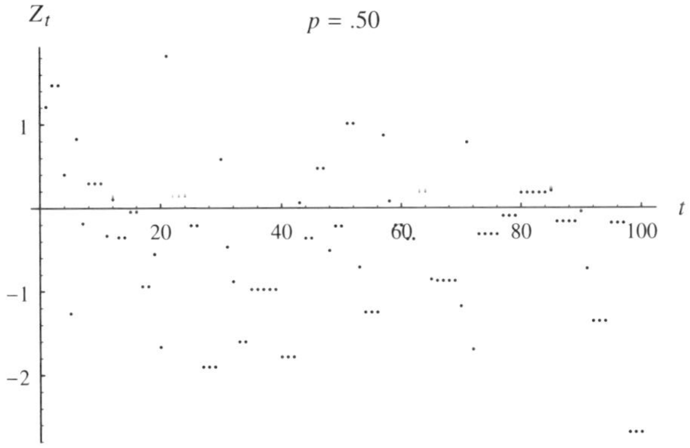

Figure 8.2: *Bernoulli*(p) *point processes with adjoined standard normal random variables.*

Therefore, the null hypothesis that $E[\Delta_k] = m_k$, $k = 1, 2, \ldots, 9$ makes sense. On the other hand, if the m_k correspond to $p = 0.25$, but the Δ_k come from $p = 0.50$, we find

$$\psi^2 = 145.808 + 34.577 + \cdots + 11.920 >> 30$$

and the hypothesis is strongly rejected. The Δ_k and m_k are far apart, and the chance they come from the same process is very small. Likewise, when the m_k correspond to $p = 0.25$, but the Δ_k are obtained with $p = 0.75$, we find

$$\psi^2 = 340.728 + 20.178 + \cdots + 8.9 >> 30$$

and again the hypothesis is rejected strongly.

HOC plots corresponding to the observed HOC in Table 8.2 are shown in Figure 8.3. For the sake of comparison, the plot of simple expected HOC from Gaussian white noise is included also. The HOC plots discriminate between the three point processes, and also set them apart from white Gaussian noise.

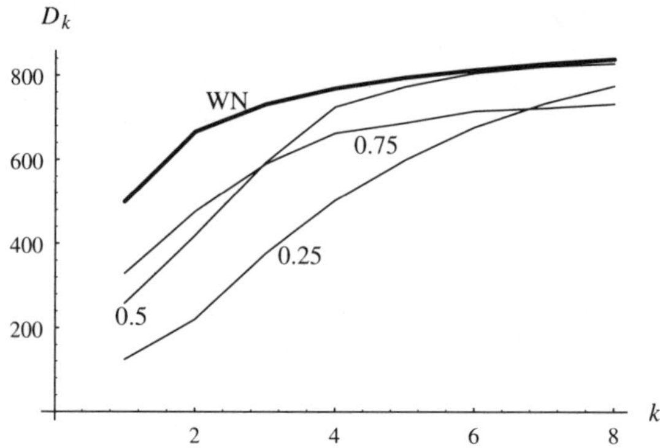

Figure 8.3: *Simple HOC from three Bernoulli processes with adjoined normal random variables. The dark line represents simple expected HOC from white noise.* $N = 1000$.

■

8.1.1 ψ^2 as a Measure of Distance from White Noise

In discriminating between two signals, a useful idea is to measure their distance from a given reference signal. In many respects, the most convenient reference is white Gaussian noise, for which the expected simple HOC are given by equation (6.9),

$$E[D_k] = (N-1)\left\{\frac{1}{2} + \frac{1}{\pi}\sin^{-1}\left(\frac{k-1}{k}\right)\right\}$$

With this, $E[\Delta_k]$ can be easily computed. The idea of using the ψ^2 statistic to measure similarity to white noise has been exploited in several applications, including applications in *nondestructive evaluation* [4].

Example 8.2: Application in NDE [4]

HOC analysis has been applied in the ultrasonic classification of adhesive joints for the purpose of nondestructive evaluation (NDE). This was done by using ψ^2 as a *distance measure from white noise*. That is, using m_k from white Gaussian noise, and observed Δ_k from ultrasonic echo signals.

In a detailed study in [4], ultrasonic echo signals were obtained from aluminum-to-aluminum bonded plate specimens representing various adhesive and cohesive bond properties. Adhesive properties relate to the physical/chemical attraction between the adhesive layer and the adherents, while cohesive properties relate to the method of curing and the chemical composition of the adhesive. The aluminum adherents were pretreated using unsealed chromic acid anodization (CAA) according to the MIL-A-8625D standard.

The ψ^2 was evaluated with $K = 8$ and $N = 512$, and m_k from white Gaussian noise. Figure 8.4 shows the probability distributions of two populations of $\psi^2/1000$ from CAA

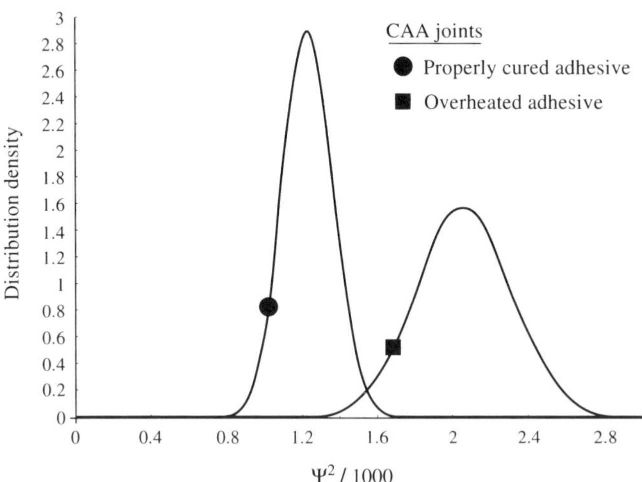

Figure 8.4: *Empirical probability distribution of $\psi^2/1000$ from CAA aluminum specimens with cured (solid •) and overheated (solid ■) adhesive layers. (Dickstein et al. (1991). Reprinted with permisson from Ultrasonics, Vol. 29, pp. 355–365.)*

aluminum specimens corresponding to (1) properly cured and (2) overheated adhesive layers. The figure illustrates the fact that ψ^2 is sensitive to the effect of curing parameters. In the same study [4], ψ^2 as a measure of distance from white noise was also found sensitive to other cohesive and adhesive factors, and in another study the statistic was found sensitive to harsh environmental factors such as high temperature and exposure to lengthy immersion in water of aluminum adhesive joints [5].

■

8.1.2 Weighted ψ^2

An improvement in the discrimination power of ψ^2 is obtained by weighing the quadratic terms in the sum. The modified statistic has the form

$$\psi^2 \equiv \sum_{k=1}^{K} w_k \frac{(\Delta_k - m_k)^2}{m_k} \tag{8.2}$$

where the weights w_k are positive and $\sum_{k=1}^{K} w_k = 1$.

Example 8.3: More on NDE [4]

The weighted ψ^2 was used in the classification of certain composite (produced from 16 ply graphite-epoxy) adhesive joints and was found superior to the unweighted ψ^2 [4]. Figure 8.5 shows the empirical probability distributions of ψ^2 values ($K = 8$, $N = 1024$, white noise as reference) corresponding to two different pretreatments (different types of gritting), type 1 and type 2. The improvement in the discrimination power of the weighted ψ^2 is apparent.

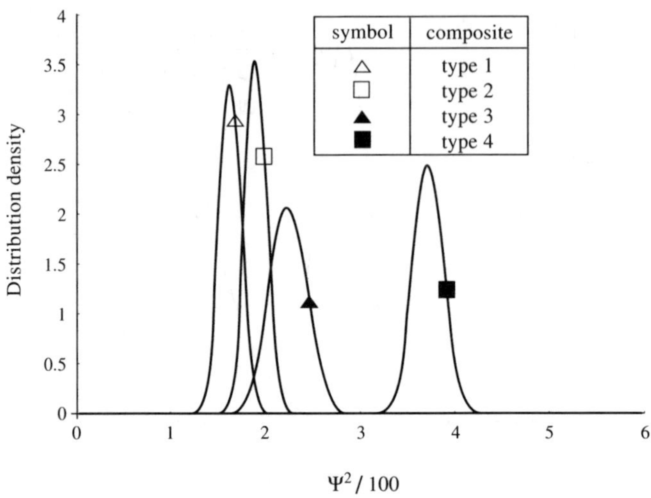

Figure 8.5: *Empirical probability distribution of unweighted* (△, □) *and weighted (solid* ▲, ■) $\psi^2/100$ *from composite joints. (Dickstein et al. (1991). Reprinted with permisson from Ultrasonics, Vol. 29, pp. 355–365.)*

In this application it was found that the weighted ψ^2 performed best by choosing $w_2 = w_3 = 0.5$, setting the rest of the weights equal to 0. This is an example where effective discrimination is achieved on the basis of D_1, D_2, D_3, only.

■

8.2 HOC PLOTS

In Section 6.2.3 we have used a HOC plot and accompanying probability limits in testing the hypothesis of white noise. The same procedure can be used in testing the hypothesis that a given time series follows a specific model. More generally, HOC plots (or paths) together with hypothesized probability limits can serve as a fast graphical means for discrimination, classification, and hypothesis testing [8], [9], [11].

When the asymptotic distribution of HOC is approximately normal, then for each *fixed* θ the cosine formula together with the variance approximations (6.2) or (6.3) can be used in the construction of, say, 95% probability limits defined by

$$E[D_\theta] \pm 1.96\sqrt{\text{Var}[D_\theta]} \tag{8.3}$$

It is important to note that the limits (8.3), as well as several of their specialized forms discussed in this section, provide an approximate probability statement for each fixed θ, and not simultaneously for all θ in some set. A crude way to obtain from (8.3) simultaneous probability statements is described in Problem 7.

8.2.1 Simple HOC Plots

Approximate probability bounds for simple HOC,

$$E[D_k] \pm 1.96\sqrt{\text{Var}[D_k]} \tag{8.4}$$

are obtained from the cosine formula together with Algorithm 6.1. See Problem 5.

The following graphical examples convey the idea of HOC plots better than words. See also Example 1.3 in Chapter 1.

Example 8.4: HOC Plots from *ARMA* Models

With $N = 1000$, Figure 8.6 shows the probability limits (8.4) corresponding to simple HOC obtained from four specific stationary $ARMA$ models hypothesized under H_0. An observed HOC plot (dashed line) falls completely within the probability limits of an $AR(1)$ with parameter 0.5. We therefore do not reject the hypothesis that the simple HOC are compatible with the hypothesis

$$H_0 : Z_t \text{ is an } AR(1) \text{ with parameter } 0.5$$

and conclude that the series from which the D_j were extracted *oscillates* to an appreciable extent as an $AR(1)$ with the indicated parameter. In all the other cases the observed HOC plot is not completely within the respective limits, and so the respective hypotheses are rejected. It is sufficient that a single D_j falls outside the limits for the hypothesis to be rejected.

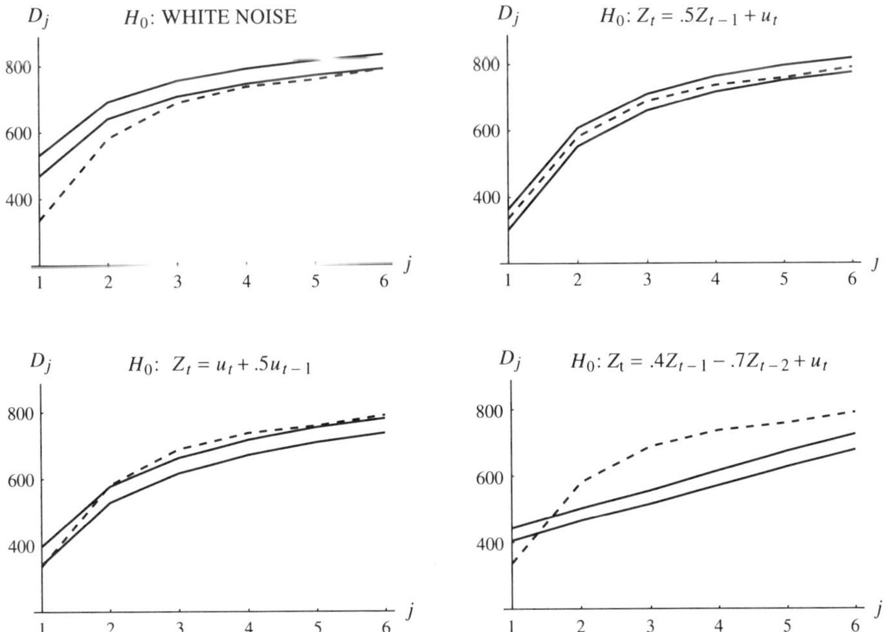

Figure 8.6: *The rate of increase in D_j (dashed line) resembles that of $AR(1)$ with parameter 0.5, and independent $u_t \sim \mathcal{N}(0,1)$. The solid lines are the probability limits under H_0.*

■

When testing specifically for white noise, it is simpler to use the approximate bounds (6.11) for simple HOC

$$(N-1)\left\{\frac{1}{2}+\frac{1}{\pi}\sin^{-1}\left(\frac{k-1}{k}\right)\right\}$$

$$\pm 1.96(N-1)^{1/2}\left\{\frac{1}{4}-\left[\frac{1}{\pi}\sin^{-1}\left(\frac{k-1}{k}\right)\right]^2\right\}^{1/2} \tag{8.5}$$

The probability bounds obtained with this expression can also serve as a quick reference for measuring deviations from white noise, as illustrated in the following example.

Example 8.5: Deviation of the Variable Star from White Noise [7]

Figure 8.7 shows the observed HOC path (dashed) with 20 D_k from the variable star data discussed in Chapter 7, and the corresponding white noise bounds (8.5) (dark lines). Except for a short intersection, the HOC plot is outside the limits, so that the variable star data are far from being white noise. On the other hand, an observed HOC plot from white noise (thin line) is comfortably within the limits as expected.

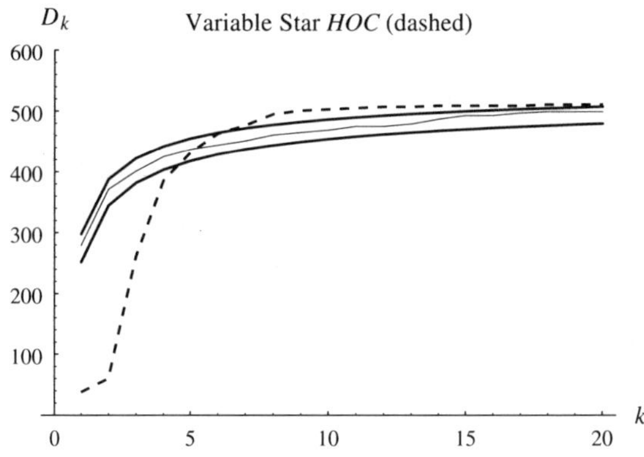

Figure 8.7: *Observed simple HOC path (dashed) from the variable star series, and 95% probability limits (thick solid) from white noise. The thin solid line is an observed HOC path from white noise.* $N = 550$.

■

8.2.2 HOC Plots from the α-Filter

It is sometimes more convenient to use the normalized HOC $\gamma(\alpha)$ and $\hat{\gamma}(\alpha)$. Then, formula (6.3) becomes

$$\text{Var}[\hat{\gamma}(\alpha)] \approx \frac{\gamma(\alpha)(1-\gamma(\alpha))}{N-1}\left\{\frac{1-2\gamma(\alpha)+\nu_1(\alpha)}{1-\nu_1(\alpha)}\right\} \tag{8.6}$$

where

$$\nu_1(\alpha) = \frac{1-2\lambda_1(\alpha)+\lambda_2(\alpha)}{2(1-\lambda_1(\alpha))}$$

and

$$\lambda_k(\alpha) = \frac{1}{2} + \frac{1}{\pi}\sin^{-1}\rho_k(\alpha), \qquad k = 1, 2$$

and $\rho_k(\alpha)$ is the kth order autocorrelation of the filtered process $\{\mathcal{L}_\alpha(Z)_t\}$, $\mathcal{L}_\alpha$ being an arbitrary parametric filter. In particular, we can use the α-filter ($AR(1)$ filter)

$$\mathcal{L}_\alpha \equiv 1 + \alpha\mathcal{B} + \alpha^2\mathcal{B}^2 + \cdots, \qquad \alpha \in (-1, 1)$$

Then, under the assumption of white Gaussian noise (see Problem 4),

$$\rho_k(\alpha) = \alpha^k, \qquad k = 0, 1, 2, \ldots \tag{8.7}$$

and the HOC from the α-filter are given by

$$\rho_1(\alpha) = \alpha, \qquad \gamma(\alpha) = \frac{1}{\pi}\cos^{-1}\alpha \tag{8.8}$$

These expressions can be substituted in (8.6) in order to obtain the 95% white noise probability limits

$$\gamma(\alpha) \pm 1.96\sqrt{\text{Var}[\hat{\gamma}(\alpha)]} \qquad \alpha \in (-1, 1) \tag{8.9}$$

that go along with (normalized) HOC plots from the α-filter. Again we emphasize that (8.9) is an approximation which holds for each *fixed* α, and not simultaneously for all α in $(-1, 1)$. See Problem 7.

Example 8.6: Tracking the Vocal Sound of a Whale [11]

This example is a continuation of Example 1.6 in Chapter 1, where HOC from the α-filter were used in creating a HOC-gram. Employing the same data set, Figure 8.8 shows the observed HOC plots (dashed lines). The figure also contains probability limits (solid lines) (8.9) corresponding to white noise, obtained with (8.7), (8.8), in order to give an idea of how far the data are from white noise. Indeed, for a strong vocal sound the HOC path is far from the white noise limits, while a signal which contains less information produces a HOC path closer to that of white noise.

In this application, also simple HOC plots were used in [11]. Although equivalent, the plots from the α-filter were found somewhat more informative. Interestingly, successful discrimination was possible on the basis of D_1, D_2, D_3, only.

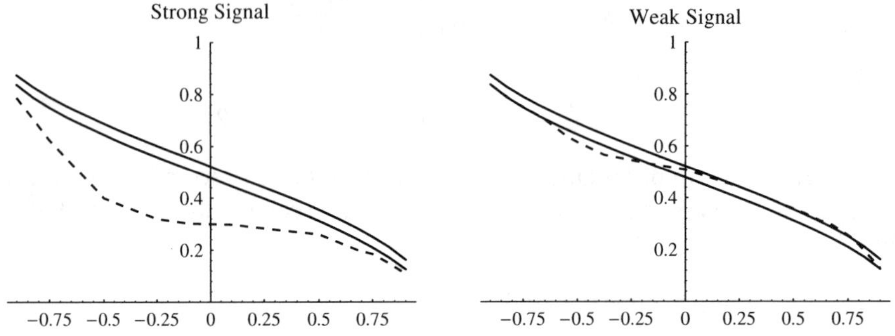

Figure 8.8: *Observed (normalized) HOC path from the α-filter, and* 95% *probability limits from white noise.*

■

8.2.3 Power Simulation

The limits (8.3) provide approximate 95% bounds for each D_θ. However, in the context of hypothesis testing, the test is based on D_θ for *all* θ in some interval or set. This means that the size of the test is not necessarily 0.05. To clarify this point, we consider simple HOC.

Suppose our test is based on $D_1, D_2, \ldots, D_6$. The hypothesis that the series oscillates in some specific fasion can be formulated by testing

$$H_0 : E[D_k] = \mu_k, \qquad k = 1, 2, \ldots, 6$$

where the μ_k are given constants obtained from a certain process.

Our test is based on $D_1, D_2, \ldots, D_6$ simultaneously, and the hypothesis is rejected if at least one D_k falls outside the probability bounds. Intuitively, a test which is based on more than a single D_k may have a higher probability than 5% of rejecting a true hypothesis, and in fact experience shows that with six D_k this probability is about 10%. At the same time, the probability of rejecting a false hypothesis (power) is higher for our simultaneous test than for a single D_k.

Power simulation results are reported in Table 8.3. The table gives the power for testing the hypothesis of white noise against various $ARMA$ alternatives. The bounds are obtained under the hypothesis of white Gaussian noise, but the D_k are generated by the indicated process. Clearly, as the alternative gets farther away from white noise, we expect a higher rate of rejection from a reliable test. The power is estimated from 50 independent series each of length 450. In each case, the table also provides a typical single binary realization

$$I_{[D_k \text{ falls outside the limits}]}, \qquad k = 1, 2, \cdots, 6$$

where $I_{[A]}$ is the indicator of the event A. Thus, 110011 means that D_1,D_2,D_5, D_6 fall outside the limits (8.4), but D_3, D_4 fall inside the limits. The binary pattern tells us which D_k "were active" in rejecting the hypothesis of white noise. The binary patterns show that D_1 need not be active, and that a rule based on six D_k is more likely to reject a false hypothesis than a rule based on D_1 only.

TABLE 8.3 Power Simulation for Testing White Noise versus the Indicated Process, and a Typical Rejection Pattern

Process	Rejection Pattern	Power
White Noise	000000	0.10
$AR(1), \phi = .05$	010000	0.26
$MA(1), \theta = .1$	100000	0.40
$AR(1), \phi = .2$	011000	0.90
$AR(1), \phi = .5$	110101	1.00
$AR(2), \phi_1 = .1, \phi_2 = -.15$	011110	0.88
$ARMA(1,1), \phi_1 = .1, \theta_1 = -.1$	110000	0.86
$ARMA(2,2), \phi_1 = .1, \phi_2 = -.4, \theta_1 = 0, \theta_2 = .3$	011111	1.00
$ARMA(2,2), \phi_1 = .1, \phi_2 = -.2, \theta_1 = .2, \theta_2 = .1$	001100	0.88

8.3 SCATTER PLOTS

In many cases a pair of simple HOC (D_j, D_k) suffices for fast discrimination. The most obvious thing to do then, is to plot D_j versus D_k and examine the resulting scattergram. If classification is of concern, it is often advantageous to construct a simple decision rule of the form $w_1 D_j + w_2 D_k + w_3 = 0$ in deciding upon class membership [20]. Many other variants of this are possible too, such as scattergrams of pairs $(D_1 D_2, D_2 D_3)$.

Example 8.7: Scattergrams Using Simple HOC

Let $Z_t = 0.8 Z_{t-1} + \epsilon_t$, where ϵ_t is white Gaussian noise with mean 0 and variance 1. Define the periodically correlated process (see Problem 5 in Chapter 1)

$$Y_t = \cos\left(\frac{2\pi t}{10}\right) \times Z_t$$

Figure 8.9 shows the graphs of Z_t and Y_t. Although Y_t appears to be stationary, it is only ostensibly so. It is interesting to see the extent to which the simple HOC were affected by the cosine multiple. This is illustrated in Figure 8.10, which shows a scattergram of points (D_1, D_2) obtained from 100 independent realizations, together with a linear decision function. The two processes give rise to distinctly different clusters of pairs (D_1, D_2), and hence we have a useful linear decision rule. The same is repeated with pairs (D_2, D_3) in Figure 8.11, where we see another, fairly clear, separation between two clusters corresponding to the two processes, respectively. A rather sharp separation is obtained from the scaled pairs $(S_1, S_2) \equiv (10^{-4} D_1 D_2, 10^{-4} D_2 D_3)$ shown in Figure 8.12. On the other hand, based on experiments, neither the pairs (D_1, D_2) nor $(D_1 D_2, D_2 D_3)$ appear to be sufficiently useful for discrimination between Z_t and the additive periodically correlated process $\cos(2\pi t/10) + Z_t$.

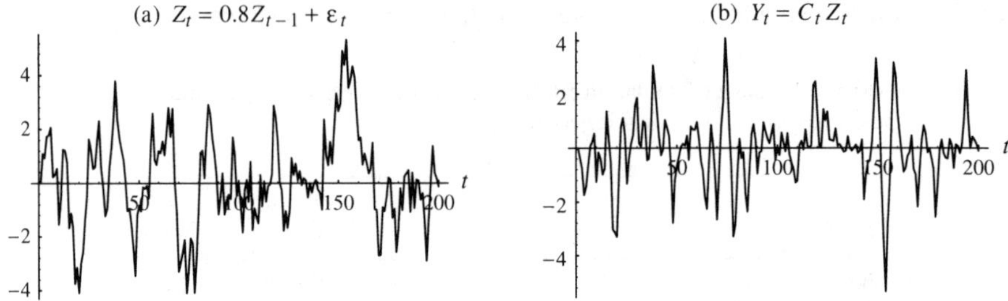

Figure 8.9: *(a) An $AR(1)$ stationary process Z_t. (b) A periodically correlated process $Y_t = C_t \times Z_t$.*

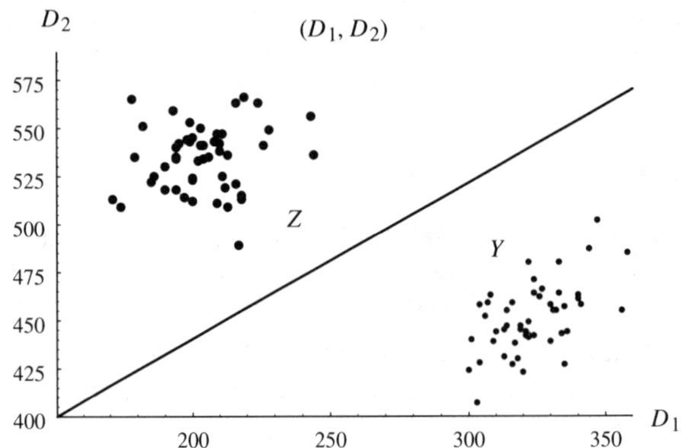

Figure 8.10: *Scattergram of 100 points (D_1, D_2) from independent realizations of stationary and periodically correlated processes. $N = 1000$.*

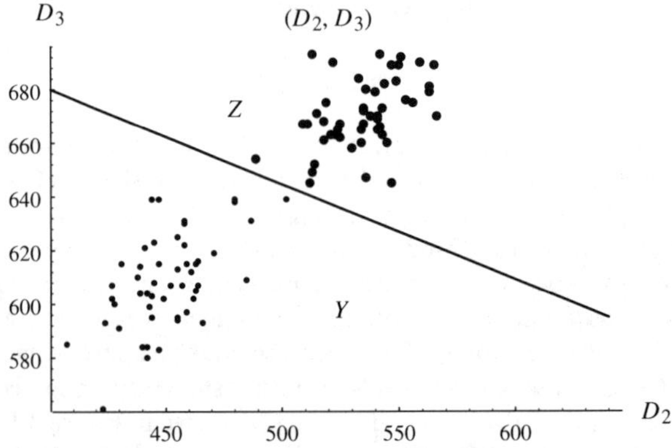

Figure 8.11: *Scattergram of 100 points (D_2, D_3) from independent realizations of stationary and periodically correlated processes. $N = 1000$.*

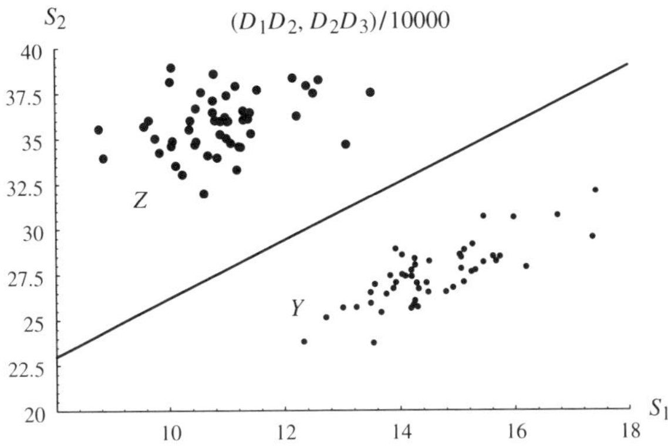

Figure 8.12: *Scattergram of 100 scaled points* $(10^{-4}D_1D_2, 10^{-4}D_2D_3)$ *from independent realizations of stationary and periodically correlated processes.* $N = 1000$. ■

8.4 ONLINE TRACKING

So far the series size N was held fixed. But as was pointed out earlier, information about the oscillatory evolution of a time series can be obtained from the observed zero-crossing rate as a function of N, an example of which we saw in Section 4.2.4 in regard to an explosive process. To a variable degree, all the above methods can be ramified by allowing a variable N [3], [10]. For example, one can measure similarity to white noise by comparing the observed zero-crossing rate (ZCR) with the probability bounds (properly normalized) (8.5) by allowing N to vary [3]. This is depicted in Figure 8.13, which shows the observed zero-crossing rate $D_1/(N-1)$ for $N = 25, 50, \ldots, 2000$, for several different processes in relation to the normalized white noise bounds. Only the dashed line obtained from the ZCR of a white noise realization is completely within the bounds, while the ZCR plots obtained from other processes (nonwhite) are outside the bounds. A straightforward extension in the spirit of this book is to fix θ and let N vary in $D_\theta/(N-1) = D_{\theta,N}/(N-1)$. This is referred to as a *HOC process*.

When the process is nonstationary, the notion of spectrum as known from the theory of stationary processes loses its usual meaning and needs to be redefined. Since nonstationarity can be manifested in many ways, there is no single definition that encompasses the multitude of all different possibilities. In this connection we mention the theory of evolutionary spectra developed in [18], which addresses a certain aspect of nonstationarity as expressed by a time dependent spectrum. It is clear, however, that a process may be nonstationary and still have a "fixed spectrum"—a periodically correlated process is a particular instance of this. Without getting into an endless discussion of a complex problem, it seems though that regardless of the type of nonstationarity, the instantaneous zero-crossing rate is a much simpler concept that can serve in lieu of a precisely defined spectrum. Even simpler than this, we can always consider the zero-crossing rate observed within a moving data window or observed adaptively giving more weight to more current data as a means for summarizing a time

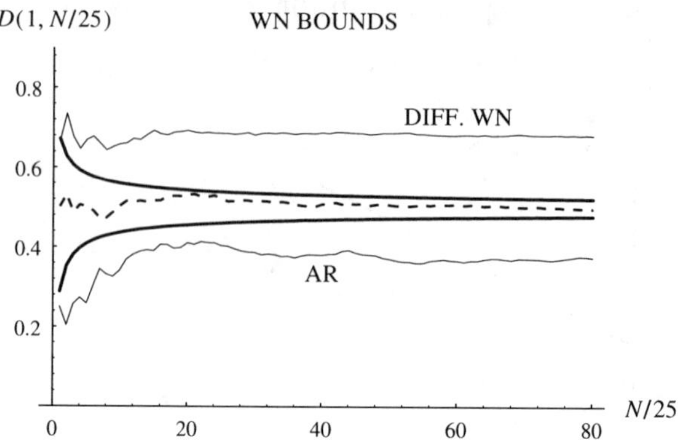

Figure 8.13: *The zero-crossing rate (dashed line) of white noise enters the bounds (thick lines) for white noise, while the rates of a differenced white noise (upper thin line), and an* $AR(2)$ *process (bottom thin line) are outside the bounds.*

dependent oscillation pattern. The adaptive method which we describe next is computationally quite appealing [15].

Fix a positive integer L, and define a *localized* normalized zero-crossing rate $\hat{\gamma}(t)$ by the sum

$$\hat{\gamma}(t) = \beta \sum_{s=t-L+1}^{t} w^{t-s} d_s \tag{8.10}$$

where $0 < w < 1$ is a discount factor, d_s is the zero-crossing indicator, and β is a positive scale defined by

$$\beta \equiv \begin{cases} \dfrac{1-w}{1-w^L}, & \text{if } 0 < w < 1 \\ \dfrac{1}{L}, & \text{if } w = 1 \end{cases}$$

This definition becomes less enigmatic by putting in (8.10) $d_s = 1$ and $t = 0$. As time unfolds, the discount factor gives more weight to the current zero-crossing rate, but it discounts that of the past. A useful feature of (8.10) is that it can be computed adaptively using the recursion

$$\hat{\gamma}(t) = w\hat{\gamma}(t-1) + \Delta\hat{\gamma}(t) \tag{8.11}$$

where

$$\Delta\hat{\gamma}(t) = \beta(d_t - w^L d_{t-L}) \tag{8.12}$$

An interpretation of $\hat{\gamma}(t)$ is that when a dominant frequency exists locally, $\pi\hat{\gamma}(t)$ traces it in time. In general, $\pi\hat{\gamma}(t)$ traces a weighted average of an "evolving spectrum."

As an application we consider the utterance[2] of "zero" and "one" together with the respective $\hat{\gamma}(t)$'s shown in Figures 8.14 and 8.15, respectively. In these cases $\hat{\gamma}(t)$ points to two different oscillation patterns, and may be viewed as a type of "empirical spectrum."

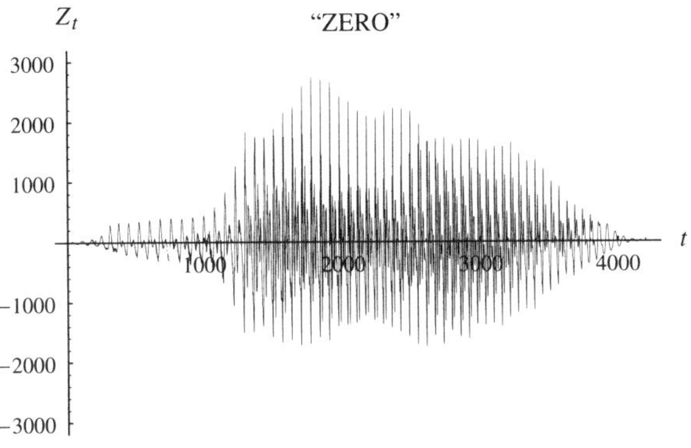

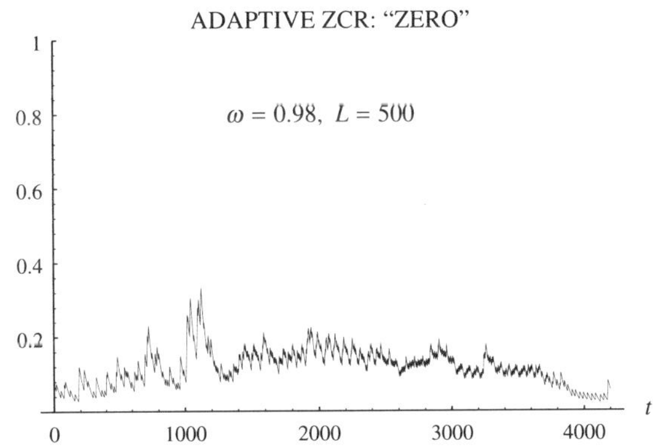

Figure 8.14: *Utterance of "zero" and its adaptive zero-crossing rate.*

The idea of observing the zero-crossing rate as a function of time and in combination with the CM method is particularly useful for adaptive instantaneous frequency tracking, a topic we mention very briefly next [6], [15].

Using the rate $\hat{\gamma}(t)$ together with a recursive filter applied iteratively as new data arrive at time t, we obtain a mechanism for adaptive frequency tracking. There are several ways to apply such a scheme in practice, one of which is to obtain at time t the filtered zero-crossing indicator d_t needed for (8.12) from Y_t and Y_{t-1},

$$d_t = I_{[Y_t \neq Y_{t-1}]}$$

where

[2]D. Burton kindly furnished the data.

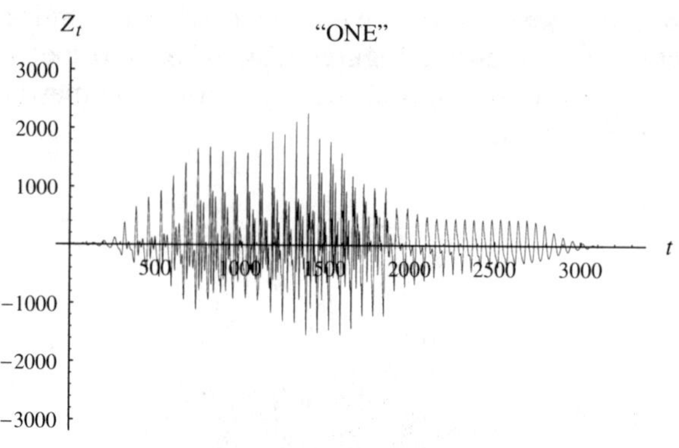

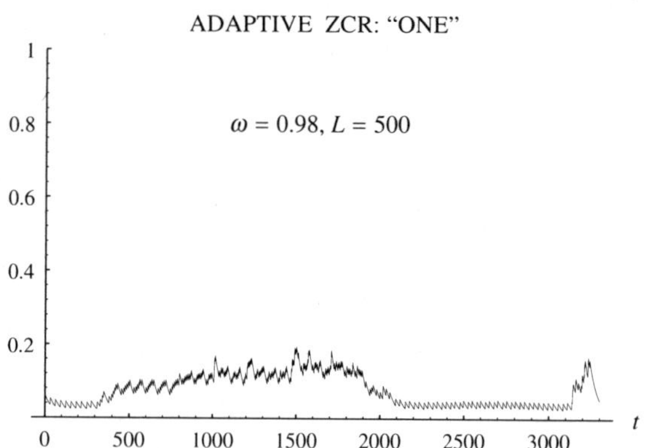

Figure 8.15: *Utterance of "one" and its adaptive zero-crossing rate.*

$$Y_t = \sum_{i=1}^{p} a_i(t) Y_{t-i} + \sum_{j=1}^{q} b_j(t) Z_{t-j}$$

Figure 8.16 shows how this algorithm traces a frequency that undergoes an abrupt change from 1 to 2 radians per unit time for a single sinusoid plus white noise using the alpha filter

$$Y_t = \alpha(t) Y_{t-1} + Z_t$$

where $\alpha(t) = \cos(\pi\hat{\gamma}(t-1))$. Since the HK algorithm is not allowed to converge here (a single iteration), the method necessitates a relatively high signal-to-noise ratio. We can see that as L decreases the adaptive estimator detects (learns) the change in frequency faster at a cost of a higher variance. Clearly, bandpass filters can do better than the alpha filter, but require more precise initial guesses. Also, an adaptive first order sample autocorrelation can be used in frequency tracking in much the same way as the adaptive zero-crossing rate, with the possible advantage of a reduced variance.

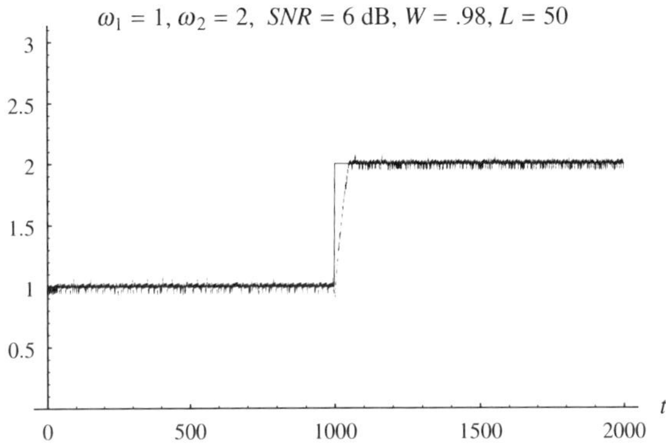

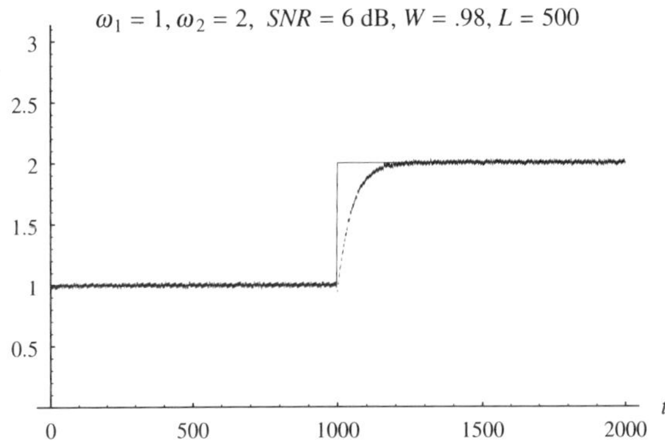

Figure 8.16: *A frequency is being traced by an adaptive zero-crossing rate together with a recursive filter. A greater L leads to a smoother estimate.*

More precise frequency tracing can be achieved by nonadaptive procedures, but at a price—in general, they are much slower to learn of sudden frequency changes and their implementation can be very time-consuming. As an example of a nonadaptive procedure, consider the HK or CM algorithm applied to overlapping stretches of data [16], [21]. For reasonable signal-to-noise ratio, we may expect near convergence within each stretch (i.e., data window) with a few iterations—provided no change takes place—and hence improved estimation. On the other hand, each iteration requires filtering of the entire data window which slows down the procedure considerably.

Figure 8.17 shows the trace obtained by an application of the HK algorithm—with five iterations of the alpha filter—to overlapping data windows of size $L = 500$, using the same data set as before. The window was shifted to the right one time unit at the end of each fifth iteration. The improvement in precision achieved at the price of a relatively long learning period is apparent. As for implementation, the adaptive procedure was by far faster.

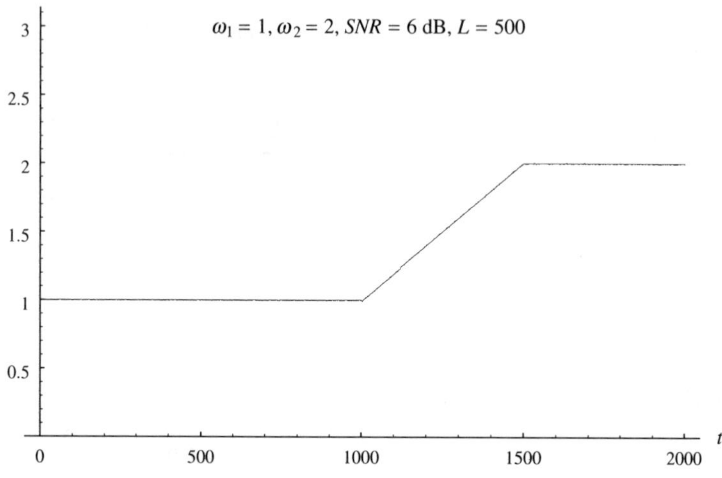

Figure 8.17: *A frequency is being traced by a nonadaptive zero-crossing rate using the HK algorithm—five iterations—on moving overlapping data windows of length* $L = 500$.

8.5 PROBLEMS AND COMPLEMENTS

1. Show that for a stationary ergodic process, the Δ_k are asymptotically nonnegative. That is, show

$$\frac{\Delta_k}{N} \to \begin{cases} P(X_t^{(k)} \neq X_{t-1}^{(k)}), & \text{if } k = 1 \\ P(X_t^{(k)} \neq X_{t-1}^{(k)}) - P(X_t^{(k-1)} \neq X_{t-1}^{(k-1)}), & \text{if } k = 2, \ldots, K-1 \\ P(X_t^{(K-1)} = X_{t-1}^{(K-1)}), & \text{if } k = K \end{cases}$$

2. Define the $K \times 1$ vectors, $\boldsymbol{\Delta} = (\Delta_1, \ldots, \Delta_K)'$, $\hat{\mathbf{p}} = 1/(N-1)\boldsymbol{\Delta}$, $\mathbf{p} = E[\hat{\mathbf{p}}]$, and the $K \times K$ diagonal matrix $\mathbf{C} = \text{diag}(p_1, \ldots, p_K)$. Show that

$$\psi^2 = (N-1)(\hat{\mathbf{p}} - \mathbf{p})'\mathbf{C}^{-1}(\hat{\mathbf{p}} - \mathbf{p})$$

3. [12] Let $\{Z_t\}$ be a Bernoulli(p) process with adjoined standard normal random variables.

   (a) Show that $\{Z_t\}$ is strictly stationary with autocovariance

$$\text{Cov}(Z_0, Z_t) = (1-p)^t$$

   (b) Find $E[\Delta_j]$, $j = 1, 2, 3$.
   ($E[\Delta_1] = (N-1)p/2$, $E[\Delta_2] = (N-1)p(1-p/3)$.)

4. Consider the α-filter, and assume the process is white Gaussian noise. Show that

$$\rho_k(\alpha) = \alpha^k, \qquad k = 0, 1, 2, \ldots$$

(Hint: Use $\mathcal{J}_\alpha(\rho)_n$ in Section 5.6.)

5. **(a)** Show that for $N = 1000$, and

$$(\rho_1, \rho_2, \ldots, \rho_7) = (0.5, 0.25, 0.125, 0.0625, 0.03125, 0.015625, 0.0078125)$$

the output from Algorithm 6.1 is

k	$E[D_k]$	Var$[D_k]$
1	333.000	248.296
2	579.850	195.717
3	684.687	162.259
4	739.669	144.061
5	773.816	132.051
6	797.436	123.217

(b) From the table construct 95% probability limits.
(c) Consider the simple HOC ($N = 1000$) values $300, 550, 709, 750, 780, 800$. Do these HOC resemble the expected HOC in the table?

6. Consider the lynx data in Problem 1 in Chapter 7. Construct a HOC plot to test the hypothesis that the data follow the model $Z_t = \beta\cos((2\pi/10)t + \phi) + \epsilon_t$ under appropriate conditions on ϕ and ϵ_t.
7. *Bonferroni Inequality [17, p. 8]*. Suppose that for $\alpha_k \in (0, 1)$,

$$P(D_{\theta_k} \in I_k) = 1 - \alpha_k, \qquad k = 1, \ldots, K$$

Show that in general

$$P(D_{\theta_1} \in I_1, \ldots, D_{\theta_K} \in I_K) \geq 1 - (\alpha_1 + \cdots + \alpha_K)$$

(Hint: If A_k are events, then $P(\text{All } A_k \text{ are true}) = 1 - P(\text{At least one } A_k \text{ is false})$, and $P(A_1 \bigcup A_2) \leq P(A_1) + P(A_2)$.)
8. Use (8.5) to construct probability bounds for the HOC process $D_{2,N}/(N-1)$.

REFERENCES

[1] Anderson, T. W. *An Introduction to Multivariate Statistical Analysis* (2nd ed.), New York: Wiley, 1984.

[2] Bartlett, M. S. *Stochastic Processes* (3rd ed.), Cambridge, England: Cambridge Univ. Press, 1978.

[3] Berger, M., B. Kedem, J. A. O'Connell, and J. F. Troendle. "A method for online testing by HOC processes," unpublished, 1987.

[4] Dickstein, P. A., J. K. Spelt, and A. N. Sinclair. "Application of a higher order crossing feature to non-destructive evaluation: A sample demonstration of sensitivity to the condition of adhesive joints," *Ultrasonics*, Vol. 29, No. 5, pp. 355–365, Sept. 1991.

[5] Dickstein, P. A., J. K. Spelt, A. N. Sinclair, and Y. Bushlin. "Investigation of non-destructive monitoring of the environmental degradation of structural adhesive joints," *Mater. Eval.*, Vol. 49, No. 12, pp. 1498–1505, Dec., 1991.

[6] Dragošević, M., S. Stanković, and M. Čarapić. "An approach to recursive estimation of time-varying spectra," *Proc. ICASSP 82 Conf.*, Vol. 3, (Paris, France, May 1982), pp. 2080–2083.

[7] Kedem, B. "Search for periodicities by axis-crossings of filtered time series," *Signal Process.*, Vol. 10, No. 2, pp. 129–144, March 1986.

[8] Kedem, B. "Higher-order crossings in time series model identification," *Technometrics*, Vol. 29, No. 2, pp. 193–204, 1987.

[9] Kedem, B. "A fast graphical goodness of fit test for time series models," in *Time Series and Econometric Modelling*, I.B. MacNeill and G.J. Umphrey, Eds., pp. 65–76, Dordrecht, the Netherlands: Reidel, 1987.

[10] Kedem, B. and D. Martin. "Strictly oscillatory processes," Tech. Rep. TR-87-01, Department of Mathematics, Univ. of Maryland, College Park, 1987.

[11] Kedem, B., and T. Li. "Higher order crossings from a parametric family of linear filters," Tech. Rep. TR-89-47, Department of Mathematics, Univ. of Maryland, College Park, 1989.

[12] Kedem, B. and E. Slud. "Higher order crossings in the discrimination of time series, I, II ," Tech. Rep. TR-79-66 and TR-79-81, Department of Mathematics, Univ. of Maryland, College Park, 1979.

[13] Kedem, B. and E. Slud. "On goodness of fit of time series models: An application of higher order crossings," *Biometrika*, Vol. 68, No. 2, pp. 551–556, Aug. 1981.

[14] Kedem, B. and E. Slud. "Time series discrimination by higher order crossings," *Ann. Stat.*, Vol. 10, No. 3, pp. 786–794, Sept. 1982.

[15] Li, T. and B. Kedem. "Adaptive frequency tracking by zero-crossing counts," Tech. Rep. TR91-26, Department of Mathematics, Univ. of Maryland, College Park, 1991.

[16] Lopes, S. *Spectral Analysis in Frequency Modulated Models*, Doctoral Dissertation, Department of Mathematics, Univ. of Maryland, College Park, 1991.

[17] Miller, R. G., Jr. *Simultaneous Statistical Inference* (2nd ed.), New York: Springer-Verlag, 1981.

[18] Priestley, M. B. *Spectral Analysis and Time Series*, Vol. 2, London: Academic Press, 1981.

[19] Reed, G. W. *Some Properties and Applications of Higher Order Crossings*, Doctoral Dissertation, Department of Mathematics, Univ. of Maryland, College Park, 1983.

[20] Tou, J. T. and R. C. Gonzalez. *Pattern Recognition Principles*, Reading, Mass.: Addison-Wesley, 1974.

[21] Yakowitz, S. "Automatic signal tracking and broadband FM detection in a noisy channel," informal report, March 1990.

9

Prediction of Level-Crossings

Given a time series $\{Y_t\}$, $t = 1, 2, 3, \ldots$, in numerous practical situations one is interested in the prediction of the future level-crossing event

$$\{Y_t \geq r\}$$

from past information about $Y_1, Y_2, \ldots, Y_{t-1}$, and knowledge of some auxiliary stochastic data represented by a (column) vector $\{\mathbf{Z}_t\}$ related to $\{Y_t\}$. The auxiliary data represent other time series which may influence the evolution of $\{Y_t\}$. It is customary to refer to $\{\mathbf{Z}_t\}$ as *covariate* data, and to the components of $\{\mathbf{Z}_t\}$ as *time dependent covariates*. The prediction problem is addressed in what follows by appealing to ideas from the theory of modern statistical inference. This theory, developed during the past twenty or twenty-five years, is both beautiful and highly useful. However, we shall not be concerned with a very rigorous formal treatment. Rather, we shall be content with some regularity conditions, and a rather informal discussion concerning the validity of our results. We note that quite a few of the concepts in this chapter were not discussed in earlier chapters, but will be defined as needed. The present development follows closely the work in [24].

It is convenient to refer to the event $\{Y_t \geq r\}$ as "success," and introduce the indicator binary process,

$$X_t \equiv I_{[Y_t \geq r]} = \begin{cases} 1, & \text{if } Y_t \geq r \\ 0, & \text{if } Y_t < r \end{cases}$$

We formulate the prediction problem as the estimation of the one step conditional probability

$$P(X_t = 1|\mathcal{F}_{t-1})$$

where $\mathcal{F}_{t-1}$ represents all that is known to the observer at time $t-1$ about $X_{t-1}, X_{t-2}, \ldots$, and about past covariate information $\mathbf{Z}_{t-1}, \mathbf{Z}_{t-2}, \ldots$. The sophisticated reader can recognize that $\mathcal{F}_{t-1}$ stands for the σ-field, $\mathcal{F}_{t-1} = \sigma(X_{t-1}, X_{t-2}, \ldots, \mathbf{Z}_{t-1}, \mathbf{Z}_{t-2}, \ldots)$. No harm is caused, however, if $\mathcal{F}_{t-1}$ is thought of as the past "history" or "information" as embodied in the past data $X_{t-1}, X_{t-2}, \ldots, \mathbf{Z}_{t-1}, \mathbf{Z}_{t-2}, \ldots$. It follows that $\mathcal{F}_{t-1} \subset \mathcal{F}_t$ because more is known at time t than at time $t-1$.

In general, the binary process may be stationary or nonstationary, and the explanatory components of $\{\mathbf{Z}_t\}$ may consist of functions of past values of $\{X_t\}$. For example, this situation is encountered in predicting whether the runoff of a river crosses a critical threshold r, given past rainfall and runoff data [26]. Here runoff is the underlying process Y_t, rainfall is the process R_t, and the covariate data vector $\mathbf{Z}_{t-1}$ consists of past rainfall and runoff data. Some possibilities are

$$\mathbf{Z}_{t-1} = (1, X_{t-1}, Y_{t-1}, R_{t-1}, R_{t-1}R_{t-2})'$$

or

$$\mathbf{Z}_{t-1} = (1, X_{t-1}, Y_{t-1}Y_{t-2}, R_{t-1}X_{t-7}, R_{t-2}Y_{t-4})'$$

The inclusion of "1" is nothing but a mathematical convenience.

Our approach to the problem is through a useful *parametric model*, referred to as a *logistic regression model*, and *partial likelihood parametric inference* concerning a *time-invariant* vector parameter $\boldsymbol{\beta}$. This enables prediction and/or hypothesis testing concerning the strength of dependence between $\{X_t\}$ and the explanatory vector $\{\mathbf{Z}_t\}$.

9.1 Partial Likelihood

The notions of likelihood and maximum likelihood estimation are among the most important underpinnings of modern statistical theory. Historically, the theory was developed for independent observations, but it has been known for quite some time that the theory can be extended to dependent data, provided the dependence structure is such that the past can be discounted in some sense. Ideally, the past is discounted so that more and more information is accumulated regarding a fixed number of time-invariant parameters. A particular way of doing this is through *conditional inference*. In the context of binary and categorical data, this has recently been done by assuming *conditional independence* ([9], [15]) or by assuming *conditional* m*th-order Markov dependence* given time dependent covariates ([10], [12], [17]; see also [13], [16]). For example, if $X_1, \cdots, X_N$ is a binary time series, conditional independence given auxiliary information, AI, implies the factorized likelihood,

$$p_\theta(X_1, X_2, \ldots, X_N|AI) = \prod_{j=1}^{N} p_\theta(X_j|AI)$$

while the conditional first order Markov property gives the factorization

$$p_\theta(X_2, \ldots, X_N|X_1, AI) = \prod_{j=2}^{N} p_\theta(X_j|X_{j-1}, AI)$$

There is a subtlety here stemming from the requirement that the auxiliary information must be known in full *throughout* the period of observation. Thus, in the context of time dependent covariates, if the primary data of interest depend on future covariates, the conditional likelihood cannot be written down without further assumptions on the dependence structure. Partial likelihood (abbreviated PL) bypasses this problem because it allows inference from incoming *past* information as time unfolds. To be more specific, suppose the nested σ-fields $\mathcal{G}_0 \subset \mathcal{G}_1 \subset \mathcal{G}_2$ are generated by the covariates, and define $\mathcal{F}_t = \sigma(\mathcal{G}_t, \{X_s, s \leq t\})$. Then also $\mathcal{F}_0 \subset \mathcal{F}_1 \subset \mathcal{F}_2$. A conditional likelihood function for a parameter θ is given by conditioning on $\mathcal{G}_2$,

$$p_\theta(X_1, X_2, X_3|\mathcal{G}_2) = p_\theta(X_1|\mathcal{G}_2)p_\theta(X_2|X_1, \mathcal{G}_2)p_\theta(X_3|X_1, X_2, \mathcal{G}_2)$$

A partial likelihood, on the other hand, is given by

$$p_\theta(X_1|\mathcal{F}_0)p_\theta(X_2|\mathcal{F}_1)p_\theta(X_3|\mathcal{F}_2)$$

To bring our point across, focus on the first factor in these products. We can see that $p_\theta(X_1|\mathcal{G}_2)$ depends on future auxiliary information, as opposed to $p_\theta(X_1|\mathcal{F}_0)$ where only past auxiliary information and past data count.

To summarize, the key idea behind PL is that PL permits sequential conditional inference: data are processed as they arrive in time, taking into account all that is known to the observer at the time, including auxiliary information. Hence, time ordering is an essential element. The logistic regression model studied in this chapter renders these ideas clear.

Partial likelihood was introduced by Cox (in 1972 and 1975) [7], [8], and its theoretical justification, including large sample theory, has been developed in [22] and [25]. In the context of survival theory and counting processes, the theory has been developed in [1], [2], [11], [21]. The general definition given next follows [23, ch. 6].

Let $\mathcal{F}_k$, $k = 0, 1, 2, \ldots$, be an increasing sequence of past histories (σ-fields), and let $X_1, X_2, \ldots$ be a sequence of random variables on some common probability space, such that X_k is $\mathcal{F}_k$-measurable. This means that events of the form $\{X_k \leq x\}$, for all real x, belong to $\mathcal{F}_k$. Let $p_k(x_k;\theta)$ be the probability density of X_k given $\mathcal{F}_{k-1}$. The *partial likelihood function* relative to θ and the data $\{X_t\}$ is given by the product

$$\mathrm{PL}(\theta; \mathbf{X}_N) \equiv \mathrm{PL}(\theta; X_1, \ldots, X_N) = \prod_{k=1}^{N} p_k(X_k;\theta) \tag{9.1}$$

We think of $\mathcal{F}_{k-1}$ as the history (σ-field) generated by past X_t, $t \leq k-1$, and past (also present if possible) covariate information. In this respect PL generalizes the notions of *both* likelihood and conditional likelihood. PL reduces to the usual *likelihood* when the auxiliary covariate information is absent, and it becomes *conditional likelihood* when all the auxiliary covariate information is known throughout the period of observation. Clearly, there is nothing in our formulation that prevents the X_t from being independent.

As we shall see, as long as we can write the product (9.1), no matter what the underlying philosophy or reasoning is, rigorous inference about θ is possible using asymptotic results from martingale (defined below) theory.

9.2 The Logistic Model

Logistic regression models have been used for years by statisticians, econometricians, and psychometricians. Early references are [4], [6], and [18], while [19] gives a comprehensive review of current literature. Recently, the following logistic regression model (also known as the *logit* model) has drawn attention in connection with time series applications,

$$p_t(\boldsymbol{\beta}) \equiv P_{\boldsymbol{\beta}}(X_t = 1|\mathcal{F}_{t-1}) = \frac{1}{1+\exp[-\boldsymbol{\beta}'\mathbf{Z}_{t-1}]} \tag{9.2}$$

where $\boldsymbol{\beta}$ is a (column) vector parameter of the same dimension as $\mathbf{Z}_{t-1}$. For convenience we assume that the first coordinate of $\mathbf{Z}_{t-1}$ is 1, so that the intercept-coefficient β_0 is constant, and that $\mathbf{Z}_{t-1}$ contains past values of X_t. The model has been considered in various forms in [9], [10], [12], [16], [24]. A typical choice for $\boldsymbol{\beta}'\mathbf{Z}_{t-1}$ could be

$$\boldsymbol{\beta}'\mathbf{Z}_{t-1} = \beta_0 + \beta_1 X_{t-1} + \beta_2 Y_{t-1} + \beta_3 Y_{t-2} + \beta_4 W_{t-2}$$

or

$$\boldsymbol{\beta}'\mathbf{Z}_{t-1} = \beta_0 + \beta_1 W_t + \beta_2 X_{t-1} + \beta_3 X_{t-1} W_t + \beta_4 W_{t-2} Y_{t-7}$$

and so on, given an underlying one-dimensional process $\{Y_t\}$ whose level-crossings are of interest, and an auxiliary process $\{W_t\}$. We can see that when $\boldsymbol{\beta}'\mathbf{Z}_{t-1}$ includes linear functions of $X_{t-1}, X_{t-2}, \ldots$, the model (9.2) becomes a form of *binary autoregression*.

Since X_t is binary, (9.2) implies that

$$p_t(x_t; \boldsymbol{\beta}) = P_{\boldsymbol{\beta}}(X_t = x_t|\mathcal{F}_{t-1}) = [p_t(\boldsymbol{\beta})]^{x_t}[1 - p_t(\boldsymbol{\beta})]^{1-x_t}$$

The corresponding partial likelihood is simply the product

$$\mathrm{PL}(\boldsymbol{\beta}) = \prod_{t=1}^{N} p_t(X_t; \boldsymbol{\beta}) = \prod_{t=1}^{N} [p_t(\boldsymbol{\beta})]^{x_t}[1 - p_t(\boldsymbol{\beta})]^{1-x_t} \tag{9.3}$$

The maximizer $\hat{\boldsymbol{\beta}}$ of PL($\boldsymbol{\beta}$) is called the *maximum partial likelihood estimator* (MPLE) of $\boldsymbol{\beta}$. A remarkable fact about $\hat{\boldsymbol{\beta}}$ is that under rather mild regularity conditions on the large sample behavior of the covariate process $\{Z_t\}$ as $N \to \infty$, it is a numerically stable estimator of $\boldsymbol{\beta}$, which is consistent and asymptotically normal with easily estimated covariance matrix.

We have modeled $p_t(\boldsymbol{\beta})$ in (9.2) in terms of $F(\boldsymbol{\beta}'\mathbf{Z}_{t-1})$, where $F(x) = 1/(1+\exp(-x))$ is a standard logistic distribution function. However, other "link" functions F can be used [19]. A particularly attractive link is defined by $F \equiv \Phi$, where Φ is the standard normal distribution function. In this case we obtain what is known as the *probit* model,

$$\tilde{p}_t(\boldsymbol{\beta}) \equiv P_{\boldsymbol{\beta}}(X_t = 1|\mathcal{F}_{t-1}) = \Phi(\boldsymbol{\beta}'\mathbf{Z}_{t-1}) \tag{9.4}$$

We shall not pursue the probit model (9.4). We only remark that all our results with the logistic regression model (9.2) have their analogs under the probit model [24]. See Problem 5.

9.2.1 Logistic Autoregression

How can time series models of the type (9.2) arise? Here is an example. Let $\{Y_t\}$, $t = 0, \pm 1, \pm 2, \ldots$, be an autoregressive process of order p,

$$Y_t = \gamma_0 + \gamma_1 Y_{t-1} + \cdots + \gamma_p Y_{t-p} + \lambda \epsilon_t$$

where λ is a constant, and the ϵ_t are independently and identically distributed (IID) random variables logistically distributed, $\epsilon_t \sim f(x) = e^x/(1+e^x)^2$, with mean 0 and variance $\pi^2/3$. See Figure 9.1 for a comparison between $f(x)$ and the density of $\mathcal{N}(0, \pi^2/3)$. Now fix a threshold $r \in (-\infty, \infty)$, and define a binary time series by clipping Y_t at this threshold,

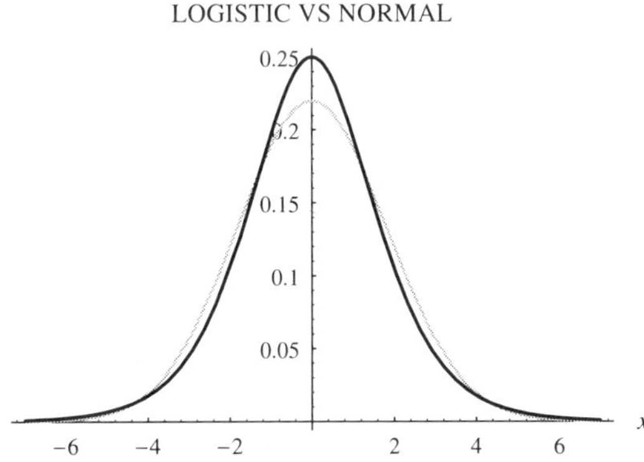

Figure 9.1: *Standard logistic density (black line) compared with $\mathcal{N}(0, \pi^2/3)$ density (gray line).*

$$X_t \equiv I_{[Y_t \geq r]} = \begin{cases} 1, & \text{if } Y_t \geq r \\ 0, & \text{if } Y_t < r \end{cases}$$

Then $X_t = I_{[Y_t \geq r]}$ satisfies (9.2) for each fixed r, with

$$\mathbf{Z}_{t-1} = (1, Y_{t-1}, Y_{t-2}, \ldots, Y_{t-p})'$$

and

$$\boldsymbol{\beta} = \frac{1}{\lambda}(\gamma_0 - r, \gamma_1, \ldots, \gamma_p)'$$

That is,

$$\begin{aligned} p_t(\boldsymbol{\beta}) &\equiv P_{\boldsymbol{\beta}}(X_t = 1 | \mathcal{F}_{t-1}) \\ &= \frac{1}{1 + \exp[-(\gamma_0 - r + \gamma_1 Y_{t-1} + \cdots + \gamma_p Y_{t-p})/\lambda]} \end{aligned} \tag{9.5}$$

In other words, an $AR(p)$ model with logistic errors implies (9.2) for all r. Conversely, it is easy to see that if (9.5) holds for $X_t = I_{[Y_t \geq r]}$ for all r, then Y_t is an $AR(p)$ process with logistically distributed noise. However, the model (9.5) for $X_t = I_{[Y_t \geq r]}$ with a *fixed* known r, is obviously much less restrictive than the assumption that Y_t is an $AR(p)$ process with logistic errors, and there can be situations where the data seem to obey (9.5) for $X_t = I_{[Y_t \geq r]}$ with some fixed r, but not an $AR(p)$ model. Our rainfall-runoff example discussed at the end of the chapter appears to be an illustration of this.

9.2.2 Some Typical Calculations

The logistic regression model (9.2), lends itself to calculations essential for the large sample theory in the next section. We mention briefly some key facts that are being used time and again in what follows.

1. Prediction:
$$p_t(\boldsymbol{\beta}) = E[X_t | \mathcal{F}_{t-1}]$$
2. Martingale difference property:
$$E[\mathbf{Z}_{t-1}(X_t - p_t(\boldsymbol{\beta})) | \mathcal{F}_{t-1}] = \mathbf{0}$$
Hence, $E[\mathbf{Z}_{t-1}(X_t - p_t(\boldsymbol{\beta}))] = \mathbf{0}$.
3. Orthogonality: For $s < t$,
$$E[\mathbf{Z}_{s-1}\mathbf{Z}'_{t-1}(X_s - p_s(\boldsymbol{\beta}))(X_t - p_t(\boldsymbol{\beta})) | \mathcal{F}_{s-1}] = \mathbf{0}$$
Hence, $E[\mathbf{Z}_{s-1}\mathbf{Z}'_{t-1}(X_s - p_s(\boldsymbol{\beta}))(X_t - p_t(\boldsymbol{\beta}))] = \mathbf{0}$.
4. Getting the factor $p_s(\boldsymbol{\beta})(1 - p_s(\boldsymbol{\beta}))$ under the integral:
$$\begin{aligned} E[\mathbf{Z}_{s-1}\mathbf{Z}'_{s-1}(X_s - p_s(\boldsymbol{\beta}))^2] &= E[\mathbf{Z}_{s-1}\mathbf{Z}'_{s-1}E[(X_s - p_s(\boldsymbol{\beta}))^2 | \mathcal{F}_{s-1}]] \\ &= E[\mathbf{Z}_{s-1}\mathbf{Z}'_{s-1}p_s(\boldsymbol{\beta})(1 - p_s(\boldsymbol{\beta}))] \end{aligned}$$

Notice that E really means $E_{\boldsymbol{\beta}}$, an expectation when the parameter value is $\boldsymbol{\beta}$, unless stated otherwise, and that $\mathbf{0}$ means a vector in 2 but a matrix in 3.

9.3 Large Sample Theory

Denote the MPLE by $\hat{\boldsymbol{\beta}}$. In this section we study the asymptotic properties of $\hat{\boldsymbol{\beta}}$ as the sample (series) size N increases. As mentioned earlier, $\hat{\boldsymbol{\beta}}$ is obtained by maximizing the partial likelihood (9.3) with respect to the $\boldsymbol{\beta}$. This is done by differentiating PL($\boldsymbol{\beta}$) in (9.3) with respect to the vector $\boldsymbol{\beta}$.

Let the dimension of $\boldsymbol{\beta}$ be d, $\boldsymbol{\beta} \in \mathcal{R}^d$, and let ∇ denote the gradient operator:

$$\nabla \equiv \left(\frac{\partial}{\partial \beta_1}, \frac{\partial}{\partial \beta_2}, \cdots, \frac{\partial}{\partial \beta_d} \right)'$$

Hence, for a smooth function $f(\boldsymbol{\beta})$,

$$\nabla' f(\boldsymbol{\beta}) \equiv \left(\frac{\partial f}{\partial \beta_1}, \frac{\partial f}{\partial \beta_2}, \cdots, \frac{\partial f}{\partial \beta_d} \right)$$

and the Hessian matrix of second order partial derivatives is given by

$$\nabla \nabla' f(\boldsymbol{\beta}) = \left(\frac{\partial^2 f}{\partial \beta_j \beta_k} \right)$$

With this notation,

$$\nabla p_t(\boldsymbol{\beta}) = \mathbf{Z}_{t-1} p_t(\boldsymbol{\beta})(1 - p_t(\boldsymbol{\beta}))$$

and the *score vector* (d-dimensional) is defined by

$$\mathbf{S}_N(\boldsymbol{\beta}) \equiv \nabla \log \text{PL}(\boldsymbol{\beta}) = \sum_{s=1}^{N} \mathbf{Z}_{s-1}(X_s - p_s(\boldsymbol{\beta}))$$

The *score vector process*, $\mathbf{S}_t(\boldsymbol{\beta})$, $t = 1, \ldots, N$, is defined by the partial sums,

$$\mathbf{S}_t(\boldsymbol{\beta}) \equiv \sum_{s=1}^{t} \mathbf{Z}_{s-1}(X_s - p_s(\boldsymbol{\beta}))$$

The score process, being the sum of martingale differences, is easily seen to be a martingale with respect to the filtration $\mathcal{F}_0 \subset \mathcal{F}_1 \subset \mathcal{F}_2 \subset \cdots$. That is, $E[\mathbf{S}_t(\boldsymbol{\beta})|\mathcal{F}_{t-1}] = \mathbf{S}_{t-1}(\boldsymbol{\beta})$. Clearly, $E[\mathbf{S}_t(\boldsymbol{\beta})] = \mathbf{0}$. Next we define

$$\mathbf{I}(\boldsymbol{\beta}) \equiv \nabla \nabla'(-\log \text{PL}(\boldsymbol{\beta})) = \sum_{t=1}^{N} \mathbf{Z}_{t-1} \mathbf{Z}'_{t-1} p_t(\boldsymbol{\beta})(1 - p_t(\boldsymbol{\beta}))$$

The quantity $\mathbf{I}(\boldsymbol{\beta})/N$ is the *sample information matrix per observation* for estimating $\boldsymbol{\beta}$. It is easily seen that $\mathbf{I}(\boldsymbol{\beta})$ is the sum of conditional covariance matrices,

$$\mathbf{I}(\boldsymbol{\beta}) = \sum_{s=1}^{N} \mathrm{Var}_{\beta}[\mathbf{Z}_{s-1}(X_s - p_s(\boldsymbol{\beta}))|\mathcal{F}_{s-1}]$$

Since $\mathbf{S}_t(\boldsymbol{\beta})$ is a martingale, we refer to $\mathbf{I}(\boldsymbol{\beta}_0)$ as the *cumulative conditional variance-covariance matrix* for $\mathbf{S}_N(\boldsymbol{\beta})$ at $\boldsymbol{\beta} = \boldsymbol{\beta}_0$.

The large sample properties of the MPLE $\hat{\boldsymbol{\beta}}$ are studied with the aid of $\mathbf{S}_t(\boldsymbol{\beta})$, and $\mathbf{I}(\boldsymbol{\beta})$. This theory was developed by [1], [3], [25]. We follow the general development given in [23, chs. 6 and 7]. The approach taken in these references for proving consistency and asymptotic normality of MPLEs is based on the martingale Central Limit Theorem for $\mathbf{S}_N(\boldsymbol{\beta})/\sqrt{N}$, the almost sure concavity of the random function $\mathrm{PL}(\boldsymbol{\beta})$, $\boldsymbol{\beta} \in \mathcal{R}^d$ (see Problem 1), and the stability of the sample information matrix $\mathbf{I}(\boldsymbol{\beta})/N$. The following assumption is a trick that leads to some notable elegance of the exposition [24].

Assumption A.1. There is a probability measure ν on $\mathcal{R}^d$ for which $\int_{\mathcal{R}^d} \mathbf{z}\mathbf{z}'\nu(d\mathbf{z})$ is positive definite, such that under (9.2) with $\boldsymbol{\beta} = \boldsymbol{\beta}_0$,

$$\frac{1}{N}\sum_{t=1}^{N} I_{[\mathbf{Z}_{t-1}\in A]} \xrightarrow{P} \nu(A), \qquad N \to \infty$$

for all d-dimensional sets $A \subset \mathcal{R}^d$.

Assumption **A.1** implies that for every continuous function $g : \mathcal{R}^d \to \mathcal{R}$,

$$\frac{1}{N}\sum_{t=1}^{N} g(\mathbf{Z}_{t-1}) \xrightarrow{P} \int_{\mathcal{R}^d} g(\mathbf{z})\nu(d\mathbf{z})$$

as $N \to \infty$. From this convergence, $\mathbf{I}(\boldsymbol{\beta})/N$ has a limit, say $\boldsymbol{\Lambda}(\boldsymbol{\beta})$, given by

$$\boldsymbol{\Lambda}(\boldsymbol{\beta}) = \int_{\mathcal{R}^d} \frac{e^{\boldsymbol{\beta}'\mathbf{z}}}{(1+e^{\boldsymbol{\beta}'\mathbf{z}})^2}\mathbf{z}\mathbf{z}'\nu(d\mathbf{z}) \tag{9.6}$$

The matrix $\boldsymbol{\Lambda}(\boldsymbol{\beta})$ is called the *information matrix per observation* for estimating $\boldsymbol{\beta}$. By **A.1**, it is positive definite and hence also nonsingular for every $\boldsymbol{\beta}$. Now, $\nabla \log \mathrm{PL}(\hat{\boldsymbol{\beta}}) = \mathbf{0}$. Thus by expanding $\nabla \log \mathrm{PL}(\hat{\boldsymbol{\beta}})$ using Taylor series to one term about $\boldsymbol{\beta}_0$, and ignoring higher order terms, we obtain the useful approximation up to terms asymptotically negligible in probability,

$$\begin{aligned}
\sqrt{N}(\hat{\boldsymbol{\beta}} - \boldsymbol{\beta}_0) &\approx (-\frac{1}{N}\nabla\nabla' \log \mathrm{PL}(\boldsymbol{\beta}_0))^{-1}\frac{1}{\sqrt{N}}\nabla \log \mathrm{PL}(\boldsymbol{\beta}_0) \\
&= (\frac{1}{N}\mathbf{I}(\boldsymbol{\beta}_0))^{-1}\frac{1}{\sqrt{N}}\mathbf{S}_N(\boldsymbol{\beta}_0) && (9.7) \\
&\approx \boldsymbol{\Lambda}^{-1}(\boldsymbol{\beta}_0)\frac{1}{\sqrt{N}}\mathbf{S}_N(\boldsymbol{\beta}_0) && (9.8)
\end{aligned}$$

Notice that $\mathbf{I}(\boldsymbol{\beta}_0)$ behaves asymptotically as $N\boldsymbol{\Lambda}(\boldsymbol{\beta}_0)$, so that $(\mathbf{I}(\boldsymbol{\beta}_0))^{-1}\mathbf{S}_N(\boldsymbol{\beta}_0)$ converges in probability to 0. Moreover, since $\mathbf{S}_N(\boldsymbol{\beta}_0)$ is a martingale,

$$\frac{1}{\sqrt{N}}\mathbf{S}_N(\boldsymbol{\beta}_0) \xrightarrow{D} \mathcal{N}(\mathbf{0}, \boldsymbol{\Lambda}(\boldsymbol{\beta}_0))$$

It follows that the asymptotic covariance matrix of $\boldsymbol{\Lambda}^{-1}(\boldsymbol{\beta}_0)(1/\sqrt{N})\mathbf{S}_N(\boldsymbol{\beta}_0)$ is $\boldsymbol{\Lambda}^{-1}(\boldsymbol{\beta}_0)\boldsymbol{\Lambda}(\boldsymbol{\beta}_0)\boldsymbol{\Lambda}^{-1}(\boldsymbol{\beta}_0) = \boldsymbol{\Lambda}^{-1}(\boldsymbol{\beta}_0)$. Thus, appealing to Slutsky's theorem and the CLT for martingales, we have:

Theorem 9.1. [24] *The MPLE $\hat{\boldsymbol{\beta}}$ is almost surely unique for all sufficiently large N, and as $N \to \infty$,*

(i) $\hat{\boldsymbol{\beta}} \xrightarrow{P} \boldsymbol{\beta}_0$

(ii) $\sqrt{N}(\hat{\boldsymbol{\beta}} - \boldsymbol{\beta}_0) \xrightarrow{D} \mathcal{N}(\mathbf{0}, \boldsymbol{\Lambda}^{-1}(\boldsymbol{\beta}_0))$

(iii) $\sqrt{N}(\hat{\boldsymbol{\beta}} - \boldsymbol{\beta}_0) - \dfrac{1}{\sqrt{N}}\boldsymbol{\Lambda}^{-1}(\boldsymbol{\beta}_0)\mathbf{S}_N(\boldsymbol{\beta}_0) \xrightarrow{P} 0$

We turn next to the question of goodness of fit. In any attempt to fit (9.2), it is important to examine the logistic regression *residuals* $X_t - p_t(\hat{\boldsymbol{\beta}})$ in order to judge the quality of fit. This, however, can be done in many different ways. A recent survey of tests of fit including graphical techniques can be found in [19, ch. 5].

One useful way to test goodness of fit is to *classify* the responses X_t according to mutually exclusive events defined in terms of the covariates $\mathbf{Z}_{t-1}$, and then check for each category the deviation of the number of positive responses from its conditional expected value [20]. More precisely, let $C_1, \ldots, C_k$, constitute a partition of $\mathcal{R}^d$. For $j = 1, \ldots, k$, define

$$M_j \equiv \sum_{t=1}^{N} I_{[\mathbf{Z}_{t-1} \in C_j]} X_t$$

and

$$E_j(\boldsymbol{\beta}) \equiv \sum_{t=1}^{N} I_{[\mathbf{Z}_{t-1} \in C_j]} p_t(\boldsymbol{\beta})$$

Put $\mathbf{M} \equiv (M_1, \ldots, M_k)'$, $\mathbf{E}(\boldsymbol{\beta}) \equiv (E_1(\boldsymbol{\beta}), \ldots, E_k(\boldsymbol{\beta}))'$. The goodness of fit can be tested with the help of quadratic forms of the type

$$(\mathbf{M} - \mathbf{E}(\hat{\boldsymbol{\beta}}))'\mathbf{V}(\mathbf{M} - \mathbf{E}(\hat{\boldsymbol{\beta}}))$$

where $\mathbf{V}$ is a suitable $k \times k$ matrix.

For testing the *hypothesis* that $\boldsymbol{\beta} = \boldsymbol{\beta}_0$, we can use statistics of the form

$$\sum_{j=1}^{k}(M_j - E_j(\boldsymbol{\beta}_0))^2 W_j$$

or

$$\sum_{t=1}^{N}(X_t - p_t(\boldsymbol{\beta}_0))^2 W_t$$

using appropriate normalizations and weights W.

The theory underlying the preceding goodness of fit and hypothesis testing statistics is contained in the next theorem. The theorem follows readily from Theorem 9.1 and several applications of the multivariate Martingale Central Limit Theorem as given in [1, appendix II].

Theorem 9.2. [24] *Let $C_1, \ldots, C_k$, be a partition of $\mathcal{R}^d$. Then we have as $N \to \infty$:*

(i)

$$\sqrt{N}((\mathbf{M} - \mathbf{E}(\boldsymbol{\beta}_0))'/N, (\hat{\boldsymbol{\beta}} - \boldsymbol{\beta}_0)')' \xrightarrow{D} \mathcal{N}(\mathbf{0}, \boldsymbol{\Sigma})$$

where $\boldsymbol{\Sigma}$ is a square matrix of dimension $d + k$,

$$\boldsymbol{\Sigma} = \begin{pmatrix} \mathbf{A} & \mathbf{B}' \\ \mathbf{B} & \boldsymbol{\Lambda}^{-1}(\boldsymbol{\beta}_0) \end{pmatrix}$$

Here $\mathbf{A}$ is a diagonal $k \times k$ matrix with the jth diagonal element given by

$$\sigma_j^2 \equiv \int_{C_j} \frac{e^{\boldsymbol{\beta}_0'\mathbf{z}}}{(1 + e^{\boldsymbol{\beta}_0'\mathbf{z}})^2} \nu\,(d\mathbf{z})$$

The matrix $\boldsymbol{\Lambda}^{-1}(\boldsymbol{\beta}_0)$ is the limiting $d \times d$ inverse of the information matrix, and the jth column of $\mathbf{B}$ is given by

$$\boldsymbol{\Lambda}^{-1}(\beta_0) \int_{C_j} \frac{e^{\boldsymbol{\beta}_0'\mathbf{z}}}{(1 + e^{\boldsymbol{\beta}_0'\mathbf{z}})^2} \mathbf{z}\nu\,(d\mathbf{z}) = \int_{C_j} \frac{e^{\boldsymbol{\beta}_0'\mathbf{z}}}{(1 + e^{\boldsymbol{\beta}_0'\mathbf{z}})^2} \boldsymbol{\Lambda}^{-1}(\boldsymbol{\beta}_0)\mathbf{z}\nu\,(d\mathbf{z})$$

(ii) As $N \to \infty$,

$$\frac{(\mathbf{E}(\hat{\boldsymbol{\beta}}) - \mathbf{E}(\boldsymbol{\beta}_0))}{\sqrt{N}} - \sqrt{N}\mathbf{B}'\boldsymbol{\Lambda}(\boldsymbol{\beta}_0)(\hat{\boldsymbol{\beta}} - \boldsymbol{\beta}_0) \xrightarrow{P} \mathbf{0}$$

(iii) As $N \to \infty$, the asymptotic distribution of the statistic

$$\chi^2(\boldsymbol{\beta}_0) \equiv \frac{1}{N} \sum_{j=1}^{k} \frac{(M_j - E_j(\boldsymbol{\beta}_0))^2}{\sigma_j^2}$$

is χ_k^2.

Another useful test statistic is defined for $a \geq 0$ by

$$W_a(\boldsymbol{\beta}_0) \equiv \frac{\sum_{t=1}^{N} \left\{[(X_t - p_t(\boldsymbol{\beta}_0))^2/(v(t))^a] - (v(t))^{1-a}\right\}}{\left\{\sum_{t=1}^{N} (v(t))^{1-2a}(1 - 4v(t))\right\}^{1/2}} \tag{9.9}$$

where $v(t) \equiv p_t(\boldsymbol{\beta}_0)(1 - p_t(\boldsymbol{\beta}_0))$ for all t. Then it can be shown that [24],

$$W_a(\boldsymbol{\beta}_0) \xrightarrow{D} \mathcal{N}(0, 1)$$

as $N \to \infty$. It should be noted that the $\mathcal{N}(0, 1)$ approximation may not be very reliable when values $p_t(\boldsymbol{\beta}_0)$ are close to 0 or 1, unless $a \in [0, 1/2]$. This point will be touched upon again in our rainfall-runoff example.

We end this section with well-known results concerning the partial likelihood ratio statistic. First notice that by expanding $\log \mathrm{PL}(\boldsymbol{\beta}_0)$ to two terms about $\hat{\boldsymbol{\beta}}$,

$$\log \mathrm{PL}(\boldsymbol{\beta}_0) \approx \log \mathrm{PL}(\hat{\boldsymbol{\beta}}) + \mathbf{S}_N(\hat{\boldsymbol{\beta}})(\boldsymbol{\beta}_0 - \hat{\boldsymbol{\beta}}) - \frac{1}{2}(\hat{\boldsymbol{\beta}} - \boldsymbol{\beta}_0)'\mathbf{I}(\boldsymbol{\beta}_0)(\hat{\boldsymbol{\beta}} - \boldsymbol{\beta}_0)$$

and that $\mathbf{S}_N(\hat{\beta}) = \nabla \log \mathrm{PL}(\hat{\boldsymbol{\beta}}) = \mathbf{0}$. It follows from (ii) of Theorem 9.1 that as $N \to \infty$,

$$2[\log \mathrm{PL}(\hat{\boldsymbol{\beta}}) - \log \mathrm{PL}(\boldsymbol{\beta}_0)] \approx N(\hat{\boldsymbol{\beta}} - \boldsymbol{\beta}_0)'\boldsymbol{\Lambda}(\boldsymbol{\beta}_0)(\hat{\boldsymbol{\beta}} - \boldsymbol{\beta}_0)$$

is asymptotically χ^2_d. One can use this last statistic for testing the hypothesis $H_0 : \boldsymbol{\beta} = \boldsymbol{\beta}_0$. With a little more effort we can also test the hypothesis H_0 that c ($c \leq d$) of the components of $\boldsymbol{\beta}$ are equal to specific values, for example, that c of the components are equal to 0. To do that, maximize $\mathrm{PL}(\boldsymbol{\beta})$ with respect to the remaining unspecified $d - c$ parameters, and denote the maximum by $\mathrm{PL}(\boldsymbol{\beta}^*)$. Clearly, $\mathrm{PL}(\hat{\boldsymbol{\beta}}) \geq \mathrm{PL}(\boldsymbol{\beta}^*)$. Then under H_0 (see [23, theorem 6.5]), as $N \to \infty$,

$$2[\log \mathrm{PL}(\hat{\boldsymbol{\beta}}) - \log \mathrm{PL}(\boldsymbol{\beta}^*)] \xrightarrow{D} \chi^2_c \tag{9.10}$$

9.3.1 Asymptotic Relative Efficiency

How efficient are maximum partial likelihood estimates compared with the usual maximum likelihood estimates? While no general answer to this question is possible, we can give an answer in the case of an $AR(p)$ process as discussed in Section 9.2.1. In this special case, a fully specified model for $\{X_t, \mathbf{Z}_t\}$ is readily available, and a comparison is possible via the information matrices corresponding to partial and full likelihoods.

Specifically, set $r = 0, \lambda = 1, \gamma_0 = 0$, and consider the stationary $AR(p)$ process, $Y_t = \beta_1 Y_{t-1} + \cdots + \beta_p Y_{t-p} + \epsilon_t$. Here, $\boldsymbol{\beta} = (\beta_1, \ldots, \beta_p)'$, $X_t = I_{[Y_t \geq 0]}$, and $\mathbf{Z}_{t-1} = (Y_{t-1}, Y_{t-2}, \ldots, Y_{t-p})'$, and

$$\epsilon_t = Y_t - \boldsymbol{\beta}'\mathbf{Z}_{t-1} \tag{9.11}$$

are iid logistic random variables with density $f(x) = e^x/(1+e^x)^2$. The variable ϵ_t is independent of $\mathcal{F}_{t-1} = \sigma(\mathbf{Z}_s, s < t)$.

Under the assumption of stationarity, let $\mathbf{Z}$ be distributed as $\mathbf{Z}_{t-1}$. Also let ϵ be distributed as ϵ_t, independently of $\mathbf{Z}$. The notation E_{β_0} is used to denote mathematical expectation when $\boldsymbol{\beta} = \boldsymbol{\beta}_0$.

The partial likelihood information matrix derived in (9.6) and corresponding to (9.2) and (9.3), is given by

$$\boldsymbol{\Lambda}^{\mathrm{PL}}(\boldsymbol{\beta}_0) \equiv \boldsymbol{\Lambda}(\boldsymbol{\beta}_0) = E_{\boldsymbol{\beta}_0}\left[\frac{e^{\boldsymbol{\beta}_0'\mathbf{Z}}}{(1+e^{\boldsymbol{\beta}_0'\mathbf{Z}})^2}\mathbf{Z}\mathbf{Z}'\right] = E_{\boldsymbol{\beta}_0}[f(\boldsymbol{\beta}_0'\mathbf{Z})\mathbf{Z}\mathbf{Z}'] \quad (9.12)$$

By fixing $\mathbf{Z}_0 = (Y_1, \ldots, Y_p)'$, the likelihood $L(\boldsymbol{\beta})$ based on $Y_{p+1}, \ldots, Y_N$, is given by

$$L(\boldsymbol{\beta}) = \prod_{t=p+1}^{N} f(\epsilon_t) \quad (9.13)$$

where ϵ_t is given in (9.11). The information matrix is obtained in exactly the same manner as in the partial likelihood case, as the limit of the information about $\boldsymbol{\beta}$ per observation. This information matrix, equal to the inverse of the asymptotic covariance matrix for the maximum likelihood estimator of $\boldsymbol{\beta}$ when the true parameter value is $\boldsymbol{\beta}_0$, is given by

$$\boldsymbol{\Lambda}^{L}(\boldsymbol{\beta}_0) \equiv 2E_{\boldsymbol{\beta}_0}\left[\frac{e^{\epsilon}}{(1+e^{\epsilon})^2}\mathbf{Z}\mathbf{Z}'\right] = 2E_{\boldsymbol{\beta}_0}[f(\epsilon)\mathbf{Z}\mathbf{Z}'] = \frac{1}{3}E_{\boldsymbol{\beta}_0}[\mathbf{Z}\mathbf{Z}'] \quad (9.14)$$

upon noting that $\int_{-\infty}^{\infty} f(x)f(x)\,dx = 1/6$.

Since $f(x) \leq 1/4$, it follows immediately from (9.12) and (9.14), that for every vector $\mathbf{b} \in \mathcal{R}^p$,

$$\mathbf{b}'\boldsymbol{\Lambda}^{\mathrm{PL}}(\boldsymbol{\beta}_0)\mathbf{b} \leq \frac{3}{4}\mathbf{b}'\boldsymbol{\Lambda}^{L}(\boldsymbol{\beta}_0)\mathbf{b} \quad (9.15)$$

Thus, any scalar parameter derived from $\boldsymbol{\beta}$ linearly can be estimated with *asymptotic relative efficiency* (ARE) at best $3/4$ via the partial likelihood logistic regression method as compared with a complete maximum likelihood $AR(p)$ analysis. The worst ARE is obtained for $p_t(\boldsymbol{\beta}_0)$ that spend most of their time close to 1 or 0, that is, the case where prediction is very good!

9.4 Applications of Logistic Regression

9.4.1 Threshold Exceedances by Rain Rate

The following rainfall example helps to demonstrate the intuitive notion that prediction of X_t becomes more and more difficult when the underlying process, from which the binary data are defined, gets closer to the threshold level.

Unlike the case with rain gauges where direct rainfall measurements are possible, when it comes to satellite observations, rain rate is inferred indirectly from *covariates* such as microwave temperature and radar reflectivity. Because the relationship between rain rate and these covariates is not fully understood at present, and is in fact experimental in nature, the estimation of rainfall from space measurements poses a great challenge to meteorologists. It has, however, been observed that knowledge of whether rain rate exceeds a fixed threshold at time t at points in space may suffice under some conditions for this problem. This remarkable fact is the basis for the so-called "threshold method" (see [14]). Taking this into account, logistic regression has been applied to observed rain rate and simulated microwave temperature, in order to determine the exceedance probability of a fixed threshold by rain rate. If the exceedance probability is more than 50%, the corresponding rain rate observation is classified as being above the threshold (see [5]).

Let R_t denote a time series of true mm/hour rain rate observed at a fixed location every 15 minutes, and denote by $V6_t, V37_t$, and $H6_t, H37_t$, microwave brightness temperature of the vertical (V) and horizontal (H) polarization at 6 and 37 GHz. These are our time dependent observations. In an ideal situation these data should be available from satellite measurements. However, in the absence of real data, they were generated from R_t using nonlinear empirical relationships [5]. The binary time series is defined as $X_t = I_{[R_t \geq r]}$, r being a fixed threshold measured in mm/hour. We consider two logistic regression models of the form (9.2) corresponding to two different thresholds 5,3, and using two different sets of covariates, $\{V6, H6\}$ and $\{V6, H6, V37, H37\}$, respectively.

A word about notation. In the present application, $X_t = I_{[R_t \geq r]}$ is predicted from complete knowledge of $\{V6, H6, V37, H37\}$ *at time t*, so that now in accordance with our previous notation the "past" $\mathcal{F}_{t-1}$ is generated by the "future" $\{V6_t, H6_t, V37_t, H37_t\}$. The notation is somewhat unfortunate, but from the mathematical point of view there is no real reason why it should be changed. A quick remedy is to simply replace in (9.2) $\mathcal{F}_{t-1}$ by $\{V6_t, H6_t\}$, or $\{V6_t, H6_t, V37_t, H37_t\}$, whichever applies. For example, with $\{V6_t, H6_t\}$ and $r = 5$, the logistic model (9.2) becomes

$$\begin{aligned} p_t(\boldsymbol{\beta}) &\equiv P_{\boldsymbol{\beta}}(X_t = 1 | V6_t, H6_t) \\ &= \frac{1}{1 + \exp[-(\text{constant} + \beta_1 \times V6_t + \beta_2 \times H6_t)]} \end{aligned}$$

Table 9.1 shows cases of observed X_t versus its logistic prediction $p_t(\hat{\boldsymbol{\beta}})$, where $\hat{\boldsymbol{\beta}}$ was estimated from *another* time series and *other* corresponding covariates obtained at a different location. We can see that when R_t is "far" from the given threshold, the prediction is perfect, resulting in $p_t(\hat{\boldsymbol{\beta}})$ near 1 or 0. As R_t approaches the threshold level, the prediction becomes less decisive, as is indeed expected, but still in this example it always assigns more than 50% chance to the true events.

TABLE 9.1 Prediction of $X_t = I_{[R_t \geq r]}$ from $p_t(\hat{\beta})$ in Rain Rate Time Series

$r = 5$,	$V6, H6$		$r = 3$,	$V6, H6, V37, H37$	
R_t	X_t	$p_t(\hat{\beta})$	R_t	X_t	$p_t(\hat{\beta})$
1.855	0	0.000	3.067	1	0.659
11.754	1	1.000	2.956	0	0.287
6.672	1	1.000	3.564	1	0.999
7.337	1	1.000	3.414	1	0.999
5.319	1	1.000	3.604	1	0.999
4.392	0	0.000	3.388	1	0.999
5.210	1	0.999	2.742	0	0.026
3.742	0	0.000	2.675	0	0.003
3.128	0	0.000	2.513	0	0.000
4.992	0	0.233	1.022	0	0.000
4.607	0	0.000	5.833	1	1.000

Source: Chiu and Kedem (1990).

9.4.2 Application to Rainfall-Runoff Data

In our next application, the logistic regression model (9.2) is applied in the analysis of daily rainfall-runoff data obtained by the National Weather Service in the Bird Creek, Ohio, watershed, and described in [26]. The data were collected in intervals of 13–15 weeks during each of the years 1939–64. In what follows, we regard daily runoff Y_t as the response variable, and rainfall R_t as the explanatory variable [24]. Since flooding is of interest, it is natural to try to understand the relationship between level exceedances $X_t = I_{[Y_t \geq r]}$ and the explanatory variables. Our primary goal is to illustrate how the model (9.2) can be estimated and pass goodness of fit tests.

The 26 years of data were split into a testing set (the 10 years 1939–48, consisting of 1031 rainfall-runoff pairs), and a training set (the 16 years 1949–64, consisting of 1691 rainfall-runoff pairs). The model (9.2) was fitted to the training data, and then the quality of fit was judged from *both* the training data and the testing data. Since the models we contemplate involve explanatory variables defined from $(Y_{t-1}, \ldots, Y_{t-4}, R_t, \ldots, R_{t-3})$, we lose 4 observations per year which lead to 991 rainfall-runoff pair observations in the testing set, and 1627 in the training set. The threshold values chosen are $r = 1, 3$ cubic feet/second. During the training period 1949–64, there were 401 and 87 positive responses (i.e., level-upcrossings) corresponding to levels $r = 1, 3$, respectively. The corresponding respective numbers for the testing period 1939–48 were 244, 56.

Our first results are given in Table 9.2. The table gives the estimated values, from the 1949–64 training data set, of the components $\hat{\beta}_i$ of the MPLE $\hat{\boldsymbol{\beta}}$, and the standardized values $\hat{\beta}_i / \sqrt{\text{Var}(\hat{\beta}_i)}$, corresponding to $X_t = I_{[Y_t \geq r]}$, $r = 1, 3$. In each case, the 10-dimensional covariate vector is

$$\mathbf{Z}_t' = (1, R_t, Y_{t-1}, R_t Y_{t-1}, R_{t-1}, R_t R_{t-1}, R_{t-2}, R_{t-1} R_{t-2}, Y_{t-2}, Y_{t-1} Y_{t-2})$$

To assess model adequacy for the logistic model (9.2) fitted to the 1949–64 rainfall-runoff data, we calculated the goodness of fit statistics $\chi^2(\hat{\boldsymbol{\beta}})$ (described in Theorem 9.2, and $W_a(\hat{\boldsymbol{\beta}})$, replacing $\boldsymbol{\beta}_0$ with $\hat{\boldsymbol{\beta}}$, where the MPLE $\hat{\boldsymbol{\beta}}$ was obtained from the 1949–64 data. Thus, the goodness of fit statistics $\chi^2(\hat{\boldsymbol{\beta}})$ and $W_a(\hat{\boldsymbol{\beta}})$ were evaluated from both the training data set (1949-64), and from the testing data set (1939–48), using the *same* $\hat{\boldsymbol{\beta}}$ obtained from the 1949–64 training data. Also σ_j^2 was estimated from the data to which the χ^2 test was being applied, using the estimator

$$\hat{\sigma}_j^2 \equiv \frac{1}{N} \sum_{t=1}^{N} I_{[\mathbf{Z}_{t-1} \in C_j]} p_t(\hat{\boldsymbol{\beta}})(1 - p_t(\hat{\boldsymbol{\beta}}))$$

TABLE 9.2 Logistic Regression Parameter Estimates $\hat{\beta}_i$ Obtained from the 1949–64 Rainfall-Runoff Data

		$r = 1$		$r = 3$	
i	Covariate	$\hat{\beta}_i$	$\frac{\hat{\beta}_i}{\sqrt{\mathrm{Var}(\hat{\beta}_i)}}$	$\hat{\beta}_i$	$\frac{\hat{\beta}_i}{\sqrt{\mathrm{Var}(\hat{\beta}_i)}}$
0	Intercept	−6.31	−13.60	−6.25	−13.90
1	R_t	161.20	6.90	99.70	6.60
2	Y_{t-1}	4.38	9.60	1.05	5.80
3	$R_t Y_{t-1}$	58.50	1.40	64.40	3.90
4	R_{t-1}	70.90	3.00	−50.50	−1.90
5	$R_t R_{t-1}$	−2509.00	−1.50	3392.00	2.30
6	R_{t-2}	−63.00	−4.30	−23.50	−1.20
7	$R_{t-1} R_{t-2}$	−2124.00	−1.20	-2498.00	−2.20
8	Y_{t-2}	0.24	0.64	0.29	2.40
9	$Y_{t-1} Y_{t-2}$	0.31	1.40	−0.04	−2.70

For the statistic $\chi^2(\hat{\boldsymbol{\beta}})$, the partition cells $C_j \subset \mathcal{R}^{10}$ are defined as the intersections of all sets satisfying the conditions that R_t is in one of the three intervals $[0, 0.004]$, $(0.004, 0.008]$, or $(0.008, \infty)$, $R_{t-1} + R_t$ is in one of the intervals $[0, 0.01]$, $(0.01, 0.02]$, or $(0.02, \infty)$, and Y_{t-1} is in $[0, 0.5]$, $(0.5, 1]$, or $(1, \infty)$. This gives $k = 3^3 = 27$ partition cells. However, because $\boldsymbol{\beta}_0$ is replaced by $\hat{\boldsymbol{\beta}}$, $\chi^2(\hat{\boldsymbol{\beta}})$ is (asymptotically) stochastically smaller than χ_{27}^2 on the training data (due to having estimated $\boldsymbol{\beta}_0$ from the same data), and stochastically larger than χ_{27}^2 on the testing data (since $\hat{\boldsymbol{\beta}}$ is approximately independent of $(\mathbf{M} - \mathbf{E}(\boldsymbol{\beta}_0))$ for the testing data). An indication of this is provided by noting that for the training data [24],

$$E[\chi^2(\hat{\boldsymbol{\beta}})] \approx k - 2\,\mathrm{tr}\,\mathbf{B}'\boldsymbol{\Lambda}(\boldsymbol{\beta}_0)\mathbf{B}\mathbf{A}^{-1} + \mathrm{tr}\,\mathbf{B}'\boldsymbol{\Lambda}(\boldsymbol{\beta}_0)\mathbf{B}\mathbf{A}^{-1}$$

$$= k - \sum_{j=1}^{k}(\mathbf{B}'\boldsymbol{\Lambda}(\boldsymbol{\beta}_0)\mathbf{B})_{jj}/\sigma_j^2$$

while for the testing data the middle term $-2\,\mathrm{tr}\,\mathbf{B}'\boldsymbol{\Lambda}(\boldsymbol{\beta}_0)\mathbf{B}\mathbf{A}^{-1}$ vanishes due to approximate independence of the training and testing data, and hence

$$E[\chi^2(\hat{\boldsymbol{\beta}})] \approx k + \mathrm{tr}\,\mathbf{B}'\boldsymbol{\Lambda}(\boldsymbol{\beta}_0)\mathbf{B}\mathbf{A}^{-1}$$

$$= k + \sum_{j=1}^{k}(\mathbf{B}'\boldsymbol{\Lambda}(\boldsymbol{\beta}_0)\mathbf{B})_{jj}/\sigma_j^2$$

The estimated value of $\sum_{j=1}^{k}(\mathbf{B}'\boldsymbol{\Lambda}(\boldsymbol{\beta}_0)\mathbf{B})_{jj}/\sigma_j^2$ was 5.3 for $r = 1$, and 3.6 for $r = 3$. The difference is attributed to the very different response probabilities $p_t(\hat{\boldsymbol{\beta}})$ corresponding to different levels r.

Unlike the case with $W_a(\boldsymbol{\beta}_0)$, the statistics $W_a(\hat{\boldsymbol{\beta}})$, $a = 0, 1$, are at best only very roughly asymptotically normal. In addition, some care is needed in interpreting the results due to division by $[p_t(\hat{\boldsymbol{\beta}})(1 - p_t(\hat{\boldsymbol{\beta}}))]^a$. This is so because many of the logistic probabilities $p_t(\hat{\boldsymbol{\beta}})$ are either quite close to 0 or 1, a tendency already seen in the previous example. For this reason, $W_0(\hat{\boldsymbol{\beta}})$ is more reliable than $W_1(\hat{\boldsymbol{\beta}})$.

The values of $\chi^2(\hat{\boldsymbol{\beta}}), W_0(\hat{\boldsymbol{\beta}}), W_1(\hat{\boldsymbol{\beta}})$ are given in Table 9.3, leading to interesting conclusions. The table indicates that the logistic model (9.2) with $X_t = I_{[Y_t \geq r]}$ for $r = 1, 3$, is quite adequate for both the training and testing data because the three statistics admit relatively small values, except for $W_1(\hat{\boldsymbol{\beta}})$ for the test data with $r = 3$ (for comparison, the 97.5th percentile of χ_{27}^2 is 43.2, and use the appropriate modifications needed for the training and test data; for $\mathcal{N}(0, 1)$ the 97.5th percentile is 1.96). However, as remarked above, the fact that the predicted response probabilities $p_t(\hat{\boldsymbol{\beta}})$ are often very close to 0 or 1, makes the statistic $W_1(\hat{\boldsymbol{\beta}})$ less reliable and more difficult to interpret than the sum of the *unnormalized* squared residuals $W_0(\hat{\boldsymbol{\beta}})$.

TABLE 9.3 Goodness of Fit Statistics on Training and Testing Data for the Logistic Model (9.2) with Coefficients Displayed in Table 9.2

	Training Data (1627 obs.)		Test Data (991 obs.)	
Statistic	$r = 1$	$r = 3$	$r = 1$	$r = 3$
$\chi^2(\hat{\boldsymbol{\beta}})$	42.40	18.90	53.10	18.70
$W_0(\hat{\boldsymbol{\beta}})$	−0.93	0.10	−0.73	1.64
$W_1(\hat{\boldsymbol{\beta}})$	0.28	−0.26	0.01	16.50

We end this example by noting that the rainfall-runoff data do not seem to follow an $AR(p)$ model with logistic errors. In this sense we have an example where logistic regression seems more adequate than linear autoregression for the prediction of level-upcrossings.

9.5 Problems and Complements

1. By fixing the values of the covariates, show that $\log \text{PL}(\boldsymbol{\beta})$ is strictly concave in $\boldsymbol{\beta}$ (a $d \times 1$ vector). Conclude that if the MPLE $\hat{\boldsymbol{\beta}}$ exists, it is unique. (Hint: A function $f : \mathcal{R}^d \to \mathcal{R}$ with continuous second order partial derivatives is strictly concave if $\nabla\nabla' f$ is negative definite.)
2. Define an $N \times d$ matrix $\mathbf{Z}$ by $\mathbf{Z} = (\mathbf{Z}_0, \mathbf{Z}_1, \ldots, \mathbf{Z}_{N-1})'$, and the $N \times 1$ vectors $\mathbf{X} = (X_1, X_2, \ldots, X_N)'$, $\mathbf{p}(\boldsymbol{\beta}) = (p_1(\boldsymbol{\beta}), p_2(\boldsymbol{\beta}), \ldots, p_N(\boldsymbol{\beta}))'$. Argue that the Newton–Raphson algorithm for obtaining the MPLE $\hat{\boldsymbol{\beta}}$ is given by

$$\boldsymbol{\beta}^{j+1} = \boldsymbol{\beta}^j + (\mathbf{Z}'\mathbf{V}(\boldsymbol{\beta}^j)\mathbf{Z})^{-1}\mathbf{Z}'(\mathbf{X} - \mathbf{p}(\boldsymbol{\beta}^j))$$

where $\mathbf{V}(\boldsymbol{\beta})$ is the $N \times N$ diagonal matrix

$$\mathbf{V}(\boldsymbol{\beta}) = \text{Diag}(p_1(\boldsymbol{\beta})(1 - p_1(\boldsymbol{\beta})), \ldots, p_N(\boldsymbol{\beta})(1 - p_N(\boldsymbol{\beta})))$$

and where $\boldsymbol{\beta}^0$ is an initial guess. (Hint: $\boldsymbol{\beta}^{j+1} = \boldsymbol{\beta}^j + \mathbf{I}^{-1}(\boldsymbol{\beta}^j)\mathbf{S}_N(\boldsymbol{\beta}^j)$.)
3. Show that for $\boldsymbol{\beta}$ close to $\hat{\boldsymbol{\beta}}$, $\mathbf{Z}'\mathbf{X} \approx \mathbf{Z}'\mathbf{p}(\boldsymbol{\beta})$
4. *Partial likelihood ratio [23]*. Consider the PL as defined in (9.1). Among the important properties of PL is that if θ_0 is the true parameter and θ is any other admissible parameter value, then

$$\frac{\text{PL}(\theta; \mathbf{X_k})}{\text{PL}(\theta_0; \mathbf{X_k})} = \frac{\text{PL}(\theta; X_1, \ldots, X_k)}{\text{PL}(\theta_0; X_1, \ldots, X_k)} = \prod_{j=1}^{k} \frac{p_j(X_j; \theta)}{p_j(X_j; \theta_0)}$$

is a *martingale* with respect to the filtration $\mathcal{F}_k$. Show that the partial likelihood ratio $\text{PL}(\theta; \mathbf{X_k})/\text{PL}(\theta_0; \mathbf{X_k})$ is indeed a martingale.
5. *Asymptotic relative efficiency [24]*. Consider the stationary $AR(p)$ process, $Y_t = \beta_1 Y_{t-1} + \cdots + \beta_p Y_{t-p} + \epsilon_t$, and let $X_t = I_{[Y_t \geq 0]}$, $\mathbf{Z}_{t-1} = (Y_{t-1}, Y_{t-2}, \ldots, Y_{t-p})'$. Assume that the ϵ_t are iid standard normal random variables with distribution function Φ. Define the probit model

$$\tilde{p}_t(\boldsymbol{\beta}) \equiv P_{\boldsymbol{\beta}}(X_t = 1|\mathcal{F}_{t-1}) = \Phi(\boldsymbol{\beta}'\mathbf{Z}_{t-1})$$

   (a) Follow the discussion in Section 9.3.1 to show that

$$\boldsymbol{\Lambda}^L(\boldsymbol{\beta}_0) = E_{\boldsymbol{\beta}_0}[\mathbf{Z}\mathbf{Z}']$$

   and

$$\boldsymbol{\Lambda}^{\text{PL}}(\boldsymbol{\beta}_0) = E_{\boldsymbol{\beta}_0}\left[\mathbf{Z}\mathbf{Z}'\frac{\phi^2(\boldsymbol{\beta}_0'\mathbf{Z})}{\Phi(\boldsymbol{\beta}_0'\mathbf{Z})(1 - \Phi(\boldsymbol{\beta}_0'\mathbf{Z}))}\right]$$

(b) Conclude that the upper bound for asymptotic relative efficiency (ARE) as described in Section 9.3.1 is $2/\pi$.

6. *Independent binary time series [24]*. Suppose there are multiple *independent* binary realizations X_t^j and corresponding vectors of covariates $\mathbf{Z}_{t-1}^j, j = 1, \ldots, m$. Define

$$p_t^i(\boldsymbol{\beta}) \equiv P_{\boldsymbol{\beta}}(X_t^i = 1|\mathcal{F}_{t-1}) = \frac{1}{1 + \exp[-\boldsymbol{\beta}'\mathbf{Z}_{t-1}^i]}$$

where $\mathcal{F}_{t-1}$ is generated by all the X_s^i, $\mathbf{Z}_s^i$, $0 \le s \le t-1$.

(a) Show that the partial likelihood is given by

$$\mathrm{PL}(\boldsymbol{\beta}) = \prod_{t=1}^{N} \prod_{i=1}^{m} [p_t^i(\boldsymbol{\beta})]^{x_t^i} [1 - p_t^i(\boldsymbol{\beta})]^{1-x_t^i}$$

(b) Show that the score vector process given by

$$\mathbf{S}_t(\boldsymbol{\beta}) = \sum_{s=1}^{t} \sum_{i=1}^{m} \mathbf{Z}_{s-1}^i (X_s^i - p_s^i(\boldsymbol{\beta}))$$

$t = 1, 2, \ldots, N$, is a martingale, and that

$$\mathbf{S}_N(\boldsymbol{\beta}) = \nabla \log \mathrm{PL}(\boldsymbol{\beta})$$

(c) Show that

$$\mathbf{I}(\boldsymbol{\beta}) \equiv \nabla\nabla'(-\log \mathrm{PL}(\boldsymbol{\beta})) = \sum_{s=1}^{N} \mathrm{Var}_{\boldsymbol{\beta}}[\sum_{i=1}^{m} \mathbf{Z}_{s-1}^i (X_s^i - p_s^i(\boldsymbol{\beta}))|\mathcal{F}_{s-1}]$$

(d) Argue that under appropriate conditions the preceding development (for $m = 1$) can be extended straightforwardly to the case of multiple independent binary time series.

7. *Expected value of quadratic forms*. Recall that the trace, denoted by tr, of a square matrix is the sum of the diagonal elements. Let $\mathbf{Y}$ be an $n \times 1$ random vector with mean vector $\mathbf{m}$ and variance-covariance matrix $\boldsymbol{\Sigma}_{\mathbf{Y}}$, and $\mathbf{Q}$ an $n \times n$ real symmetric matrix. Show that

$$E[\mathbf{Y}'\mathbf{Q}\mathbf{Y}] = \mathbf{m}'\mathbf{Q}\mathbf{m} + \operatorname{tr} \boldsymbol{\Sigma}_{\mathbf{Y}}\mathbf{Q}$$

(Hint: $\mathbf{Y}'\mathbf{Q}\mathbf{Y}$ is scalar, so that E and tr commute, and use the fact that tr $\mathbf{AB}$ = tr $\mathbf{BA}$ (i.e., the trace is invariant under cyclical permutations).)

REFERENCES

[1] Andersen, P. and R. Gill. "Cox's regression model for counting processes: A large sample study," *Ann. Stat.*, Vol. 10, pp. 1100–1120, 1982.

[2] Arjas, E. and P. Haara. "A marked point process approach to censored failure data with complicated covariates," *Scand. J. Stat.*, Vol. 11, pp. 193–210, 1984.

[3] Arjas, E. and P. Haara. "A logistic regression model for hazard: Asymptotic results," *Scand. J. Stat.*, Vol. 14, pp. 1–18, 1987.

[4] Berkson, J. "Application of the logistic function to bioassay," *J. Amer. Stat. Assoc.*, Vol. 39, pp. 357–365, 1944.

[5] Chiu, L. S. and B. Kedem. "Estimating the exceedance probability of rain rate by logistic regression," *J. Geophys. Res.*, Vol. 95, D3, pp. 2217–2227, 1990.

[6] Cox, D. R. *The Analysis of Binary Data*, London: Methuen, 1970.

[7] Cox, D. R. "Regression models and life tables," *J. R. Stat. Soc.*, B, Vol. 34, pp. 187–202, 1972.

[8] Cox, D. R. "Partial likelihood," *Biometrika*, Vol. 62, pp. 69–76, 1975.

[9] Fahrmeir, L. "State space modeling and conditional mode estimation for categorical time series," paper in *New Directions in Time Series Analysis*, Part I, Brillinger, D. et al. eds., pp. 87–109, Springer-Verlag, New York, 1992.

[10] Fahrmeir, L. and H. Kaufmann. "Regression models for non-stationary categorical time series," *J. Time Ser. Anal.*, Vol. 8, pp. 147–160, 1987.

[11] Gill, R. "Understanding Cox's regression model: A martingale approach," *J. Am. Stat. Assoc.*, Vol. 79, pp. 441–447, 1984.

[12] Kaufmann, H. "Regression models for nonstationary categorical time series: Asymptotic estimation theory," *Ann. Stat.*, Vol. 15, pp. 79–98, 1987.

[13] Kedem, B. *Binary Time Series*, New York: Dekker, 1980.

[14] Kedem, B. and H. Pavlopoulos. "On the threshold method for rainfall estimation: Choosing the optimal threshold level," *J. Am. Stat. Assoc.*, Vol. 86, pp. 626–633, 1991.

[15] Keenan, D. M. "A time series analysis of binary data," *J. Am. Stat. Assoc.*, Vol. 77, pp. 816–821, 1982.

[16] Liang, K.-Y. and S. L. Zeger. "A class of logistic regression models for multivariate binary time series," *J. Am. Stat. Assoc.*, Vol. 84, pp. 447-451, 1989.

[17] Muenz, L. R. and L. V. Rubinstein. "Markov models for covariate dependence of binary sequences," *Biometrics*, Vol. 41, pp. 91-101, 1985.

[18] Nerlove, M. and S. J. Press. "Univariate and multivariate log-linear and logistic models," Rep. R-1306, Rand Corporation, Santa Monica, Calif., 1973.

[19] Santner, T. J. and D. E. Duffy. *The Statistical Analysis of Discrete Data*, New York: Springer-Verlag, 1989.

[20] Schoenfeld, D. "Chi-squared goodness-of-fit for the proportional hazard regression model," *Biometrika*, Vol. 67, pp. 145–153, 1980.

[21] Slud, E. "Multivariate dependent renewal processes," *Adv. Appl. Probab.*, Vol. 16, pp. 347–362, 1984.

[22] Slud, E. "Partial likelihood for continuous-time stochastic processes," *Scand. J. Stat.*, Vol. 19, pp. 97–109, 1992.

[23] Slud, E. *Martingale Methods in Statistics*, 1993.

[24] Slud, E. and B. Kedem. "Partial likelihood analysis of logistic regression and autoregression," submitted to *Statistica Sinica* (revised Jan. 1993).

[25] Wong, W. H. "Theory of partial likelihood," *Ann. Stat.*, Vol. 14, pp. 88–123, 1986.

[26] Yakowitz, S. "Nearest-neighbour methods for time series analysis," *J. Time Ser. Anal.*, Vol. 8, pp. 235–247, 1987.

Appendix A

Some Data Sets

A.1 HOURLY GATE PHASE I DATA

Table A.1 gives the hourly area average rain rate, interpolated to hourly values, of the original data set—4 km gridded data from Phase I of the Global Atmospheric Research Program, Atlantic Tropical Experiment (GATE) produced by Hudlow and Patterson (M.D. Hudlow and V.L. Patterson, *GATE Radar Rainfall Atlas*, NOAA Special Report, available from U.S. Government Printing Office, Washington DC 20402, 158 pp., 1979). Source of the interpolated data: T. L. Bell, NASA/GSFC, private communication.

Here, area = 280 km $\times$ 280 km, and N =450 hours. Starting from Julian day 179 at midnight,

Column 1: Number of hours elapsed since the start of the series (Hour).

Column 2: Number of minutes elapsed since the start of the series (Minute).

Column 3: Mean rain rate in mm/hour (RR). The corresponding (simple) HOC plot is shown in Figure A.1.

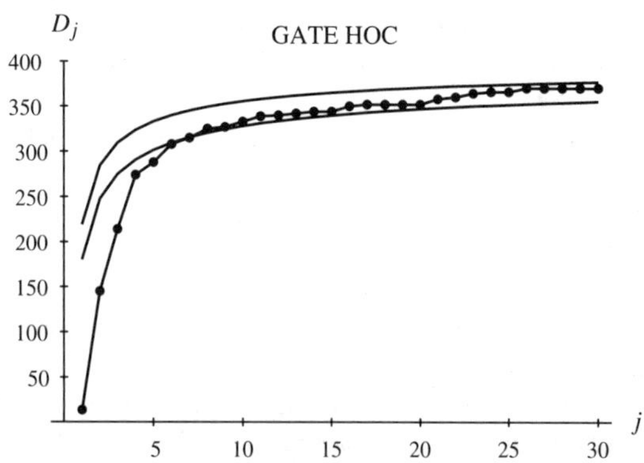

Figure A.1: *A HOC plot (simple, N=400) of area average rain rate from GATE I, and the corresponding 95% white noise bounds (solid lines). The area average rain rate is far from being white noise.*

TABLE A.1 Interpolated Hourly Rain Rate Data

Hour	Minute	RR	Hour	Minute	RR
0	15	0.4100	52	3135	0.4181
1	75	0.4869	53	3195	0.8071
2	135	0.5221	54	3255	0.8064
3	195	0.6225	55	3315	1.1274
4	255	0.1686	56	3375	0.7338
5	315	0.2499	57	3435	0.4093
6	375	0.3469	58	3495	0.1981
7	435	0.5244	59	3555	0.1809
8	495	0.8891	60	3615	0.2249
9	555	1.3874	61	3675	0.5827
10	615	1.5349	62	3735	0.5755
11	675	2.4034	63	3795	0.5864
12	735	2.8912	64	3855	0.6262
13	795	2.0011	65	3915	0.6622
14	855	1.0467	66	3975	0.6270
15	915	0.8515	67	4035	0.6258
16	975	0.7095	68	4095	0.4982
17	1035	0.5015	69	4155	0.2852
18	1095	0.2214	70	4215	0.0935
19	1155	0.1357	71	4275	0.1027
20	1215	0.3097	72	4335	0.1879
21	1275	0.4458	73	4395	0.1534
22	1335	0.2588	74	4455	0.1548
23	1395	0.1874	75	4515	0.1193
24	1455	0.1660	76	4575	0.1118
25	1515	0.1804	77	4635	0.0798
26	1575	0.3271	78	4695	0.0436
27	1635	0.4480	79	4755	0.0080
28	1695	0.4523	80	4815	0.0106
29	1755	0.5137	81	4875	0.0363
30	1815	0.4929	82	4935	0.0700
31	1875	0.6689	83	4995	0.0966
32	1935	0.5913	84	5055	0.1012
33	1995	0.5923	85	5115	0.1529
34	2055	0.6398	86	5175	0.1170
35	2115	0.6344	87	5235	0.0920
36	2175	0.5557	88	5295	0.0938
37	2235	0.4014	89	5355	0.1431
38	2295	0.4906	90	5415	0.1351
39	2355	0.7899	91	5475	0.4191
40	2415	1.2601	92	5535	0.5659
41	2475	1.0314	93	5595	0.7363
42	2535	1.2176	94	5655	0.5868
43	2595	0.8159	95	5715	1.1983
44	2655	0.6718	96	5775	1.4788
45	2715	0.6079	97	5835	1.1106
46	2775	0.3645	98	5895	1.1142
47	2835	0.2941	99	5955	1.1172
48	2895	0.2408	100	6015	0.6854
49	2955	0.0632	101	6075	0.5655
50	3015	0.0520	102	6135	0.6374
51	3075	0.2529	103	6195	0.7172

TABLE A.1 *(Continued)*

Hour	Minute	RR	Hour	Minute	RR
104	6255	0.8229	154	9255	0.0654
105	6315	0.8538	155	9315	0.1810
106	6375	1.1399	156	9375	0.1021
107	6435	2.0136	157	9435	0.1464
108	6495	2.1718	158	9495	0.2628
109	6555	2.6986	159	9555	0.2684
110	6615	3.0673	160	9615	0.4261
111	6675	2.7456	161	9675	0.4793
112	6735	1.9366	162	9735	0.2857
113	6795	1.3360	163	9795	0.2783
114	6855	0.9533	164	9855	0.3923
115	6915	0.9955	165	9915	0.2472
116	6975	0.8270	166	9975	0.1480
117	7035	0.9225	167	10035	0.1560
118	7095	0.8330	168	10095	0.0855
119	7155	0.6251	169	10155	0.0988
120	7215	0.5262	170	10215	0.0864
121	7275	0.4956	171	10275	0.0453
122	7335	0.8650	172	10335	0.0316
123	7395	0.7745	173	10395	0.0207
124	7455	0.5606	174	10455	0.0156
125	7515	0.3898	175	10515	0.0147
126	7575	0.2492	176	10575	0.0403
127	7635	0.1590	177	10635	0.0327
128	7695	0.0737	178	10695	0.0178
129	7755	0.0318	179	10755	0.0179
130	7815	0.0211	180	10815	0.0145
131	7875	0.0048	181	10875	0.0222
132	7935	0.0111	182	10935	0.0130
133	7995	0.0225	183	10995	0.0232
134	8055	0.0473	184	11055	0.0185
135	8115	0.0371	185	11115	0.0275
136	8175	0.0340	186	11175	0.0332
137	8235	0.0064	187	11235	0.0453
138	8295	0.0051	188	11295	0.0450
139	8355	0.0071	189	11355	0.0172
140	8415	0.0187	190	11415	0.0069
141	8475	0.0748	191	11475	0.0017
142	8535	0.3538	192	11535	0.0075
143	8595	0.3616	193	11595	0.0102
144	8655	0.1559	194	11655	0.0020
145	8715	0.0836	195	11715	0.0072
146	8775	0.1610	196	11775	0.0166
147	8835	0.0216	197	11835	0.0126
148	8895	0.0000	198	11895	0.0251
149	8955	0.0000	199	11955	0.0418
150	9015	0.0000	200	12015	0.0282
151	9075	0.0035	201	12075	0.0358
152	9135	0.0060	202	12135	0.0351
153	9195	0.0309	203	12195	0.0313

TABLE A.1 *(Continued)*

Hour	Minute	RR	Hour	Minute	RR
204	12255	0.0293	254	15255	1.7975
205	12315	0.0151	255	15315	2.3641
206	12375	0.0151	256	15375	2.4007
207	12435	0.0268	257	15435	1.6642
208	12495	0.0375	258	15495	1.9134
209	12555	0.0653	259	15555	1.0693
210	12615	0.0592	260	15615	0.9228
211	12675	0.0604	261	15675	0.6345
212	12735	0.1021	262	15735	0.3443
213	12795	0.1979	263	15795	0.2088
214	12855	0.2713	264	15855	0.1682
215	12915	0.3268	265	15915	0.1753
216	12975	0.3834	266	15975	0.1393
217	13035	0.4695	267	16035	0.1674
218	13095	0.7898	268	16095	0.2311
219	13155	1.0742	269	16155	0.1373
220	13215	0.9859	270	16215	0.1101
221	13275	0.9932	271	16275	0.0264
222	13335	1.3674	272	16335	0.0130
223	13395	1.4538	273	16395	0.0042
224	13455	1.2128	274	16455	0.0004
225	13515	0.8882	275	16515	0.0002
226	13575	1.2300	276	16575	0.0004
227	13635	2.5729	277	16635	0.0012
228	13695	1.9985	278	16695	0.0025
229	13755	2.9087	279	16755	0.0002
230	13815	3.1145	280	16815	0.0004
231	13875	3.6308	281	16875	0.0120
232	13935	4.4068	282	16935	0.0206
233	13995	4.4925	283	16995	0.0136
234	14055	3.1026	284	17055	0.0176
235	14115	3.0842	285	17115	0.0020
236	14175	2.6774	286	17175	0.0004
237	14235	2.2216	287	17235	0.0009
238	14295	1.1765	288	17295	0.0078
239	14355	0.9799	289	17355	0.0109
240	14415	0.8292	290	17415	0.0143
241	14475	0.5638	291	17475	0.0055
242	14535	0.3558	292	17535	0.0040
243	14595	0.1912	293	17595	0.0144
244	14655	0.1708	294	17655	0.0014
245	14715	0.2069	295	17715	0.0002
246	14775	0.2571	296	17775	0.0000
247	14835	0.4298	297	17835	0.0000
248	14895	0.7600	298	17895	0.0000
249	14955	0.8739	299	17955	0.0000
250	15015	0.9229	300	18015	0.0000
251	15075	0.9944	301	18075	0.0000
252	15135	1.0040	302	18135	0.0000
253	15195	1.2756	303	18195	0.0000

TABLE A.1 *(Continued)*

Hour	Minute	RR	Hour	Minute	RR
304	18255	0.0010	356	21375	0.0734
305	18315	0.0001	357	21435	0.0606
306	18375	0.0000	358	21495	0.0593
307	18435	0.0000	359	21555	0.0838
308	18495	0.0000	360	21615	0.1287
309	18555	0.0000	361	21675	0.2166
310	18615	0.0000	362	21735	0.1864
311	18675	0.0004	363	21795	0.1537
312	18735	0.0000	364	21855	0.1023
313	18795	0.0000	365	21915	0.1232
314	18855	0.0000	366	21975	0.1611
315	18915	0.0000	367	22035	0.1853
316	18975	0.0000	368	22095	0.2965
317	19035	0.0002	369	22155	0.4527
318	19095	0.0004	370	22215	0.4202
319	19155	0.0003	371	22275	0.4074
320	19215	0.0042	372	22335	0.6065
321	19275	0.0050	373	22395	0.6804
322	19335	0.0043	374	22455	0.7485
323	19395	0.0097	375	22515	1.2629
324	19455	0.0076	376	22575	1.4835
325	19515	0.0001	377	22635	1.6404
326	19575	0.0016	378	22695	2.1035
327	19635	0.0070	379	22755	2.2113
328	19695	0.0337	380	22815	2.2489
329	19755	0.1067	381	22875	2.4750
330	19815	0.0678	382	22935	1.8730
331	19875	0.0447	383	22995	1.4208
332	19935	0.0380	384	23055	1.0254
333	19995	0.0143	385	23115	0.6456
334	20055	0.0052	386	23175	0.5473
335	20115	0.0025	387	23235	0.8463
336	20175	0.0018	388	23295	0.7854
337	20235	0.0021	389	23355	0.7565
338	20295	0.0020	390	23415	0.5969
339	20355	0.0056	391	23475	0.5739
340	20415	0.0147	392	23535	0.7958
341	20475	0.0090	393	23595	0.6650
342	20535	0.0099	394	23655	0.5980
343	20595	0.0163	395	23715	0.7669
344	20655	0.0053	396	23775	1.0413
345	20715	0.0045	397	23835	1.0161
346	20775	0.0031	398	23895	0.9518
347	20835	0.0047	399	23955	1.0095
348	20895	0.0152	400	24015	1.0848
349	20955	0.0268	401	24075	1.1695
350	21015	0.0402	402	24135	0.8383
351	21075	0.0980	403	24195	1.0303
352	21135	0.1546	404	24255	1.0335
353	21195	0.1657	405	24315	0.8254
354	21255	0.1121	406	24375	0.7466
355	21315	0.0864	407	24435	0.6214

TABLE A.1 *(Continued)*

Hour	Minute	RR	Hour	Minute	RR
408	24495	0.5248	429	25755	0.0361
409	24555	0.4496	430	25815	0.0217
410	24615	0.3952	431	25875	0.0152
411	24675	0.6151	432	25935	0.0417
412	24735	0.7058	433	25995	0.0893
413	24795	0.7090	434	26055	0.1056
414	24855	0.6844	435	26115	0.1193
415	24915	0.7479	436	26175	0.1222
416	24975	0.7545	437	26235	0.1693
417	25035	1.0532	438	26295	0.1375
418	25095	1.0090	439	26355	0.0817
419	25155	0.8426	440	26415	0.0380
420	25215	0.8481	441	26475	0.0269
421	25275	0.2998	442	26535	0.0408
422	25335	0.2473	443	26595	0.0483
423	25395	0.1038	444	26655	0.0373
424	25455	0.0499	445	26715	0.0454
425	25515	0.0359	446	26775	0.0391
426	25575	0.0251	447	26835	0.0560
427	25635	0.0279	448	26895	0.0428
428	25695	0.0284	449	26955	0.0529

A.2 ANNUAL MEAN AIR TEMPERATURE

Table A.2 gives the annual mean air temperature (Celsius) from 1781 to 1980 at Hohenpeissenberg, Germany. The values for 1811 and 1812 are missing. (Source: Report No. 155 of the Deutschen Wetterdienstes, West Germany, 1981.) The table reads from left to right.

TABLE A.2 Annual Mean Air Temperature, C°, 1781–1980

7.2	6.2	6.7	5.0	4.7	5.2	6.9	6.2	6.2	6.9	7.0	6.8	7.3	7.6
7.1	6.7	7.6	6.3	5.8	7.2	7.0	7.2	6.2	6.7	5.1	7.8	7.5	6.1
7.8	8.8	—	—	5.5	5.4	5.8	4.6	5.9	7.0	6.5	5.5	6.8	8.0
6.3	6.7	6.5	6.5	6.7	6.9	4.3	6.0	8.1	6.2	6.4	7.7	6.9	7.5
6.4	6.4	7.7	6.4	7.5	5.7	5.9	5.2	5.4	7.2	5.6	6.4	6.0	5.6
5.1	7.0	5.4	5.9	5.5	6.6	6.9	5.7	7.0	5.0	6.5	7.1	6.9	4.6
6.4	6.7	5.7	6.8	5.8	4.6	4.5	7.1	6.3	5.9	5.4	6.2	5.8	5.6
4.2	6.0	5.3	5.9	4.9	5.9	5.8	5.7	4.5	5.0	4.6	4.7	5.2	5.7
5.8	5.7	5.3	4.9	5.9	6.8	6.1	6.4	5.6	5.7	6.6	6.8	5.5	5.8
6.0	5.7	5.2	6.1	7.2	5.8	6.4	6.2	5.8	6.5	5.3	6.7	5.4	7.1
7.2	5.3	6.3	5.8	6.0	6.7	6.3	6.7	6.0	6.9	5.2	6.3	5.4	7.5
5.9	6.4	6.9	6.6	6.1	5.2	5.2	6.0	7.4	5.7	7.2	6.7	7.5	7.4
7.4	7.1	7.0	6.2	7.1	5.6	5.8	4.9	6.8	6.4	7.5	6.5	7.7	5.5
5.7	6.5	5.5	6.7	6.9	6.2	6.0	5.9	6.7	6.4	5.9	6.6	6.6	6.6
7.0	5.8	6.3	5.7										

A.3 CANADIAN LYNX DATA

Table A.3 gives the base 10 logarithms of the annual number of lynx trapped in Canada along the Mackenzie River for the period 1821–1934 ($N = 114$). (Source: M.J. Campbell and A.M. Walker, "A survey of statistical work on the Mackenzie River series of annual Canadian lynx trappings for the years 1821–1934 and a new analysis," *J. R. Stat. Soc. A*, Vol. 140, pp. 411–431, 1977).

TABLE A.3 Base 10 Logarithms Z_t of Annual Lynx Trappings

year	Z	year	Z	year	Z	year	Z
1821	2.430	1850	2.557	1879	2.303	1907	3.264
1822	2.506	1851	2.576	1880	2.360	1908	2.538
1823	2.767	1852	2.352	1881	2.671	1909	2.582
1824	2.940	1853	2.556	1882	2.867	1910	2.907
1825	3.169	1854	2.864	1883	3.310	1911	3.142
1826	3.450	1855	3.214	1884	3.449	1912	3.433
1827	3.594	1856	3.435	1885	3.646	1913	3.580
1828	3.774	1857	3.458	1886	3.400	1914	3.490
1829	3.695	1858	3.326	1887	2.590	1915	3.475
1830	3.411	1859	2.835	1888	1.863	1916	3.579
1831	2.718	1860	2.476	1889	1.591	1917	2.829
1832	1.991	1861	2.373	1890	1.690	1918	1.909
1833	2.265	1862	2.389	1891	1.771	1919	1.903
1834	2.446	1863	2.742	1892	2.274	1920	2.033
1835	2.612	1864	3.210	1893	2.576	1921	2.360
1836	3.359	1865	3.520	1894	3.111	1922	2.601
1837	3.429	1866	3.828	1895	3.605	1923	3.054
1838	3.533	1867	3.628	1896	3.543	1924	3.386
1839	3.261	1868	2.837	1897	2.769	1925	3.553
1840	2.612	1869	2.406	1898	2.021	1926	3.468
1841	2.179	1870	2.675	1899	2.185	1927	3.187
1842	1.653	1871	2.554	1900	2.588	1928	2.723
1843	1.832	1872	2.894	1901	2.880	1929	2.686
1844	2.328	1873	3.202	1902	3.115	1930	2.821
1845	2.737	1874	3.224	1903	3.540	1931	3.000
1846	3.014	1875	3.352	1904	3.845	1932	3.201
1847	3.328	1876	3.154	1905	3.800	1933	3.424
1848	3.404	1877	2.878	1906	3.579	1934	3.531
1849	2.981	1878	2.476				

Index

A

D

E

F

G

H

N

O

P

Q

R

S

T

U

V

W

Y

Z